JOSEPH FRANZ ROCK

PHYTOGEOGRAPHY OF NORTHWEST AND SOUTHWEST CHINA

EDITED WITH INDICES BY HARTMUT WALRAVENS

ÖSTERREICHISCHE AKADEMIE DER WISSENSCHAFTEN
PHILOSOPHISCH-HISTORISCHE KLASSE
SITZUNGSBERICHTE, 799. BAND

BEITRÄGE ZUR KULTUR- UND GEISTESGESCHICHTE ASIENS

NR. 67

ÖSTERREICHISCHE AKADEMIE DER WISSENSCHAFTEN
PHILOSOPHISCH-HISTORISCHE KLASSE
SITZUNGSBERICHTE, 799. BAND

Joseph Franz Rock

Phytogeography of Northwest and Southwest China

Edited with indices by
Hartmut Walravens

Verlag der
Österreichischen Akademie
der Wissenschaften

Wien 2010 OAW

Vorgelegt von w. M. Ernst Steinkellner
in der Sitzung am 2. Oktober 2009

British Library Cataloguing in Publication data
A Catalogue record of this book is available from the British Library

Die verwendete Papiersorte ist aus chlorfrei gebleichtem Zellstoff hergestellt,
frei von säurebildenden Bestandteilen und alterungsbeständig.

ISBN 978-3-7001-6726-6

Copyright © 2010 by
Österreichische Akademie der Wissenschaften
Wien

Druck und Bindung: Prime Rate kft., Budapest

Printed and bound in the EU

http://hw.oeaw.ac.at/6726-6
http://verlag.oeaw.ac.at

Contents

Preface

Joseph Francis Rock (1884-1962) has become well-known by his many publications in the fields of botany, Southwest China, and especially, the Naxi. He can justly be called the father of Naxi studies even if single researchers preceded him with individual articles. The fact that he collected about 10,000 ritual pictographic Naxi manuscripts and translated and explained many of these with the help of *dto-mbas*, that he prepared a two volume Naxi dictionary[1] which is a kind of encyclopedia, and provided a comprehensive history and geography of the Naxi may justify this statement.[2]

Also on the botanical side, Rock's achievements are considerable. He became – after William Hillebrand[3] – the second father of Hawaiian botany, which is easily proved by his publications.[4] From the ten thousands of plants specimens that he collected especially in Southwest and Northwest China, botanical gardens all over the world profited. The herbaria at Harvard University, the U.S. National Museum (Smithsonian Institution) and the Arnold Arboretum hold most of the original collections while the Museum of Comparative Zoology acquired almost two thousand bird skins through Rock.[5]

Rock was a great book collector, and while he boasted an excellent collection in his chosen fields of study he also provided American institutions with rare items that he found. A large part of his Chinese library went to the University of Washington.

Last not least, Rock was an excellent photographer. Major collections of pictures (prints) are found at the National Geographic Society, the Library of Congress (ex CIA?), Harvard University, Royal Botanical Gardens Edinburgh and Berlin State Library.[6]

It may be useful here to briefly mention some of the major publications on Rock:

---

[1] *A ¹Na-²khi-English encyclopedic dictionary.* By J. F. Rock, Honorary Research Associate, Far Eastern Institute, University of Washington. Part 1-2. Roma: Is.M.E.O. 1963, 1972. XLII, 512, 589 pp. 57 pl. (Serie Orientale Roma 28.)

[2] Rock's publications are recorded in Joseph Franz Rock (1884-1962): *Berichte* .... 2002 (see below).

[3] Wilhelm F. Hillebrand, 1821-1886: *Flora of the Hawaiian Islands.* A description of the phanerogams and vascular cryptogams. London: Williams & Norgate; New York: Westermann 1888. Reprint: New York, London: Hafner 1965. XCVI, 673 pp. – Hillebrand lived in Hawaii from 1851 to 1871.

[4] It may suffice to mention here *The indigenous trees of the Hawaiian Islands.* By Joseph Rock, Botanist of the College of Hawaii, Consulting Botanist, Board of Commissioners of Agriculture and Forestry, Territory of Hawaii. With two-hundred and fifteen photo-engravings. Published under patronage. Honolulu 1913. 518 pp. incl. 218 pl. – *List of Hawaiian names of plants.* By Joseph Rock, Consulting Botanist, Board of Agriculture and Forestry. Honolulu: Hawaiian Gazette 1913. 20 pp. (Territory of Hawaii, Board of Agriculture and Forestry. Botanical Bulletin 2.)

[5] Outram Bangs and James L. Peters: Birds collected by Dr. Joseph Rock in Western Kansu and Eastern Tibet. *Bulletin of the Museum of Comparative Zoology at Harvard College* 68.1928, 311-381.

[6] A number of excellent reproductions are given in Rock's books, and especially also in his contributions to the *National Geographic Magazine.* There is one exhibition catalogue: Michael Aris: *Lamas, princes, and brigands: Joseph Rock's photographs of the Tibetan borderlands of China.* With the assistance of Patrick Booz, and contributions by S. B. Sutton and Jeffrey Wagner. New York: China House Gallery 1992. 141 pp.

Stephanne B. Sutton: *In China's border provinces. The turbulent career of Joseph Rock.* Illustrated with photographs. New York: Hastings House 1974. 334 pp.

H. Walravens: Joseph Franz Rock (1884-1962). Sammler und Forscher. Eine Übersicht. *Oriens Extremus* 38.1995, 209-237

Joseph Franz Rock (1884-1962): *Berichte, Briefe und Dokumente des Botanikers, Sinologen und Nakhi-Forschers.* Mit einem Schriftenverzeichnis. Herausgegeben von H. Walravens. Stuttgart: Franz Steiner Verlag 2002. 452 pp. (VOHD Supplement. 36.)

Joseph Franz Rock: *Expedition zum Amnye Machhen in Südwest-China im Jahre 1926.* Im Spiegel von Briefen und Tagebüchern. Herausgegeben von H. Walravens. Wiesbaden: Harrassowitz 2003. 237 pp. (Orientalistik Bibliographien und Dokumentationen 19.)

Joseph Franz Rock: *Briefwechsel mit E. H. Walker, 1938-1961.* Herausgegeben von H. Walravens. Wien: Österreichische Akademie der Wissenschaften 2006. 328 pp. (Österreichische Akademie der Wissenschaften. Philosophisch-historische Klasse. Sitzungsberichte 738.)

Joseph Franz Rock (1884-1962): Tagebuch der Reise von Chieng Mai nach Yünnan, 1921-1922. Briefwechsel mit C. S. Sargent, University of Washington, Johannes Schubert und Robert Koc. Herausgegeben von H. Walravens.
Wien: Österreichische Akademie der Wissenschaften 2007. 580 pp. (Österreichische Akademie der Wissenschaften. Philosophisch-Historische Klasse. Sitzungsberichte 757.)

An edition of the correspondence between Rock and his Leipzig colleague Johannes Schubert is in press and will be published soon by the Austrian Academy of Sciences.

The present publication reproduces an original work by J. F. Rock that was written for the Royal Horticultural Society by agreement with Lord Aberconway, the president of the Society. Unfortunately, after the president's death in 1953[7] the Society changed its publication policy, and Rock's book remained unpublished among his papers.

It was not designed as a scholarly catalogue of plants but as a mix between a travelogue and a description of the flora of the areas that Rock himself explored. This seemed to be the best presentation for plant lovers, not necessarily scholars, like the constituency of the Royal Horticultural Society. The work should be well illustrated, too, as a feast for the eye, or as we might say today, a bit colloquially, as a coffee-table book.

The book is extant as a typescript, with Chinese and Tibetan script written into the text by Rock himself. This is of great importance as many of the geographical names of the areas that Rock explored were not found on maps, and other explorers had too little knowledge of the respective languages to provide accurate spellings.

While thanks to Unicode the insertion of Chinese characters into English text is no problem, Microsoft Word is still (2008) not compatible with Tibetan. So the names had to be written in TextEdit but could not be imported into MS Word text. For this reason, Tibetan script is only given in the indices. There is an advantage to it: orthographic variants by the author can easily be detected. The geographic index was sorted by hand and according to the specifics of the Chinese and Tibetan languages – as there are only a limited number of syllables they were made the filing criteria, instead of sorting letter

---

[7]   Lord Aberconway, 1879-1953.

by letter. Also, commercial sorting programmes cannot handle Tibetan superscript letters, and thus words that belong together (and are found in dictionaries in the same place) would show up at different places (e.g. Do and rDo).

Rock used the Wade/Giles transcription for Chinese (with a few variants) and a kind of Wylie transcription for Tibetan, for the sake of the English reader. It did not seem necessary to change these applications even if Hanyu Pinyin has become the standard system for Chinese.

More of a challenge is the fact that Rock used both spoken and written forms of Tibetan names; when mentioned for the first time the different spellings are often given, including the Chinese transcriptions. Most of the time the colloquial form is quoted, and not always consistently. As Rock refers to local pronunciations, it seemed inappropriate to standardize his spellings. As a reference tool, the geographical index gives the *written form* of the names in order to facilitate retrieval in historical and geographical sources.

A challenge were also the illustrations. While the book was designed to have a number of plates which were kept in an envelope with the typescript, it turned out on closer inspection by the archivist in Edinburgh that only part of the material carried the stamped page numbers of the typescript. Others had «page numbers» but it is unclear to what kind of pages they refer. Some had only negative numbers. At any rate, many of the plates filed with the typescript did not match the plate references in the typescript. The editor tried as well as he could to correlate the plates with the text.

It is certainly possible that Rock never made the final plate selection for the book because the project never reached the definitive stage of printing. On the other hand, Rock may have re-arranged the illustrations and refiled or used some for other purposes. At any rate, the captions to the plates were written by Rock himself and are reproduced faithfully, including references to negatives and the field notes (diaries). Quite a number of the originally selected plates could not be identified.

A word may be added regarding Rock's photographs. There are several thousand of them extant, with negatives mostly in the possession of the National Geographic Society, perhaps also Harvard University. So far no census, or inventory, has been made. Many of the prints in the Dept. of Prints of the Library of Congress have captions, while most of the prints in Berlin and Edinburgh do not have any. The reason is simple: Rock knew his pictures very well, and so he did not need them. For researchers the situation is rather unsatisfactory, and it would be useful to have all the Rock photographs scanned and identified.

Thanks are due above all to the Royal Botanical Gardens Edinburgh that own the original typescript and the plates, for kindly making these available for publication. Especially the archivist, Leonie Paterson, worked hard to assemble and scan the available plates. The Oriental Department of the Berlin State Library, custodian of half of Rock's papers, acquired a xerox copy of the Edinburgh typescript and kindly furnished it for printing.

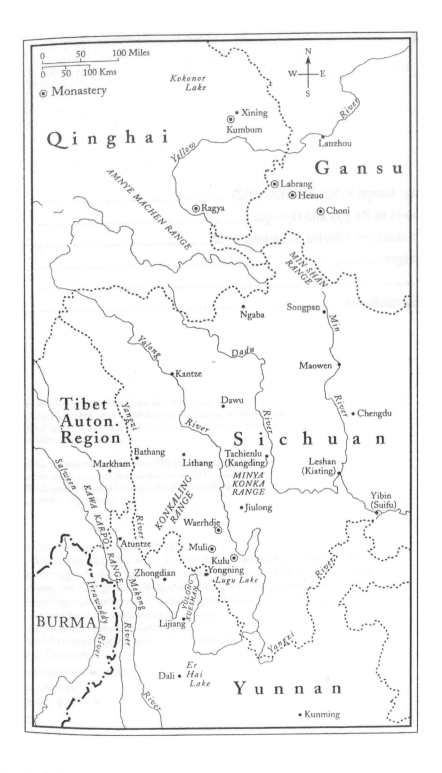

0   50   100 Miles

0   50   100 Kms

⊙ Monastery

N
W ✦ E
S

Kokonor Lake

Qinghai

⊙ Xining
⊙ Kumbum

Yellow River

• Lanzhou

Gansu

⊙ Labrang
⊙ Hezuo
⊙ Choni

AMNYE MACHEN RANGE

⊙ Ragya

MIN SHAN RANGE

• Ngaba
• Songpan

Min River

Yalong

• Kantze

Dadu River

• Maowen

Tibet Auton. Region

• Dawu

Yangzi

Sichuan

River

• Chengdu

Salween

• Markham
• Bathang

• Lithang

• Tachienlu (Kangding)

• Leshan (Kiating)

KAWA KARPO RANGE

KONKALING RANGE

MINYA KONKA RANGE

• Yibin (Suifu)

Waerhdje ⊙

• Jiulong

• Atuntze

Muli ⊙

River

Mekong

• Zhongdian

Kulu ⊙
• Yongning
Lugu Lake

Yangzi

YULONG XUESHAN

BURMA

Irrawaddy River

• Lijiang

Er Hai Lake

• Dali

Yunnan

River

• Kunming

List of Illustrations

The numbers refer to the pages in the text. Additional information is quoted when available, like negative numbers, pages in Rock's diaries (field notes) as well as dates.

27 **83**   Neg. 263; vol. VII:86; IX:62; Sept. 19, 1926
The The-wu village of Wang-tsang in the valley of the Pai-lung Chiang. Wang-tsang
kou (valley) to left of village. The trees back of the village are *Malus baccata* Borkh.
Elevation 6400 feet. Lower T'ieh-pu land, S.W. Kan-su.

28 **83**   Neg. 262; vol. VII: 29; vol. IX: 60. Aug. 31/1926
In the mouth of Wang-ts'ang kou (valley) Lower T'ieh-pu land. The streambed is lined
with *Quercus liaotungensis* Koidz., elevation 7200 feet.

29 **86**   Neg. R 334 AA; vol. VII: 83; Sept. 13, [1926?]
Abies chinensis Van Tiegh, growing in Wang-ts'ang Valley, Lower The-wu Land,
elevation 8000 feet. Trees are from 100-150 feet in height and from 3-4 feet in
diameter. Southeastern Min Shan, Kan-su.

30 **87**   Neg. 414AA, vol. VII: 80; Sept. 11, 1926.
Populus szechuanica Schneid. var. *Rockii* Rehd. A tree growing 80 feet tall in the
forests of Wan-tsang Valley in the Lower The-wu country south of the Kan-su Min
Shan. Elevation 8,000 feet. Spec. no. 14846.

31 **89**
Trail in the valley of the Pai-lung Chiang around a cliff built on posts and sticks with
the river roaring a hundred feet below. Between Pe-zhu and Ni-shih-k'a in Lower T'ieh-
pu Land.

32 **89**   <neg. 261; vol. VII: 36; IX: 60; Sept. 1, 1926
The new Wang-tsang Monastery, Wang-tsang Men-chhe Gom-pa, on the left bank of
the Pai-lung Chiang, Lower T'ieh-pu Land. Elevation 6,290 feet. Southern slopes of the
Kan-su Min Shan.

33 **92**
In the valley of To-erh (kou) or Do-ru Nang; to the left the Sa-kya Lamasery of Pe-ku
Gom-pa or Pai-ku ssu, elevation 7,400 feet. The walls of the lamasery are striped red
and white. Looking down stream. In the distance, center is visible the lamasery of Ra-
gya Gom-pa.

34 **92**
Cutting a trail through the snow on Yang-pu Shan, over 12,300 feet. For two days our
men and The-wu Tibetans of Yang-pu cut and shovelled a path over the snow-covered
slopes to the path. The Rhododendrons visible in the snow are *Rhod. przewalskii*. The
Kan-su – Ssu-ch'uan divide.

35 **95**   Neg. 270; vol. IX: 62; vol. VII: 47
Looking down Ma-ya kou from a bluff elevation 8,000 feet. The horizontal trail in
centre of picture is south of the Pai-lung Chiang. The valley opposite To-erh kou, ist
branch on the right is A-hsia kou (valley). For description of flora see Ma-ya kou
chapter.

36 **95**   Neg. R 366; vol. VII: 55. Sept. 6/26
View down Ma-ya kou (valley) from an elevation of 10,600 feet, from below the
summit of Lha-mo gün-gün pass. The trees are Abies, and *Picea Wilsonii* Mast., *P.
asperata* Mast., and *P. purpurea*, with an occasional *Juniperus squamata* forma
Wilsonii, the latter along the margin of the small meadow in lower centre of picture.

37 **95**   Neg. R 365; vol. VII:55 Sept. 6/26
At the head of Ma-ya kou (Valley), eastern Min Shan, elevation 10800 feet, below Lha-
mo gün-gün pass. The trees are *Abies Faxoniana* Rehd. & Wils., the shrubs in
foreground *Rhododendron Przewalskii* Kom. Looking south.

38 **97**   Neg. R 371; vol. VII: 59; Sept. 7, 1926
Looking northwest from He-ra in San-pa Kou (valley) towards Yor-wu-drag-kar,
elevation 9500 feet. Lower T'ieh-pu Land southern slopes of Min Shan. The shrubs are
mainly Salix, Rosa, Berberis, and scattered conifers, Picea, Abies, and Juniperus. The
limestone gorge in the distance is Yor-wu-drag-kar through which the main San-pa
stream flows.

39 **98**   Neg. AA 392; vol. VII: 94. Sept. 17/26
The upper or northern end of Do-ya-ya gorge. Here the limestone gives way to
conglomerate, very porous yet smooth in appearance; elevation 10,700 feet. Eastern
Min Shan, Yellow River–Yangtze divide. Willows to right, to left on the smooth walls
Juniperus saltuaria Rehd. & Wils.

40 **120**   Vol. VI: 53. Sept. 25, 1925
A lämmergeier (*Gypaetus barbatus grandis* Starr) shot on the eastern shores of the
Koko Nor, Ch'ing-hai.

41 **134**
The T'ung-tzu Ho, looking downstream towards its junction with the Pa-pao Ho (Hei-
kou Ho) and to the southern slopes of the Ma-lo-ho Shan about 15,000 feet in height, a
part of the Ch'i-lien Shan range. Elevation 9,000 feet.

42 **142**   Neg. 218
The Tshe Chhu (River) on the Na-mo-gen Thang (Plain) at an elevation of 12050 feet.
The encampment of the Rong-wo nomads in the middle distance, the spur in the
background is Na-mo-ri-ön-dza-de. The nomads taking their sheep out to graze after
having been in the center of their encampment for the night. Our camp was on the
terrace to the left above the river.

43 **161**   Neg. 25; vol. VI:43
The vanguard of our yak caravan making its way to the summit of the Khe-chhag Nye-
ra (pass) 13,200 feet elevation. Between La-brang and the Sog-wo A-rig encampment.
May 8th 1926

53 **168**   Neg. 148; vol. VI:149
Hermit quarters above Ra-gya Gom-pa at the foot of the cliffs (conglomerate sandstone) of Mount Khyung-ngön, elevation 10,700 feet. Trees to left *Juniperus tibetica* Kom. Nettles, *Urtica dioica* L. grow below the hermit quarters, of these the lama hermits subsisted in the summer.

54 **169**
Head of a blue sheep *Ovis Burrhel* common in the valley of the Yellow River on cliffs. At Ra-gya Gom-pa they are sacred and protected, they roam in herds quite tame and unafraid of man. This one was shot in Lung-mar valley, east of Ra-gya.

55 **172**   Neg. 85; vol. VI:95
View up the Yellow River from back of the summit of Am-nye Khyung-ngön, the sacred mountain back of Ra-gya Gom-pa, showing the central cleft of the mountain, the trees are mainly *Juniperus tibetica* Kom., and *Picea asperata* Mast. Elevation 11,406 feet.

56 **172**   Neg. 63; vol. VI:80
The Yellow River or Ma Chhu above the mouth of the Lung-mar or Red Valley, elevation 10,300 feet, looking upstream. The trees are mainly *Juniperus Przewalskii* Kom., and *Picea asperata* Mast. Three miles east of Ra-gya Gom-pa.

57 **176**   Neg. 284 (my neg. 98)   vol. VI:104-105, vol. IX:23
The Yellow River looking down stream from a bluff, elevation 10,900 feet on the spur which intersects Hao-wa and Nya-rug Valleys, west of Ra-gya. Spruces (*Picea asperata* Mast.), on the left on the northeastern slopes of the Yellow River Valley with *Juniperus Przewalskii* Kom. on the opposite slopes. *Juniperus tibetica*, right foreground.

58 **176**   Neg. 283; vol. VI: 104-105, neg. 97. June 27, 1926, downstream
The Yellow River looking down stream west of Ra-gya from a bluff on the spur which intersects Hao-wa and Nya-rug valleys, elevation 10,900 feet. *Picea asperata* Mast., to left and *Juniperus Przewalskii* Kom. on the opposite slopes. In immediate foreground right, *Juniperus tibetica* Kom. A rapid can be seen in middle distance.

59 **177**   Neg. 103; vol. VI: 109-110, upstream
The Valley of the Yellow River looking up stream, south-east from a bluff above Ta-ra-lung, between Hao-wa and Sa-khu-tu valleys, elevation 10,600 feet. The valley walls to the right are forested with *Picea asperata* Mast. The wrinkled appearance of the slopes to the left is due to overgrazing.

60 **177**   Neg. 104; vol. VI: 110; IX: 25, upstream
The Yellow River looking south, up stream, from a bluff above Sa-khu-tu ravine, elevation 10910 feet. Juniper trees in center below, in the mouth of Sa-khu-tu ravine. The narrow ridges on the left hillsides are the result of grazing by the heards of sheep of the nomads.

61 **177**   Neg. 101; vol. IX: 34 May 30/26
The Yellow River looking down stream west-north-west from a bluff called Ta-ra-lung, elevation 10620 feet. *Picea asperata* Mast. forests on the left. *Juniperus Przewalskii* Kom., in foreground. West of Ra-gya Gom-pa.

62 **177**   Neg. 100; vol. VI: 107; downstream
The Yellow River gorges as seen from a pass north of Nya-rug nang (valley), elevation 11,850 feet, looking down stream westnorthwest. The trees on the spur in foreground are *Juniperus tibetica* Kom., spruces *Picea asperata* on the slopes to the left.

63 **179**   Neg. 106; vol. VI: 116; IX: 25.
Picea asperata Mast. forests in the valley of the Yellow River as seen from a bluff 11700 feet elevation, between Ar-tsa and Tag-so valleys, looking down stream, northwest of Ra-gya Gom-pa.

64 **179**   Neg. 123; vol 9:29: June 2, 1926
Camp in the Tag-so Valley, elevation 10146 feet, immediately above its narrow defile cut into the valley walls of the Yellow River; spruces, birches, willows, Lonicera, Berberis, Ribes, etc. form the main vegetation.

65 **180**   Neg. R 265; June 24, 1926
Trunk of *Picea asperata* Mast., growing in Tag-so canyon, elevation 10150 feet, northwest of Ra-gya. Tag-so is tributary of the Yellow River.

66 **180**   Neg. 129; vol. IX: 30
Juniperus tibetica Kom. (no. 13946) on a grassy plot in Tag-so canyon; *Picea asperata* Mast. clings to the steep hillsides in the background. Elevation 11000 feet.

67 **180**   Neg. 112; vol. IX: 26 May 31/26
Vegetation in Tag-so canyon; to right *Betula japonica* Sieb. var. *szechuanica* Schneid., to left *Picea asperata* Mast., in center *Salix juparica* Goerz, elevation 10150 feet.

68 **180**   Neg. 111; vol. IX: 26; May 31, 1926
In the valley of Tag-so, looking upstream. *Picea asperata* Mast., covers the steep valley slopes, in middle foreground (pointed trees) *Betula japonica* Sieb. var. *szechuanica* Schneid. Shrubs to the left *Salix juparica* Goerz, other shrubs are *Sibiraea angustifolia* (Rehd.) Hao, *Lonicera syringantha* Max., and *Berberis Boschanii* Schneid. Elevation 10200 feet.

69 **180**   Neg. R 302; June 3, 1926. Spec. 13946
Juniperus tibetica Kom., showing trunk and stringy bark, growing in Tag-so valley where it forms pure stands at 10600 feet elevation. The tree is probably 500 years old.

70 **189**   Neg. 154; June 23/26
The Bâ valley at an elavation of 9940 feet, facing the northern slopes of the valley. The shrubs on the river-bank are *Lonicera syringantha*, back of them, *Salix Wilhelmsiana*

M. v. B., *Salix juparica* Goerz, *Hippophae rhamnoides* L.; the tussocks on the hillside above are *Caragana tibetica* Kom.

71 **192**   Neg. R 290 (AA); June 27/26
The Gyü-par Valley looking upstream, northern slopes of the Gyü-par Range, elevation 18600 feet. *Picea asperata* forest in background, the shrubs are willows, honeysuckles, Hippophae, Ribes, etc.

72 **192**   Neg. R 296 (AA); June 27/26
The Gyü-par Valley, northern slopes of the Gyü-par Range, looking upstream at an elevation of 10200 feet. *Picea asperata* on the steep slopes, willows in the streambed. A landslide exposes loess deposits with schist and shale embedded in the fine loess.

73 **192**   Neg. 169, June 28th, 1926
The large bend of the Yellow River at the mouth of the Gyü-par valley, north of the Gyü-par (Jü-par) Range, as seen from a bluff elevation 10480 feet. The rocks on the opposite bank of the river are a deep grayish blue slate. The Gyü-par stream enters the Yellow River, extreme lower right & through a narrow rocky defile.

74 **195**   Neg. 175 (R 219); June 29, 1926
Looking down a branch of the Gyu-par Valley and the main valley towards the Yellow River, northwest from an elevation of 11100 feet. The Yellow River flows at the foot of the eroded steep cliffs (center distance). The trees in foreground are *Picea asperata* Mast.

75 **195**   June 28, 1926
Caragana tibetica Kom., a tussock-forming leguminous shrub growing on slate scree overlooking the Yellow River, on the northern slopes of the Gyü-par range, elevation 10400 feet.

76 **195**   Neg. R 277; June 28, 1926
The Gyü-par valley looking south southeast from a bluff 11300 feet, showing spruce forest (*Picea asperata* Mast.) on the slopes to right, the only region in which the Gyü-par Range is forested. The triangular peak in the distance under the clouds is the second highest of the range and is called Gyü-par sher-nying.

77 **196**   Neg. R 254; June 27, 1926
One of the tallest *Picea asperata* Mast. (no. 14323) growing on the western slopes of the Gyü-par Range in the Gyü-par Valley, elevation 10200 feet. The base of the trunk is on the lower slopes of the hill. The tree is over 100 feet in height

78 **199**   Neg. 209; vol. VI: 255
Tetraogallus tibetanus Przewalskii Bianchi, shot on the rocky scree slopes at an elevation of 13200 feet, near the head of Ser-chhen Valley, north-east of Ra-gya.

79 **200**   Neg. 183
View into the Bâ valley with the Gyu-par range in the distance, as seen from a bluff
south, elevation 10400 ft, looking north north-west. Willows along the streambed in the
valley. Near the foot of the eroded loess cliffs are two Tibetan villages called Sa-og-
rong-wa. The tussock forming plants in the immediate foreground are *Caragana*
tibetica Kom.

80 **206**   Neg. 95; vol. VI: 102
Tshang-rgyur mGo-logs at their encampment west of the Yellow River, opposite Ra-
gya Gom-pa. The left one was on the point of taking a pinch of snuff, the latter consists
of the ashes of burnt yak dung. They wear one single sheepskin garment (tanned by
softening and rubbing it with butter).

81 **206**
A Tshang-gur Go-log at his encampment west of the Yellow River. The scars on his
abdomen are from self-inflicted burns resorted to as counter-irritation against
indigestion. Around his neck he wears an amulet containing charms.

82 **206**   neg. 140
The chief of the Bu-tshang Go-log, his encampment is south of the Ri-mang Go-long
and west of the Nga-ba tribe, south of the Yellow River. He wears a rich blue satin
brocade with gold designs, trimmed with otter skins, and on his head a fox skin. Around
his neck is a fine silver charm box studded with turquoise and coral.

83 **207**
Ferry across Yellow River at Ra-gya Gom-pa; twelve inflated sheepskins are use to one
raft, and as many as twelve persons are take across; horse and yak have to swim.
Elevation at river bank 9900 feet, looking downstream. William E. Simpson my
interpreter stands behind my horse.

84 **207**   Neg. 75   ; vol. VI:90
Tibetan ferry across the Yellow River at Ra-gya Gom-pa. Eight nomad women are
about to be ferried across on a raft of twelve sheepskins.

85 **209**   Neg. 197; vol. VI: 233
In the Tsha-chhen Valley with part of our escort of Ja-zâ nomad Tibetans at an
elevation of 12500 feet. Note the grove of *Juniperus tibetica* Kom. on the valley slope.
In foreground small bushes of *Potentilla fruticosa* L. var? They were not in flower.
West of Yellow River.

86 **210**   Neg. 196; vol. VI: 233
A Tibetan nomad of the Ja-zâ clan, from west of the Yellow River, their only garment is
a sheepskin, their head is shaved except for a lock in the back. He wears a silver charm
box, with turquoise, and a necklace of wooden beads. The right arm and shoulder is
always bare.

87 **210**   Neg. 198; vol. VI:237
The Am-nye Ma-chhen Range (Ma-chhen pom-ra) as seen from the summit of Am-nye Drug-gu, elevation 144350 feet, looking west. The valley of the Gur-zhung is below and west of the ridge in the immediate foreground. The Tshab Chhu Valley is beyond the ridge above the little cloud extending diagonally towards upper center of picture. *Juniperus tibetica* Kom. on the slopes.

88 **210**   Neg. 202; vol. VI: 238
The Yellow River gorges looking up stream from the mountains to the west of it, from a bluff 10500 feet elevation, below Am-nye Drug-gu, the mountain god of the Yön-zhi tribe. To right on the slopes *Picea asperata* Mast., middle distance *Juniperus tibetica* Kom. and willows, foreground *Rhododendron capitatum* Max. The small white flowers are *Leontopodium linearifolium* H.-M.

Introduction

This volume treats of the vegetation and geographic phase of the western part of Kan-su province and of the eastern and northeastern part of the province of Ch'ing-hai, the only region where forests occur and where plant growth is richer than elsewhere in these two provinces.

Of Kan-su, the area included commences at the border of Ssu-ch'uan north of Ching-ch'üan, a little west of the 105° meridian and north of the 32°30' parallel, and extends to the 102° meridian in the south, to the 100° in the northwest, and 39° parallel in the north.

That of Ch'ing-hai begins with the 34° parallel in the south to the 103° meridian (but not in an even line as that meridian crosses parts of Kan-su) to the 39° parallel in the north and 99° meridian in the west.

The richest botanical region of Kan-su and scenically the most beautiful, even of the whole of China, is without question the Min Shan in the south, and Lien-hua Shan north of it, both composed mainly of limestone. North of the latter the flora is poor, as the area is arid and becomes more so as the borders of Inner Mongolia are approached.

Of the Kan-su area in question that comprising the Min Shan to south of Sung-p'an in northwest Szechuan, has not only been mapped for the first time but also explored botanically in its entirety.

Previous botanical explorations were confined to the southeasternmost point (Farrer & Purdom), while other travelers simply passed over its eastern end.

Of Ch'ing-hai a considerable area was unexplored and this was mapped and combed botanically for the first time; I refer to the area of the gorges of the Yellow River, east and west, from the bend in the south to north of the Gyü-par range, and all the intervening grasslands. The region of the Am-nye Ma-chhen, and the western half of the Gyü-par Range the only part where forests occur and which are of botanical interest.

In the far north the Koko Nor, Potanin range and the extreme northern ranges of the Nan Shan system which include the Ch'ih-lien and T'o-lai ranges were crossed in several places and botanically explored but mainly for ligneous plants.

The exploration of all these regions which lasted from the spring of 1925 to spring 1927 were undertaken for the Arnold Arboretum of Harvard University and the Museum of Comparative Zoology of the same university, the first receiving all the botanical material including all the many seeds, of both ligneous and herbaceous plants, and the latter the birds numbering nearly 2,000 specimens.[1] Reports on these collections have been published by the said University.

The expedition started from Haiphong in Indochina by rail to K'un-ming, capital of Yün-nan, and thence by mule caravan across the whole of West China to the southwestern border of the Gobi desert. For the exploration of the different regions 60 yak, 14 horses and 18 camels as well as 40 mules were employed at one time or another depending on the mode of travel in vogue in the different parts. My assistants comprised 12 Na-khi from Li-chiang in northwest Yün-nan who joined me in K'un-ming, Yün-nan, whence we started in December 1924

---

[1] Outram Bangs and James L. Peters: Birds collected by Dr. Joseph Rock in Western Kansu and Eastern Tibet. *Bulletin of the Museum of Comparative Zoology at Harvard College* 68.1928, 311-381.

The photographs were all taken by the author and some of the section maps here reproduced for the first time were made by him. Other maps like those of the Yellow River and the region between La-brang (Hsia Ho) and the Am-nye Ma-chhen were made with the cooperation of the late W. E. Simpson.

Thanks are rendered to the Royal Horticultural Society and its President Lord Aberconway[2], for the interest taken, and to the officers of the National Geographic Society of Washington D.C., who generously made the prints which illustrate this volume.

London, England September 1952   J. F. Rock

---

[2] Henry Duncan McLaren, 1879-1953, 2nd baron Aberconway, British politician, horticulturist industrialist. He inherited Bodnant and deloped its garden; he was particularly fond of breeding rhododendrons and magnolias.

The Min Shan

The Kan-su Min Shan is a distinct range, for the greater part composed of grey, old limestone, only the extreme eastern part is conglomerate. Its bearing is from northwest to southeast, and extends from Shi-tshang-gar-sar to Hsi-ku, a distance of about 150 miles as the crow flies, perhaps a little less, but limestone outcroppings extend much further west than Shi-tshang or Hsin ssu 新寺, the new lamasery, as the Chinese call it. It was here that Dr. Tafel[1] was nearly killed by Shi-tshang Tibetans and robbed of everything he possessed while camping in front of the lamasery, he received a sword cut over the forehead the deep soar of which he carried to the end of his life in 1937, when he died of cancer.

The most conspicuous part of the range when viewed from the north, is a broad cleft about the center of the range, several thousand feet deep, known as the Shih-men or Rock Gate, flanked by enormous buttresses. This is however not the highest point of the range, the top of the Rock Gate is not more than 15,000 feet. A peculiar feature of the Min Shan is, that at a distance of about three miles north of the main backbone, a preliminary range extends its entire length, this range is lower, but is pierced by streams; in order to approach the main backbone, rock gates, often very narrow, have to be negotiated. As they are lower, they are not visible when the range is viewed from heights in the north. On the southern slopes of the range are similar rock gates which impede the traveler's progress.

Unlike the preliminary rock gates which carry streams, the main Shih-men is impassable and does not pierce the range, but is shut in by a circular rampart forming an amphitheater; the slopes are loose scree and boulders. The enclosing rampart can be approached however from the rear where talus slopes extend south from the main backbone, and these screes are the home of a wonderful alpine flora to be described later. Such unnegotiable clefts occur also further east in the same range. See: The main Shih-men or Rock gate 石門.

At the eastern end, the Min Shan turns south-southeast where it reaches its highest elevation in a huge limestone crag almost square and more or less detached from the main crest, it looms into the sky in shape like the famous Hua Shan 華山 in Shensi.[2] Below it, directly south, is a similar truncate limestone block, also detached and nearly square, the former is called Ta-ku-ma 大古嗎 or the great Ku-ma about 17,000 feet in height and the latter Hsiao-ku-ma 小古嗎 or Small Ku-ma 15,500 feet, the name is from the Tibetan sGo-ma one of the four guardians of the world as represented in lama dances, actually the square of a door. They are the sentinels to T'ieh-pu land in the southeast. It is the Hsiao Ku-ma which Farrer[3] calls the Thundercrown and is the only

---

[1]   Albert Tafel, 1876-1935, physician and explorer, best known through his massive *Meine Tibetreise* (Stuttgart: Union 1914. 2 vols.). See Albert Tafel† *Ostasiatische Rundschau* 1935, 304; Albert Tafel, 1876-1935. *Monumenta Serica* 1.1935/36, 496-498.

[2]   A good description is: *Hua Shan. The Taoist sacred mountain in West China. Its scenery, monasteries, and monks.* Foreword and 111 photos by Hedda Morrison. Introduction and Taoist musings by Wolfram Eberhard. Hong Kong: Vetch and Lee (1974). XXV, 135 pp.

[3]   Reginald Farrer, 1880-1920. Cf. *Reginald Farrer: Dalesman, planthunter, gardener.* Lancaster: Centre for North-West Regional Studies, Univ. of Lancaster 1991. X, 102 pp (Occasional papers,

part of the main Min Shan where he collected. In the extreme southeast the Min Shan connects with the long mountain chains which hem in the deep valleys of To-erh (kou) 多兒溝 and A-hsia (kou) 阿夏溝 the Tibetan rDo-ro-khu and Â-cha-khu respectively, whose streams flow southeast to north and debouch into the Pai-lung Chiang. This connection, a long grey limestone chain, is severed by the Pai-lung chiang which flows here in a very narrow trench. The mountains to the southeast diminish in height till we reach the Yang-pu Shan 陽布山, the Tibetan rTa-rgas La (Ta-ge La) 12,300 feet in height, on the northern slopes of which the To-erh kou has its source, while the A-hsia kou comes from southwest and joins the To-erh kou below the T'ieh-pu hamlet of Pa-ka or Pe-kar written dPal-mkhar. The whole system is still limestone including the high crag back of La-tzu ssu 拉子寺, the Tibetan Rwa-gziḍ-dgon-pa, the last fantastic limestone peak 14,500 feet, in this part of the Min Shan. The long valleys and gorges which extend from Yang-pu Shan north are arid and the vegetation is of an entirely xerophytic nature, while the northern slopes of the mountain and its narrow ravines extending from its summit, are rich in shrubs and trees; here occur Rhododendrons not found on the Min Shan proper, while conifer forests extend to 11,700 feet, this being the timberline. On its southern slopes however only a few fir (Abies) trees are to be found, remaining trees are mainly junipers, *Caragana jubata* and a few Rhododendrons, the timberline on the south side being at about the same height as on the north side. Yang-pu Shan is a more or less rounded mountain mass and was so deeply covered with snow that no rocks were visible, but judging from its contour and the rocks of its southern slopes, over which a trail leads into a deep gorge, are slate and schist while the main valley leading to the pass on the north side was cut in red sandstones. We have definitely left the Min Shan limestone behind.

The region from, Yang-pu Shan to Sung-p'an will be described separately, for with Yang-pu Shan the Kan-su Min Shan ends.

The Valleys of the Northern Slopes of the Kan-su Min Shan
The T'ao River or Lu Chhu flows nearly along the entire northern length of the Min Shan. The long spurs which extend from the real backbone of the Min Shan, both in the north and south are composed of schist, shale and old sandstone through which the limestone range, the actual Min Shan, has pushed itself, the contact between schist and limestone being plainly visible. The T'ao river following the course of least resistance cut through this loose material and encircles the range as previously related. However, a spur of hard limestone extends south-southeast across the valley of the Pai-lung Chiang, the latter, which otherwise flows in a broad valley, encounters here this formidable barricade through which it cut a very narrow channel, rushing madly eastward between the steep walls.

There are three long valleys extending from south to north over which passes lead to the south of the range, all the other valleys are of various lengths, some very short, culminating in grassy spurs whose northern slopes are covered with a dense impenetrable thicket of interlacing branches of *Rhododendron rufum* Bat., to almost the

University of Lancaster, Centre for North-West regional Studies 19.); *Reginald Farrer, at home in the Yorkshire Dales*. Gigglewick, Yorkshire: Castlebury 2002. 66 pp.

very edge of the grassy top of the spur; south of these spurs are narrow ravines with precipitous slopes, which extend in a more or less parallel line with the distant crags of the Min Shan. The intervening slopes of the range are deeply cut into wedge-like valleys, a veritable labyrinth of spurs, knife-edge ridges and ravines too high for tree growth.

Ch'e-pa kou 扯巴溝 the longest and westernmost valley extends north from the summit of the Min Shan. Its head is south of a pass 12,500 feet in height west of Kuang-k'e La whence it descends in a curve from southwest to northeast. Near the Tibetan village of rMe-ri-shol pronounced Me-ri-shöl or Me-ri-shü, the Chinese 買力什, it received a large affluent from the west. The valley is fully described in the journey from Kuang-k'e to Cho-ni via Ch'e-pa kou.

The next valley in size or length is K'a-cha kou 卡札溝, this being the correct name as it occurs on the Cho-ni prince's map (other travelers call it Chia-ch'ing kou or Kar Ching K'ou) which also extends to the summit pass Kuang-k'e La 12,550 feet, from south-southwest to northeast. It has several affluents one of which, Tsha-lu (kou) leads to the famous landmark, the main Rock Gate or Shih-men of the Min Shan. To the east of the K'a-cha kou, the Tibetan Kha-rgya nang are the following three valleys from west to east, La-li kou 拉力溝, Ma 馬 or Ma-erh kou 麻兒溝, and Po-yü kou 波峪溝. These do not extend to the backbone of the Min Shan but have their source in a high spur which extends from east to west with an average altitude of between 11,000 to 12,000 feet. Between these valleys are smaller ones as Shao-ni kou 勺尼溝 and P'a-lu kou 怕路溝.

The longest and largest on the east, leading to a pass over the Min Shan, is the Ta-yü kou 大峪溝, the approach to the Hsia T'ieh-pu 下鐵布 or Lower T'ieh-pu land. At the village of A-chüan 阿絹 also written A-chüeh 阿角, and this is the name which appears on the Cho-ni prince's map, (also called A-i-na 阿亦那) it received an affluent from the southwest called the Ta kou 大溝 or Great Valley, while its main stream coming from a pass called rTsa-ri-khi-kha, elevation 11,250 feet, is known as Hsiao kou 小溝 or Small Valley to where it received the Ta kou stream at A-chüeh. All these valleys are densely forested almost from their mouths to their sources, and will be described in detail as to their plant covering etc.

The Northern Affluents of the T'ao River

The valleys which debouch into the T'ao river from the north are of little botanical interest as they are bare, and have been cut into the grass-covered loess plateau which slopes down steeply into the T'ao valley. These valleys have their sources in high grass-covered passes of an average of 10,000 feet elevation, but which increase in height westwards. The mountain range, really a high plateau is marked on some maps as a continuation of the Hsi-ch'ing Shan that is Lui-chhab-rag Range, which again other travelers call the Tasurchai range. It has nothing to do with the Hsi-ch'ing Shan which, as already stated is an isolated limestone range like the Min Shan. Somewhat east of, and across from the mouth of the Ta-yü kou valley debouches the Hsin-p'u ho 新堡河. There are several affluents, their valleys bare and grassy; the one in which T'ao-chou chiu-ch'eng 洮州舊城 or old city is situated is called Shi-tshang Khog, in Tibetan, the

Chinese Chiu-t'ao-p'u kou shui 舊洮堡溝水, T'ao-chou old city is known to the Tibetans as Pag-tse, from the Chinese Pa-chai 八寨 or Eight villages of which the old city is apparently composed. Between this valley and Hsin-p'u kou are three other streams from west to east they are Yang-sheng-kou shui 羊升溝水 and Yang-yung-kou shui 羊永溝水 and Liu-shun-ch'uan-kou shui 劉順川溝水, Lin-t'an-hsien 臨潭縣 the chief magistracy of the region being situated in the latter valley.

Beyond Hsin-p'u-kou hui 新堡溝水 are two more valleys the Lo-ts'ang kou 洛藏溝 and the Kao-lou kou 高樓溝. West of T'ao-chou old city in Tibetan country are the valleys of Ang-hkhor, the Chinese Wan-kou the upper part of which belongs to the Cho-ni prince, then mTshams-rdo west of Ya-ru Gom-pa, written Yag-rug-dgon-pa, and the rJe-li nang.

Still further west is the largest affluent from the north the rDog Chhu which has its source in the grasslands of the A-mchhog country, it flows north of, and parallel to the T'ao River, to near hBo-ra monastery when it turns directly south and debouches into the T'ao river. One other tributary the rJe-tshang River which has its source north of Shi-tshang-dgar-gsar, in the rJe-lung Nye-ra (pass) flows east and becomes the Me-shi Khog, but where it enters the T'ao River is not known to me.

All these northern affluents have their origin in loess-covered, rounded spurs, clothed with grass but at their exits harbor shrubs of a xerophytic nature.

The Valleys Skirting the Southern Slopes of the Kan-su Min Shan
In a similar manner as the T'ao river encircles the Min Shan in the north, the Pai-lung Chiang skirts it on its south side with the difference, that the T'ao River has its source in another range further west, while one of the Pai-lung Chiang's sources is on the Min Shan proper, although it is the smaller of the two branches; thus it may be actually considered an affluent of the main Pai-lung chiang which has its source in Ssu-ch'uan, west of the Upper T'ieh-pu country. Ssu-ch'uan drives here a wedge between Ch'ing-hai and Kan-su, but is absolutely unadministered, being in the lawless territory of the Ta-tshang-lha-mo Tibetans who are The-wu, like their confrères in The-wu land but follow the profession of bandits when not engaged in agricultural pursuits.

The Pai-lung Chiang or White Dragon River, its Tibetan name being Chhu-klung-dkar-po or White River is known by that name after it passes the monastery of bKra-shis dgon-pa and the valley of rDo-ro-phu which debouches as stream into it on its right or south bank. From its source at sTag-tshang-lha-mo it is known as the rJe-khog, there it receives an affluent from the south called hBrong-khog after the confluence of which it is known as the Tshong-ri nang, to the south of this it receives an affluent called the Za-ri Khog and beyond the latter it is known as Pai-shui Chiang or the White Water River. This is the longest and western part of the stream. The branch which has its source south of the Min Shan, below Mount Kuang-k'e 光克 is known as the Brag-sgam-nag pronounced Drag-gam-na stream to where it enters a rock gate immediately south of Drag-gam-na whence, after entering a defile, it is called the Yi-wa kou 亦哇溝 ; it flows here in a narrow forested valley till its confluence with the Pai-shui Chiang coming from Tag-tshang-lha-mo. At this confluence the valley is broad but un-cultivated. It is a dangerous spot for here the lawless Tibetans from Tag-tshang lha-mo hide in the bushes of the streambed and attack caravans or travelers going south or up Kuang-k'e. All the land to the west of the confluence, as the ridge hemming in the stream to the west, is no-man's land.

About three miles or ten li before the Pai-shui Chiang's confluence at the meadow called Chhe-khu-kha, the White Water River receives two streams, one issuing from a valley called mTsho-ru kou the Chinese Tso-lu kou 作路溝, and the other below it coming from a valley called Tsha-ru kou the Chinese Ch'a-lu kou 茶路溝. The forested mouth of Tsha-ru kou is 7,850 feet above sea level; further up its slopes are grass-covered and partly terraced and a village is not far beyond its mouth, the former is in Ssu-ch'uan and not under the Cho-ni prince's control, and as Sung-p'an, the last town and magistracy in northwest Ssu-ch'uan could not, and I might well say even now, cannot control that area, the inhabitants are a law unto themselves. The country further south to Tag-tshang-lha-mo is undulating, and indicates loess-covered plateau which merges into the grasslands of the Yellow River. Beyond the confluence the main stream, here issuing from a defile, is called the hGro-tshug and thence gDon-hgra-sle Chhu up to the monastery of Tra-shi gomba (bKra-shis-dgon-pa). It is called both the Pai-shui Chiang White Water River and Pai-lung Chiang or White Dragon River in Chinese. The River continues in an east-southeast and easterly direction. It receives a number of affluents on both sides. Its north bank is hemmed in by the steep slopes of the Min Shan, while to the south a long, undulating low range forms its valley wall. The affluents on the south side come from valleys which extend more or less parallel to the south of the valley wall, beyond which there is another parallel range as yet unexplored, as is the country to the south of it which merges into grassland.

The upper slopes of the valley of the Pai-shui Chiang facing south are formed by the main backbone of the Min Shan which is of gray hard limestone, while the lower slopes are conglomerate, gravel, schist and loess, only the lateral spurs of the Min Shan which cut or extend across the valley south, are of the same gray hard limestone and through these the river has cut itself deep gorges which form the famous rock gates, one of the characteristics of this region.

Peaks occur also in the southern spur flanking the Pai-shui Chiang and these peaks, although lower are again limestone, like Tshwa-ri-ma-smon pronounced Tsha-ri-ma-mön, southwest of the ancient site of T'ieh Chow 疊州 but is Ssu-ch'uan; the intervening spurs and hills are still lower and are of schist and gravel covered with loess. The vegetation takes on a xerophytic character composed of low shrubs while the valley slopes facing north are forested with pines (*Pinus tabulaeformis*).

None of the geographic data is to be found either on existing maps or in the literature as the region was entirely unexplored.

The Northern Affluents of the Pai-lung or Pai-shui Chiang
There are three larger affluents and several smaller ones which have their sources on the southern slopes of the Min Shan. In the west it receives the Wa-pa River which has its source southeast of Drag-gam-na below a pass 11,350 feet above sea level; it flows in the valley of Wa-pa (kou) 哇巴溝 first southeast, and then south-southwest, and debouches opposite the high peak (Tshwa-ri-ma-smon) Tsha-ri-ma-mön, flowing in a

deep gorge. East of it is a high terrace on which are situated two monasteries, one above the other. The district or the region is here called Pa-shih-te-ka and the monasteries sPag-shis-gong-ma dgon-pa in Tibetan and in Chinese Tien-ha shang ssu 殿哈上寺 the lower and dGong-ma-nang dgon-pa the upper, in Chinese Tien-ha hsia ssu 殿哈下寺 respectively. The second large valley is in the east and extends from the pass Lha-mo-gün-gün at an elevation of 11,250 feet directly south under the name of Ma-ya kou 麻牙溝 near (west) of the village of Ma-ya into the Pai-shui Chiang.

The third is a very long valley which extends through many rock gates and defiles under various names southeast, whence its stream flows through the long valley known as bSam-pa, the Chinese San-pa kou, in the same direction out of the Cho-ni prince's domain as the White Water River which it joins. These valleys are very rocky and arid, harbour only scrub vegetation in their lower sections, while higher up they form magnificent limestone gorges forested with conifers and deciduous trees.

The Southern Affluents of the Pai-lung Chiang

The largest affluents are in the Lower T'ieh-pu or Hsia T'ieh-pu 下鐵布 country. The first is the Ta-ra nang written rTa-ra-nang the Chinese Ta-la kou 達拉溝 inhabited by the fierce and lawless Ta-ra The-wu. Although it was in the Cho-ni prince's realm, he was powerless to control them. It is a long valley which extends deep into the mountains, but like the next valley Wang-ts'ang kou 旺藏溝 further south, extends parallel to the main Pai-shui Valley towards the west, where it borders on the grasslands of Ssu-ch'uan in no-man's-land. This, like Wang-ts'ang valley, is densely forested. The longest valley in the east is Do-ro nang (rDo-ro nang) the Chinese To-erh kou 多兒溝; this valley which is joined by the Â-cha Nang (Ah-bya-nang) the Chinese A-hsia kou 阿夏溝, has its source in Yang-pu Shan 陽布山 which forms the divide between Cho-ni in Kan-su and Ssu-ch'uan. It, like A-hsia kou, is more or less arid except at its head waters where conifer forests abound. It flows from southeast to northwest and is the longest of all the Pai-shui Chiang affluents.

The T'ao River Valley and Its Vegetation

The flora of the beautiful valley of the T'ao River may be divided into several definite zones. First, the shrubs and trees and herbaceous plants confined to the banks of the stream itself, the flat areas in the valley covered by groves of tall poplars, and the herbaceous plants found in its meadows. This would comprise the plants confined to the valley floor.

There is a shrub vegetation restricted to well drained lower rocky slopes which hem in the valley; this vegetation may almost be termed a xerophytic one especially that found on the loess-covered hills which flank it on the north; a hundred feet or so above the valley floor this shrub vegetation gives way to grass-covered slopes. On to south walls of the valley however conifer forest descends to almost the valley floor or begins above the shrubby plants which hug the lower slopes. This region gives way to pine. In the mouths of the numerous valleys which debouch into the T'ao valley from the south we have first groves of large poplars but these relinquish the land to a mixed forest which increases further up the valleys.

Higher up, what one may call the middle zone, the larger fruiting spruces give way to *Picea purpurea*, this species which extends to over 10,000 feet elevation is in places associated with Abies or firs.

Above this zone, a considerable distance up the longer valleys, their place is taken by junipers, although certain junipers are often confined to the mouths of the valleys. With them occur Rhododendrons of the broad-leafed species.

The next and last zone constitutes the alpine one which must be divided into an alpine meadow flora and a rock and scree flora. In a broad sense these zones are distinct and range from 8,000 feet to nearly 14,000 feet on the main backbone of the Min Shan, entirely composed of limestone. This has mainly reference to the northern slopes of the Min Shan which in areas is less rich in species than the southern slopes which are dealt with separately.

The T'ao River enters Cho-ni territory at about 12,000 feet elevation by Shi-tshang-gar-sar where it winds through narrow rocky gorges, and at Ch'e-pa kou flows at 9,080 feet, a drop of about 3,000 feet. Shi-tshang-gar-sar lamasery is actually controlled by Cho-ni but not the land it stands on nor the Tibetan tribes living there. Cho-ni territory proper begins at Ch'e-pa kou (valley) one of the longest up which a trail leads south over the Min Shan into the The-wu land. At Ch'e-pa kou the T'ao River describes a wide, sweeping curve with broad and partly wooded islands in the river, the hills are low and bare beyond, west, of Ch'e-pa kou, but partly wooded on the southern valley wall. The valley is here comparatively broad and cultivated by the Tibetans of Ma-ru village situated at the foot of the hills facing north, i.e., south of the T'ao River.

Between the mouth of the Ch'e-pa kou and Cho-ni the River describes many curves around projecting spurs of the Min Shan, with alluvial fans at the mouths of emerging valleys where usually villages are situated and cultivation is carried on. The drop in the River between Ch'e-pa kou and Cho-ni amounts to nearly 1,000 feet.

Beyond Cho-ni, to the mouth of Ta-yü kou, the easternmost valley through which a trail leads over the Min Shan into Hsia or Lower The-wu land the river falls another 500 feet, the distance being short in comparison to that between Ch'e-pa and Cho-ni. South of the T'ao River the mountains rise steeply from the valley floor and are intersected by many valleys all forested to the very heads, while on the north bank the mountains are low and grass-covered, except at their bases which harbour a xerophytic scrub vegetation.

Meadows are encountered everywhere in the T'ao Valley and in the summer are studded with beautiful flowers. It may be remarked here that while the T'ao River in Cho-ni territory flows between 9,000 and 8,000 feet elevation, the Pai-lung Chiang which skirts the Min Shan to the south flows nearly 2,000 feet lower than the T'ao, the climate therefore being considerably warmer.

Spring does not commence in Cho-ni till June, while the T'ao River remains frozen from November till April, ice bridges over which bullock carts can cross occurring till then.

The immediate mountain spurs which hem in the T'ao River are composed mainly of shale, schist, gneiss and covered with loess especially those of northern valley wall, while limestone is found at the head of the valleys where the spurs composed of the above mentioned geological formation adjoin the limestone, pushed-up through this loose material.

The Vegetation Along the T'ao River Bank West of Cho-ni
The greater part of the vegetation along the actual stream and its immediately adjoining flat areas is composed mainly of willows and poplars. Of the former x *Salix taoensis* Goerz, a new hybrid, *Salix Wilhelmsiana* M.B., *Salix wallichiana* And., and *Salix sibirica* Pallas, are the prevailing species while of the latter *Populus Simonii* Carrière, forms pure stands; it is represented by trees of all ages from small saplings to huge trees occupying the large flats found in the valley near Cho-ni and beyond Shao-ni kou 勺尼 溝, also throughout the valley and especially on the alluvial fans at the mouths of affluents. *Myricaria dahurica* Ehrenb., grows in the sand and gravel and among the rocks with *Viburnum mongolicum* a shrub 4 feet tall with inconspicuous pale green flowers. *Scorzonera austriaca* Willd., a yellow flowered composite prefers the sandy moist bank with *Linum nutans* L. and *Potentilla anserina* L., a spreading herb, all three first described from Europe. The flat rosette forming *Aconitum gymnandrum* Max., a deep rich purple flowered species when growing in waste places is often 3 feet tall, while here in the sand it adopts the above mentioned lowly habit. The labiate *Nepeta macrantha* with deep purplish-blue flowers first known from the Altai mountains is partial to the river bank.

Scattered here and there with willows on both banks of the river we find the shrubs *Lycium chinense* Mill., the 4 foot tall, yellow *Rosa xanthina* Lindl., and *Cotoneaster multiflorus* Bge. 15 feet tall and white flowered, the latter two invade also drier, rocky situations along the river. *Cotoneaster adpressus* Bois, with red flowers and flat spreading habit is common on the sandy bank as well as on rocks at 8,300 feet elevation west and east of Cho-ni, while the black-stemmed and white-flowered *Cotoneaster acutifolius* Turcz. var. *villosulus* Rehd. & Wils., grows to a height of 8-10 feet; it was first collected in Hu-pei (Hupeh) province; while the former was described from plants grown from seeds introduced from China.

Very common along the river not only here but elsewhere along streams, brooks, etc., is the buckthorn *Hippophaë rhamnoides* L., which often forms impenetrable thickets, becomes a small tree when growing isolated or, on high mountain slopes, is reduced to a low bush less than one foot in height.

Shrubs in general are found both along the stream as well as above the rocky banks and to these belong *Caragana maximovicziana* Kom., yellow flowered, and reaching a height of 3-4 feet often forming low thickets, and the much rarer *Caragana densa* Kom. *Berberis dasystachya* Max., 4-5 feet tall with pale green leaves, ascends however also into the pine forest which covers the dry steep slopes of a tributary valley, with *Berberis parvifolia* Sprague, a shrub 2-4 feet and brilliant red fruits, a very handsome species. *Rosa omeiensis* Rolfe, first described from Ssu-ch'uan which has here white flowers and forms dense bushes 5-6 feet tall. *Prunus padus* L., var. *commutata* Dipp., attains the size of a small tree up to 22 feet, its flowers are white and its fruits black, it is common everywhere in the T'ao River valley, flowers in May and fruits in September to

October. *Corylus sieboldiana* Bl. var. *mandschurica* Schneid., the Manchurian hazelnut with bristly edible nuts, occurs here and there along the banks overhanging the streambed with *Ostryopsis davidiana* Dcne., belonging like the former to the birch family and found first in Mongolia by Père David, it reaches 4-5 feet in height. The in Kan-su and Ch'ing-hai widely distributed *Lonicera syringantha* adorns also the banks of the T'ao and scents the air with its fragrance. The shrubby cherries *Prunus setulosa* Batal., and *Prunus stipulacea* Max., both endemic in Kan-su, the former with purplish-red flowers and the latter with pink flowers are shrubs which branch from the base; the latter has dull green leaves and ovoid oblong red drupes. It ascends also into Abies and Rhododendron forest in the lateral valleys both south and north of the Min Shan, while the former is confined to the valleys near the streams. Not previously recorded from Kan-su and originally a native of Yün-nan is the otherwise here common *Spiraea canescens* G. Don var. *glaucophylla* Fr., with cream-colored flowers, 4-5 feet in height, and *Potentilla fruticosa* L. var. *dahurica* Ser., a low shrub with white flowers.

Other shrubby Lonicera, are *L. aemulans* Rehd., and *L. heteroloba* Batal., the first with rich yellow and the second very ornamental with dark red flowers usually prefer the banks of streams but the latter also ascends into the spruce and fir forests. *L. chrysantha* Turcz. var. *longipes* Max., which disports orange yellow flowers is confined to the valley where it forms shrubs 4-5 feet tall together with *Lonicera szechuanica* Batal., 4-5 feet tall, *Potentilla sericea* L. and *Potentilla Potaninii* Wolf. both shrubs with golden yellow flowers. *Spiraea longigemmis* Max., with its creamy-white, fragrant flowers occurs both on the banks of the T'ao as well as in the forests higher up, and the same holds good of the most ornamental of all the shrubs found in this region, the mock orange *Philadelphus pekinensis* Rupr. var. *kansuensis* Rehd., its beautiful, large white, fragrant flowers open in July and perfume the pure air and embellish the landscape; it is the only representative of the genus, but is one of the most common shrubs all over the Min Shan extending from river banks into dense forests of tributary valleys, and is at home in both damp and arid situations.

Of conifers confined to the streambed of the T'ao is *Juniperus formosana* Hay., a small tree 10 to 15 feet tall it has glaucous leaves and globose bluish-black fruits which ripen end of October; it does however extend higher into spruce forest or their outskirts.

This accounts for the shrubby vegetation on the banks of the T'ao River, whose valley is of great scenic beauty, the river flowing often in the center of the valley, then again at the foot of high cliffs, or it branches and leaves islands wooded with the mentioned willows ands other shrubs in the middle of the valley floor. Of great beauty are the old, venerable *Populus Simonii* Car., trees whose groves resemble large open parks, especially attractive in the autumn when the yellow leaves create a delightful contrast to the deep green of the spruces of the hillsides.

Interspersed throughout the T'ao River valley are lovely juicy green meadows all with a flora of their own.

Meadow-Flora of the T'ao River Valley
While the meadow flora of the T'ao Valley at 8,500 feet is not particularly rich it makes up in color of the species found. *Iris ensata* Thunb., with flowers ranging from white to

blue and yellowish-green is the most common herbaceous plant in the T'ao Valley, it grows in meadows and along roads displacing grass but not to the extent it does in the north among the parallel ranges of the Nan Shan where it covers all flat expanses in the valley flora. With it grows *Iris gracilis* Max., a small lavender blue species but more confined to grassy slopes; the leaves of the former are broad and glaucous.

Ranunculus affinis R. Br. var. *flabellatus* Franch., is common with grasses near Choni, while *Ranunculus pulchellus* G. A. Mey., and *R. flammula* L., are partial to swampy meadows as are *Caltha palustris* L., and *Caltha scaposa* Hook. f. & Thoms., the latter first known from Sikkim as is the widely spread, pink to lavender-blue flowered *Primula sibirica* Jacq. Other open meadow plants loving sunshine that came to our notice were the columbine *Aquilegia ecalcarata* Max., with reddish purple flowers, and *Ajuga calantha* Diels really a lovely labiate with dark green appressed leaves forming a rosette, in the centre of which rises a beautiful little bouquet of deep blue purplish flowers reminding of violets; it is confined more to the drier meadows and extends up the lateral valleys, but is not so very common nor gregarious. *Pedicularis muscicola* Max., with brilliant purple flowers formed beautiful cushions in the rich green meadows associated with the white flowered scrophulariaceous herb *Legotis brachystachys* Max., various Polygonum, *Thalictrum alpinum* L., and *Pedicularis cheilanthifolia* Schrenk var. *isochila* Max., with yellow, and *Pedicularis bonatiana* Li, with wine-colored flowers. The crucifer *Torularia humilis* (C. A. Mey.) O. E. Schulz, *Glaux maritima* L., both common and gregarious the scrophulariaceous *Euphrasia hirtella* Jord., and *Carex atrofusca* Schk., a widely distributed species extending from here to the Koko Nor and the arctic regions of Europe and North America, made up the rest of the turf-loving plants.

Those which prefered gravelly, well drained areas, were the papaveraceous *Hypecoum erectum* L. var. *lectiflorum* (Kar. & Kir.) Max., *Triglochin maritimum* L., a juncacious plant also extending into moist meadows where it forms thick carpets, and a lank weedy, yellow-flowered Corydalis as yet not determined no 12314.

In meadows but in the shade of willows thrived *Gentiana leucomelaena* Max., a white to pale blue erect species, *Ajuga lupulina* Max., its whitish-blue flowers hidden under greenish, cream-colored bracts, the leguminous *Oxytropis taochensis* Kom., a spreading, rosette forming plant with blue to purplish flowers, a solanaceous herb of the genus Anisodus with large greenish-yellow flowers as yet not determined no 12337, and the leguminous *Thermopsis lanceolata* R. Br. one foot or less in height, yellow flowered and most common under willows and poplars. All these plants frequent an altitude of 8,500 feet but some, as already remarked ascend higher.

The valley slopes to the north or left of the river, especially in the eastern and western ends of the valley, harbour a more xerophytic type of vegetation than those to the south of the river and so we find in the dry slate and shale rubble *Lonicera Ferdinandi* Franch., first known from Mongolia, *Caragana maximovicziana* Kom., *Berberis dasystachya* Max., *B. vernae* Schn., *B. parvifolia* only 2-4 feet high, *Rosa wilmottiae* Hemsl., *Berberis Silvataroucana* Schn., the latter preferring the fine loess slopes, as does *Juniperus formosana* Hay., rather than rocky situations. Here also belongs *Betula japonica* Sieb. var. *szechuanica* Schneid., a tree 60 feet tall, with a trunk

2-3 feet in diameter and a silvery grey to blackish bark. Here also we find *Ribes glaciale* Wall., and *Sibiraea angustata* (Rehd.) Hao.

On the loess banks and plains especially on the south bank of the T'ao River are a number of rosaceous trees of great beauty, they are especially common near the mouth of La-li kou 拉力溝 where the forests descend to near the river. Of special interest is the new *Crataegus kansuensis* Wils., (see Plate **1**) with deeply lobed leaves and deep globose tomato-red fruits; it reaches a height of 20 feet and is quite ornamental as are *Malus kansuensis* Schneid., of the same size as the former, but with velvety tomentose, slightly lobed leaves and dull red oblong fruits, extending also up the valleys to 9,500 feet, and *Malus toringoides* Hughes a fairly large tree over 40 feet high with small oblong cherry-like fruits. It occurs north and south of the Min Shan and especially here, that is west of Cho-ni, extending up K'a-cha kou (valley) where it grows on the outskirts of pine forest (*Pinus tabulaeformis* Carr.), and there reaches a height of 75 feet. Otherwise it forms groves with the above and a new hybrid *Malus kansuensis* x *M. toringoides* Rehd. hybr. nov. Near by *Prunus Padus* L. var. *commutata* Dipp., forms lovely woods, in the shade of which in July, *Delphinium tatsiense* Fr., more than four feet tall flourished. Another associate though less common here proved to be *Pyrus pashia* Ham., a tree 40 feet tall with fruits one inch in diameter, red and yellow on stout peduncles. It certainly is different here from the forms which are met in Yün-nan and Burmah and I doubt that these northern types belong to the same species. Most common of all is the spiny compact shrub, 6-8 feet in height *Prunus tangutica* (Bat.) Koehne, with dry dehiscent velvety fruits, the Tangut peach; this species is a most desirable stock plant, drought resisting, with a long tap root, it could be used as a stock for peaches in arid regions; it can stand both heat and intense cold. With it occurs *Prunus tomentosa* Thunb., as common as the former; it was first known from Japan, but occurs also in Yün-nan and extends north to Inner Mongolia.

On shady banks of the southern slopes of the valley grow the white-bracted *Hydrangea Bretschneideri* Dipp., a shrub or small tree, first known from northern China and Mongolia, and the caprifoliaceous, *Triosteum pinnatifidum* Max., which loves deep shade along streams with *Paris polyphylla* Sm., also known from Yün-nan, but here less common. *Spiraea longigemmis* Maxim., grows here scattered but prefers the open grassy slopes.

Between wooded areas on the valley slopes, grassy embankments are taken up with a herbaceous vegetation of *Dracocephalum tanguticum* Max., *Thalictrum baicalense* Turcz., *Iris gracilis* Max., at 9,000 feet and higher, *Serratula centauroides* L., a native of Siberia belonging to the Compositae, *Silene repens* Patr., and *Pedicularis rudis* Max., a yellow-flowered species. Others found on the banks of the T'ao are *Oxytropis ochrantha* Turcz., *Cynanchum inamoenum* (Max.) Loesn., a yellow flowered asclepiad, the umbelliferous *Pleurospermum Franchetianum* Hemsley, and *Dracocephalum imberbe* Bge., originally known from the Altai mountains; this latter species often grows in tussocks on gravelly slopes above the river, its deep bluish purple flowers and tussock forming habit would make it an attractive addition to a rock garden.

Later in the summer appear other herbaceous plants as *Vicia unijuga* A. Br., *Potentilla sericea* L., the composite *Stereosanthus Souliei* Franch., *Geranium* aff.

pratense L., *Spiraea longigemmis* Max., a pubescent form of the species, *Potentilla bifurca* L., a native of Siberia, *Sanguisorba canadensis* L., and others.

The rocky hillsides of the T'ao valley are taken up with *Prunus tangutica, Pr. tomentosa* Thbg., the lovely *Lonicera heteroloba* Batal., with deep red, pendant flowers which also grows on the outskirts of forests, *Sorbus Koehneana* Schn., a white flowered species first collected in Hupeh and the pinkish-flowered *Clematis gracilifolia* Rehd. & Wils., entwining shrubs. *Rhamnus leptophylla* Schn., a gnarled stiff shrub with greenish flowers occupies the scrub thickets as does the new variety var. *scabrella* Rehd., with larger leaves. In open scrub thrives *Asparagus trichophyllus* Turcz., *Asp. brachyphyllus* Turcz., both originally known from North China, *Euonymus nanoides* Loes. & Rehd., a west Szechuan plant, the crucifer *Eruca sativa* Lam. var. *lativalvis* (Beiss.) Goss., which prefers however the finer sandy slopes, and *Torularia humilis* proles *Piasezkii* f. *grandiflora* O. E. Schulz also a crucifer, but confined to the dry scrub-covered slopes of the valley. Near Cho-ni, *Primula conspersa* Balf. f. & Purd. its flowers a loud red-purple and pale green glabrous leaves first collected by Purdom[4] and belonging to the section Gemmifera, also found on Lien-hua Shan q.v., brightens the boggy meadows, while *Scutellaria amoena* Wright, with lavender flowers flourishes everywhere on the drier grassy slopes.

An undescribed *Thalictrum* (no 12567) occupies with the yellow flowered *Pedicularis alaschanica* Max., *typica* Li, a pink *Aster* sp? and *Asperula odorata* L., a rubiaceous herb with pale blue flowers, the rocky slopes at Cho-ni.

The earliest to flower in May on grassy banks and slopes are the liliaceous *Gagea pauciflora* Turcz., *Iris tenuifolia* Pall., with blue flowers, *Oxytropis falcata* Bge., a deep pinkish-purple flowered legume, *Chrysosplenium sphaerospermum* Max., a yellow saxifrage and *Androsace Mariae* Kan. var. *tibetica* (Max.) H.-Mzt., which forms compact cushions. It is very attractive on account of its pale pink to white flowers and habit of growth, it extends also to gravelly areas and into scrub forest, with *Primula stenocalyx* Max., a purplish blue species. The yellow *Astragalus scaberrimus* Bge, also encroaches on to the sandy banks of the river proper, as does the Edelweiss *Leontopodium nanum* (Hook. f. & Thoms.) H.-Mzt. *Prunus salicina* Thunb., first known from southern China grows here and there with the new willow *Salix cereifolia* Goerz, opening its white flowers in May, and fruiting in October; the drupes are pruinose and edible. The giant poplar *Populus Simonii* Carr., over 80 feet tall appears bright red in May due to its staminate catkins; it is the *P. Przewalskii* of Max., described in 1882, while it was first described as above in 1867. Associated with it is *Prunus stipulacea* Max., also May flowering, a shrub 6-8 feet, often becoming a small tree 15 feet tall. It is associated with willows and birches, its fruits, borne singly, of a purplish red and bitter to the taste, form in late September or October; it is a delightful plant on account of its lovely pink flowers.

An undescribed *Iris* sp? (no 12138) with dark purple flowers abounded also on the grassy slopes about Cho-ni at 8,500 feet and another equally undescribed *Iris* sp? (no 12851) with reddish flowers forms clumps, on grassy slopes but only between the

---

[4]   See Heriz Smith: William Purdom (1880-1921): a Westmorland planthunter in China. *Hortus* 10 (no. 38:2).1996, 49-62.

valleys of Po-yü kou and Ta-yü kou to the east of Cho-ni, extending to 9,000 feet elevation.

On the western slopes of the T'ao River valley on the outskirts of forests we find the yellow flowered *Salvia Roborowskii* Max., and the erect growing deep-purplish blue flowered *Pulsatilla ambigua* Turcz., resembling a Clemati*s*.

Elsewhere in meadows in the valley occur the composites *Leontopodium Smithianum* H.-Mzt., and *Anaphalis Hancockii* Max., with the very common *Stellera chamaejasme* L., at 8,500 feet elevation. *Senecio argunensis* Turcz., formed tussocks with a species of *Codonopsis* not yet described (no 13196) whose deep red flowers open in August.

Lonicera microphylla Willd., a fragrant yellow flowered shrub 4-6 feet, a native also of eastern Siberia is common on the northern bank of the T'ao at 8,600 feet with *Berberis mouillacana* Schneid., on loess hills, formerly known from Szechuan. It extends also to the Koko Nor and Nan Shan facing Mongolia. *Potentilla fruticosa* var. *parviflora* Wolf., with tiny leaves and golden yellow flowers, was especially common near the native ferry across the T'ao opposite K'a-cha kou (valley).

Geranium eriostemon Fisch., and the lovely blue *Corydalis curviflora* var. *pseudo-Smithii* Fedde, open their flowers in July, while at the same time *Lonicera coerulea* L. var. *edulis* Reg., produces its elongate, horn-shaped, pruinose fruits. It is fairly common on the south bank of the river opposite Cho-ni; its fruits are used in the making of a delicious preserve. Cultivated in the lamasery of Cho-ni because of their ornamental value are the small-leaved *Syringa microphylla* Diels, a shrub or tree 8-20 feet tall with lavender to pink flowers, the lilac *Syringa oblata* Lindl. var. *affinis* Lingelsh., with white flowers. *Paeonia suffruticosa* the true species whence all the other horticultural forms are derived was planted in beds in the lamasery compound. It is a beautiful shrub 4-5 feet tall (in cultivation in England it reaches 9 feet from seed introduced by the author); its petals cream-colored with dark purple blotches at the base. It also occurs wild in the mountains of the The-wu country and Wu-tu Hsien whence the cultivated plants come from, seeding prolifically (see Col. Stern's paper ...)[5]

The Mountain Flora of the T'ao Valley

Close to the scrub vegetation at the foot of the valley slopes follow the forests composed mainly of conifers. Immediately after the scrub forest is a transition type which thrives above the former on the outskirts of the conifer forest. The steep slopes of the spurs which drop into the river valley are mostly schist with here and there limestone outcroppings. On these slopes occur various species of Picea as *Picea asperata* Mast., the most common one which, in its various stages of growth, one would never recognize as one and the same species. I am still of the opinion that Wilson[6] who

---

[5]   Frederick Claude Stern: *A study of the genus Paeonia*. With 15 illustrations in colour by Lilian Snelling and Stella Ross-Craig. London: Royal Horticultural Society 1946. VIII, 155 pp.

[6]   Ernest Henry Wilson, 1876-1930, was in China 1907-1910 and in Japan 1914-1915 and 1917-19 as botanical explorer and collector; as of 1919 he was deputy director of the Arnold Arboretum. See Edward Irving Farrington: *Ernest H. Wilson, plant hunter*. With a list of his most important introductions and where to get them. Boston: The Stratford Co 1931. XXI, 197 pp., 34 pl.; Alfred Rehder: Ernest Henry Wilson. *Journal of the Arnold Arboretum* 11.1930, 181-192 (with portrait and list of publication).

worked up the conifers found them too difficult, and lumped them. Most systematists working from dried material are apt to do this without a knowledge of the living plants; one case in point is the work done lately on Meconopsis where one of the most distinct species peculiar to Yün-nan was relegated to, or united with, one with which it has only the color of the flowers in common. I shall come back to these species when discussing the alpine plants of the Li-chiang Snow-range.

The conifers occuring on the spurs north of the T'ao River extend below their summits which are much lower than the mountains to the south of the depressions between them. On the southern slopes the conifer forest extends high up the spurs to 11,500 feet, above which Rhododendrons take over to near the top at 12,000 feet elevation.

Juniperus formosana Hay., a small tree ascends into the lower forests up to 9,500 feet from the valley floor, and there it is associated with *Salix taoensis* Goerz. Above them we come to *Picea Wilsonii* Mast., and *Picea asperata* Mast., the former 60-80 feet tall with drooping branches, the latter species exceeds here the former in height. Unlike in the far north where *P. asperata* forms pure stands, here the spruces are associated with Abies as *A. sutchuenensis* Rehd. & Wils., which joins the spruces at about 10,000 feet, while the spruces themselves descend to 8,000 feet across from Cho-ni. On the outskirts of Picea forest at 10,000 feet or lower are various shrubs as the white flowered *Viburnum veitchii* C. H. Wright, the rosaceous *Maddenia hypoleuca* Koehne, the new *Salix alfredi* Goerz, *Lonicera tangutica* Max., 4-5 feet tall with pale pink flowers suspended on long pedicels and leaves hairy on the margins; this species loves the borders of meadows in forest clearings, in its shade occurs often the berberidaceous pink-flowered *Podophyllum emodi* Wall., first known from the Himalaya. On these clearings we also encounter, but at the 9,000 foot level, *Senecio campestris* D.C., species of Gentianella of the section Crossopetalum, the deep yellow *Trollius pumilus* Don, the crassulaceous *Sedum aizoon* var. *scabrum* Max., whose flower buds are red, then *Lithospermum officinale* L., and *Anemone rivularis* Ham., a white flowered species and the blue boraginaceous *Lappula redowskii* (Horn.) Greene.

Along small streams in the forests we find *Daphne Giraldii* Nitsche, with yellow flowers, and 2-3 feet in height, also *Geranium Pylzowianum* Max., and *Anaphalis lactea* Max., the latter two also in open meadows in forest clearings, while on shady banks occurs *Lathyrus pratensis* L., and the wine colored *Allium Przewalskianum* Reg., which thrives on grassy slopes on the outskirts of forests. The brilliant orange to firy red *Lilium tenuifolium* Fisch., up to 2 feet tall loves the banks of rivulets and margins of forests. Among limestone outcroppings at 8,500 feet we find *Clematis aethusifolia* Turcz., first discovered in Mongolia, climbing over shrubs or trailing on the ground and spreading over considerable areas.

In the spruce forest occur *Clematis fargesii* Franch., with large white flowers, *Aquilegia oxysepala* Trautv. & Mey., here a rather rare plant, with the common shrub *Ribes Meyeri* Max., *Rosa omeiensis* Rolfe, a shrub 4 feet with scandent branches, and the beautiful *Philadelphus pekinensis* Rupr. var. *kansuensis* Rehd., at 9,800 to 10,000 feet elevation; in its shade and those of willows in moss, the dark rose red *Paeonia veitchii* Lynch, less common than *P. anomala* L., brightens the otherwise somber forest.

At the 10,000 feet level *Picea purpurea* Mast., a tall slender species with small purplish black cones forms almost pure stands but often mixed with, or growing on the margins either above or below it, we find the more than 30 feet tall *Juniperus saltuaria* Rehd. & Wils.

At 11,000 feet south of Cho-ni, on the northern slopes of the Min Shan, within Abies forest or on the grassy upland slopes, we meet with the broad-leaved *Rhododendron rufum* Batal., a very slow growing species its flowers a lovely rose red to white with pinkish tinge reaching a height of 15 feet. Those introduced by me into cultivation, now over 27 years ago are only about six feet high and have as yet not flowered. Thus the trees encountered on the upper slopes of the Min Shan must indeed be several hundred years old. Its constant associate in this region is the lower shrubby, 5-6 feet tall, *Rhododendron Przewalskii* Maxim., its flowers white and densely packed into a compact inflorescence. It is not a handsome species and is probably closely related to if not identical with *Rhod. agglutinatum* Balf. & Forr., but Millais' *Rhod. kansuense* is probably identical with it.

It is easily distinguishable from afar by its paler green foliage and bright yellow petioles, its leaves are usually horizontal giving the bushes a flat, round appearance.

On meadows among Rhododendrons and firs the leguminous *Thermopsis alpina* Ledeb., flourishes, its rich yellow flowers lending color to the landscape as early as May, before the Rhododendrons have burst into bloom or at least *Rhod. Przewalskii* Maxim. *Rhododendron anthopogonoides* Max., a shrub 2-3 feet with white tubular flowers and very aromatic leaves reminding on the odor of Eucalyptus, is prevalent on the margins of spruce forest at 10,000 - 11,000 feet, while *Rhod. capitatum* Max., with deep purplish blue flowers is confined to open grassy hillsides at the same elevation, with *Lonicera hispida* Pall. Here also belongs *Sorbus tapashana* Schn., a tree reaching 20 feet, with large umbels of white flowers, and *Spiraea alpina* 2-3 feet in height, flowers creamy white, but dark red in bud. In the mossy spruce forest or among rocks and on trees, the ranunculaceous white flowered herb *Souliea vaginata* (Max.) Franch., is at home, at 11,000 feet above sea-level. In the dense shade of the spruce forest in moss or under Berberis grows *Primula polyneura* Franch., like its relative in Yün-nan *Pr. lichiangensis* both wine-colored to purple flowered.

In the lateral valleys which extend from the crest of the range north, into the T'ao River Valley a similar vegetative covering, but still more varied and richer in alpines is to be found especially at the heads of the three long valleys which merge into the passes leading over the range. We shall now leave the T'ao River Valley and enter the most interesting and at the same time the richest of all the valleys, namely K'a-cha kou 卡札溝; R. C. Ching[7] calls it Kar-ching K'ou which is the guttural Kan-su pronunciation of the first character; instead of the syllable cha he erroneously wrote ching, and for kou = valley he writes k'ou which is incorrect for it means mouth and not valley. The Kan-su people use also the word 谷 ku for valley, more so than kou. Walker in his paper on

[7] R. C. Ching, 1898-1986, (Ch'in Jen-ch'ang) 秦仁昌: A botanical trip in the Ho Lan Shan, Inner Mongolia 賀蘭山植物採集記略. *Bulletin of the Fan Memorial Institute Biol. Bot. Series* 10.1941, 257-265 – See also *Selected papers of Ching Ren Chang* 秦仁昌論文記. China Science Press Press 1988. X, 366 pp.

the plants collected by Ching[8] romanised the first character ch'iao which is however
not the reading for the character as here employed. On the Chinese General Staff Land
Survey map the name is erroneously given as K'a-ch'e kou 卡車溝.

The K'a-cha kou or K'a-cha Valley and Its flora

The name of the valley is derived from two prominent villages situated some 12 miles
up the valley, namely Kha-rgya ya-ru or Upper Kha-rgya, and Kha-rgya ma-ru or Lower
Kha-rgya pronounced Kha-dja (gya).

The first village at its mouth on the right bank of the Kha-gya stream is Ta-tzu-to 達
子多 situated at an elevation of 8,400 feet. The stream receives three large affluents on
the right (east), one of which the Tsha-lu leads up to the main shih-men 石門 or rock
gate the prominent landmark of the Range.

The K'a-cha is one of the most beautiful valleys of the northern slopes of the Min
Shan. In many places it is quite broad and there villages are found, the inhabitants
belonging to the Cho-ni tsu 卓尼族 or Cho-ni tribe who inhabit the T'ao River valley
as well as those of its affluents. Their villages are situated at the foot of the valley spurs,
the houses have flat roofs like those of Cho-ni and are of tamped earth. [Plate 2]

The mouth of K'a-cha kou or Ku is broad and where it debouches into the T'ao
valley there is a primitive ferry, roughly timbered and leaky. Rafts of spruce logs are
also floated down from here to Cho-ni, saving considerable time. These rafts are taken
apart and the logs sold on arrival in Cho-ni. The names of the villages are Tibetan and
have been transcribed phonetically into Chinese by the Cho-ni and Chinese officials as
in the T'ao-chou t'ing chih 洮州聽志 or the Gazetteer of T'ao chou now called Lin-
t'an 臨潭.

The K'a-cha Valley is 30 miles or 100 li in length from its mouth to the summit of
the pass Kuang-k'e and extends in a south-southwesterly direction from its mouth; Ta-
tzu-to is beautifully situated at the mouth of the valley where there is a large grove of
old poplars, *Populus Simonii* Carr., with undergrowth of willows as *Salix taoensis*
Goerz, a new hybrid, *Salix sibirica* Pall., *Salix wallichiana* Anderss., the latter a tree 15-
20 feet, the former two, shrubs; immediately beyond the willows are associated with
Prunus setulosa Batal., *Cotoneaster multiflorus* Bge., which loves the drier scrub
forests, *Viburnum glomeratum* Max., *Juniperus formosana* Hay., here a tree up to 40
feet in height.

In the shade under the willows in rich alluvial soil grows the greenish-yellow Aroid
Arisaema consanguineum Schott f. *latisectum* Engl., the lovely *Primula polyneura*
Franch., a yellow, still undetermined *Corydali*s no. 12436, and *Aquilegia ecalcarata*
Max. *Lonicera trichosantha* Bur. & Fr., with horizontally spreading branches and
yellow flowers and *Rhamnus leptophylla* Schn., extended further up the valley with
Berberis mouillacana Schn., into the spruce forest where *Rubus amabilis* Focke spread
its rambling branches bearing white flowers. *Geranium Pylzowianum* Max., prefers the
banks of the (Kha-gya) stream as does *Pedicularis torta* Max., a yellow flowered
species which, with *Pedicularis muscicola* a brilliant mass of purple when in flower,

[8]   Egbert H. Walker: Plants collected by R. C. Ching in Southern Mongolia and Kansu Province, China.
Contributions from the US. National Herbarium 28.1941, 563-675, XIII, pl. 21-27.

forms magnificent cushions several feet square in adjoining meadows; it is worthy of cultivation as it remains in flower for several months.

The western spur which rises steeply from the valley floor is densely forested with *Pinus tabulaeformis* Carr., trees more than 50 feet tall whom the people here call the Huang-sung shu [黃松樹] or the Yellow pine. Among rocks in the pine forest we meet with the fern *Polystichum molliculum* Sieb. var. *szechuanica* Schneid., *Malus toringoides* Hugh., *Prunus padus* L. var. *commutata* Dipp., *Lonicera microphylla* Will., a native also of Siberia, *Betula albo-sinensis* var. *septentrionalis* occurs here also, it extends however into the eastern part of the Min Shan at 10,000 feet along sandy streams. *Picea asperata* is here common, very variable, or more than one species is here represented, for some reason *Picea Meyeri* and *Picea Schrenkiana* have been merged with *P. asperata*. [Plate 3]

In the crevices of cliffs along the Kha-gya stream grows the large bluish-purple-flowered *Clematis macropetala* Ledeb., a climber often 20 feet long, while at a narrowing of the valley due to limestone outcroppings, *Juniperus distans* Florin has taken possession of the boulders (see Plate 4), the left valley slopes also being covered here with this juniper. There are several small villages nestled against the forested valley slopes, the last village being Kha-gya ma-ru south of the confluence of the No-mo-na stream, an affluent from the east; beyond there is no more habitation, all is wilderness. Ten miles or more from the mouth of the valley and the last village are beautiful meadows through which the Kha-gya stream flows. In these moist meadows are a multitude of herbaceous plants, the most conspicuous being the scarlet *Meconopsis punicea* Max., its long, curly scarlet petals gracefully waving like little banners in the wind. These meadows are bordered by a magnificent forest of spruces, with willows on the outskirts, under which the red poppy (*M. punicea*) is also often found as is the yellow *Corydalis straminea* Max. One of the most conspicuous plants here is *Caragana jubata* Poir., which covers the barer hillsides with thousands of its erect growing columnar stems, arrayed like an army of soldiers, some reaching a height of 12 feet, it is quite a feature of the landscape. It has here no lateral branches and is long, cylindrically-shaped, tapering to a point at the apex; its flowers are pink and recall a sweet pea. On the summit ridges at 14,000 feet it forms densely pubescent, prostrate branching shrubs.

Bordering the meadows we find *Lonicera hispida* Pall., a shrub 4-5 feet with yellow flowers, *Daphne tangutica* Max., forming globose bushes of 1-2 feet, white flowered, and climbing over bushes the graceful *Clematis gracilifolia* Rehd. & Wils., bearing purplish white flowers. On rock and crevices of boulders dwells the purple-flowered *Primula stenocalyx* Max., favouring elevations of 9,600 to 10,000 feet; it was first collected by Przewalski[9] in the Ta-t'ung alps in 1873. In swampy meadows occurs the

---

[9] Nikolaj Mihajlovič Prževal'skij, 1839-1888, Russian explorer. See *Russkie voennye vostokovedy. Biobibliografičeskij slovař*. Moskva 2005, 193-196; Vasilij Mihajlovič Gavrilenko: *Russkij putešest-vennik N. M. Prževal'skij*. Moskva: Mosk. Rabočij 1974. 142 pp.; Donald Rayfield: *The life of Nikolay Przhevalsky (1839-1888) explorer of Central Asia*. London; P. Elek 1976. XII. 221 pp.; Emil Bretschneider: *History of European botanical discoveries in China*. St. Petersburg: Academy of Sciences 1898, 959-992.

tiny *Primula gemmifera* Batal., along the Kha-gya stream, and with it the loud purple flowered *Primula conspersa* Balf. f. & Purdom.

In the drier meadows flourish *Ajuga calantha* Diels, already described previously, and in the shade under willows the lovely pink flowered *Paeonia anomala* L., common in all the valleys of the Min Shan. Associated with willows are *Viburnum mongolicum* (Pall.) Rehd., *Spiraea canescens* G. Don var. *glaucophylla* Fr., a white flowered shrub with horizontally spreading branches, *Lonicera tangutica* Max., of similar habit, *Lonicera heteroloba* Batal., with dark red flowers, and reaching a height of 8 feet; in the shade of these occurs an undescribed variety of the cream-colored *Primula chionantha* Balf. f. & Forr. (no 12470). In rock crevices and on limestone boulders along the stream thrives a beautiful Incarvillea with large purple flowers and pinkish white throat, as yet undescribed (no 12476), and on grassy banks the yellow flowered and black-spotted *Fritillaria cirrhosa* Don var. *ecirrhosa* Franch., the Chinese Pei-mu [貝母], the bulbs of which are highly priced medicinally and sell for about U.S.$ 20.00 a lb.

The trail in the valley was lined with bushes of *Sorbaria arborea* Schn. var. *glabrata* Rehd., the labiate *Elsholtzia densa* Benth., a native of India, and *Hedysarum esculentum* Ledeb., 3 feet tall, also common on cliffs along the stream; over these bushes twined *Clematis aethusifolia* Turcz., woody at the base, bearing small single, yellowish flowers. Along the stream grew a species of *Dicranostigma* as yet not determined, no. 13144, and on the roadside the campanulaceous *Adenophora* aff. *gracilis* Nannf., and *Adenophora liliifolioides* Pax. & K. Hoffm.; in adjoining meadows blossomed the silvery-leaved pale yellow, *Oxytropis ochrocephala* Bge., with *Pedicularis armata* Max., the Edelweiss *Leontopodium Dedekensi* (Burr. & Fr.) Bv., the white flowered *Galeopsis tetrahit* L., a native of Europe, the larkspur *Delphinium Henryi* Franch., the lavender pink carnation *Dianthus superbus* L., is a species of *Saussurea* with purple flowerheads no 13156, the white flowered *Parnassia setchuensis* Franch., and *Anaphalis margaritacea* (L.) Benth. & Hook., a native also of western America and Kamchatka, and the white flowered herb *Achillea ptarmica* L., a native of Europe. All flowering in July.

Other luxuriant, moist turf plants at 9,500 feet and beyond were *Ligularia sagitta* Max., *Gentiana gracilipes* Turrill, a dark blue, purplish species, and among rocks the handsome purplish-red, large flowered *Incarvillea compacta* Max. which extended to 10,000 feet elevation. Higher up on the outskirts of *Picea purpurea* forests, the maple *Acer tetramerum* Pax var. *betulifolium* Rehd., with *Cornus macrophylla* Wall., *Viburnum glomeratum* Max., and the new variety *scabrella* Schn., formed the deciduous copice. On the valley slopes and stream flourished luxuriantly *Salix plocotricha* Schneid., with a x *Salix taoensis* Görz, *Caragana brevifolia* Kom., with thin leaves and yellow flowers; and *Berberis diaphana* Max. with single, red, turbinate fruits. *Lonicera deflexicalyx* Batal., its orange colored fruits borne on short pedicels, *Lonicera chrysantha* Turcz. var. *longipes* Max., and the rosaceaous *Sorbus Prattii* Koehne with small leaflets and white fruits the size of a pea, grew together with *Sorbus Koehneana* Schneid., also white-fruited and possessing silvery white leaflets. *Euonymus Giraldii* Loes. var. *ciliatus* Loes., with large winged fruits and red arillus, a shrub 3-4 feet, and *Rosa Sweginzowii* Koehne, with large leaflets glaucous beneath and large pyriform, spiny red fruits borne on spiny peduncles, occurred here also.

Rosa omeiensis Rolfe, its sessile fruits orange red and *Lonicera szechuanica* Batal., and as its name relates, first found in the southern, neighboring province of Ssu-ch'uan, concluded the list of shrubby plants. Here and there the beautiful *Aconitum volubile* var? sent its white slender vines, with trusses of pendant purplish-red flowers over shrubs.

Near the mouth of Tsha-lu valley on a meadow called Chhu-chhui-dzong, elevation 9,450 feet bloomed *Ligularia virgaurea* Max., *L. sagitta* Max., both yellow-flowered and *Delphinium tongolense* Franch., a lovely blue-flowered species; in the shade of bordering willows *Aconitum laeve* Royle, and *Pedicularis labellata* Jacq., hid their lovely blooms.

The Flora in Upper K'a-cha kou

The meadow Chhu-chhui-dzong, elevation 9,450 feet, is surrounded by the shrubby bluish purple-flowered *Rhododendron capitatum* Max., 3-4 in height forming round bushes, a very gregarious plant often forming large colonies, with it one finds *Lonicera coerulea* L. var. *edulis* Reg., but the latter does not ascend beyond this elevation while the former extends often to the summit of the ridges.

The trees one encounters here are mainly *Picea purpurea* Mast., while some of the shrubs previously enumerated ascend higher up the valley. *Meconopsis punicea* Max., becomes more common, it occurs in the open alpine meadows as well as in the shade of willows.

Beyond the meadow the valley narrows, the valley slopes become very rocky indicating that the main Min Shan, entirely composed of limestone, has been reached. The slopes are here covered with dead timber. Along the stream the ground is boggy and moss covered (a species of Hypnum). In this marshy meadow we encounter the blue *Corydalis curviflora* var. *pseudo-Smithii* Fedde, *Salvia Przewalski* purple-flowered, *Pedicularis oederi* var. *heteroglossa* Prain, *Polygonum hookeri* Meisn., and *Trollius pumilus* Don.

At 9,800 feet where the valley narrows further we meet with the first large individual *Rhododendron rufum* Batal.; the grassy steep hillsides or rather mountain slopes being covered with *Rhododendron capitatum* Max.; *Rhod. rufum* Batal. increases in numbers and trees 15-20 feet tall with rose colored flowers spotted deep purple or carmine are common with its associate *Rhod. Przewalskii* Max., a shrub 6-8 feet, which E. H. Walker mistook for *Rhod. agglutinatum* Balf. f. & Forr., which it resembles, and may be identical. In his enumeration of R. C. Ching's plants he records this species from this valley but makes no mention of *Rhod. Przewalskii* which is the second common, large-leaved species on the Min Shan. Its leaves are covered on the undersurface with a deciduous tomentum. Its flowers are white or pink and spotted purple.

The cherry *Prunus stipulacea* Max., here a shrub 6-8 feet was still in flower at 10,000 feet elevation, it fruits in July and produces rich red oval drupes. Seeds of this and all the other Prunus were introduced by me as were those of nearly all the ligneous plants and many herbaceous ones of the Min Shan and other mountain ranges. See List of Seeds of ligneous and herbaceous plants collected in 1925 and 1926 by J. F. Rock and distributed by the Arnold Arboretum, in *Journ. Arn. Arb.* 14.1933, 43-52.

On the grassy slopes we encountered here the first yellow poppy *Meconopsis integrifolia* Franch. above 10,000 feet elevation. The flowers here are much larger than those found in Yün-nan, the hair of the leaves are not reddish but light brown. Here also occur two species of juniper *Juniperus distans* Florin, 25 feet tall, with large black pruinose fruits, and *J. saltuaria* Rehd. & Wils. with small globose shining black fruits, the latter does also not grow taller than about 25 feet at this elevation; it, unlike the former which grows on the limestone over 1,000 feet lower, ascends to below the summit pass where it is associated with *Rhododendron rufum* Bat. near the 12,000 feet level.

There is also a third species *Juniperus squamata* var. *Fargesii* Rehd. & Wils., with small oval black fruits and bright green foliage; this species reaches a height of 25-30 feet; it descends into the The-wu land to the same elevation 10,000 feet south of the Min Shan.

Looking up the valley one beholds enormous limestone crags covered with snow, these enclose the valley and are the outer rim of the summit crest of the Min Shan over which Kuang-k'e pass 光克山口 leads south into the upper The-wu land.

From 10,000 feet on *Caragana jubata* Poir., becomes very common on the hillsides and along the bank of the stream, they stand erect like Carnegia cactus. *Picea purpurea* Mast., the purple spruce, *Salix sibirica* Pall., with yellow staminate catkins grow scattered over the rocky slopes in company with the two broad leaved rhododendrons. *Salix sibirica* extends also into most of the lateral ravines along brooks.

There are several forkings of the valley whose names we could not learn, these lateral ravines terminate in grass-covered spurs. North of the last ravine on the left bank of the Kha-gya stream which here flows in a deep rocky chasm, is an alpine meadow where The-wu's coming from south of the range often camp, this meadow is called Li-se-thang, and is 11,000 feet above the sea. Northeast is a huge limestone bluff with talus slopes about 13,000 feet in height, and beyond another rocky peak about 15,000 feet in height which flanks the pass Kuang-k'e 光克 on the east (see Plate **5**); it is the main pass which leads south of the Min Shan into the The-wu land. The slopes opposite Li-se-thang were one mass of white and pink *Rhododendron rufum* Bat., with Picea forest, while the crags were covered with snow. On the alpine meadows we found *Rhododendron capitatum* Max., variable in color from deep lavender to bluish purple forming colonies, with masses of a farinose form of *Primula chionantha* Balf. f. & Forrest, with pale cream-colored flowers and darker eye, the upper stem, pedicels and calyces a mealy white. Along a very narrow, shallow watercourse grew *Salix sibirica* Pall., a shrub 8-10 feet in height with yellow catkins, and *Salix myrtillacea* and, of lower stature than the foregoing; *Rhododendron rufum* Bat. was the predominant tree here reaching 15-18 feet in height with here and there a *Rhododendron thymifolium* Max., a shrub 2-3 feet with small purple-lavender flowers. The ranunculaceous white flowered *Souliea vaginata* (Max.) Franch. with *Primula flava* Max., with white farinose undersurface of leaves and sulphur-yellow flowers grew on the sandy banks at the foot of crags or on the crags themselves. On the moist alpine meadows at 1,000 feet below there flourished the yellow flowered *Pedicularis oederi* var. *heteroglossa* Prain, the deep ultramarine *Corydalis curviflora* var. *Smithii* Fedde, and *Meconopsis punicea* Max., with *Mecon. integrifolia* nodded their gorgeous blooms all over the open glades. *Caragana jubata* Poir. was still very common. *Abies Faxoniana* Rehd. & Wils., forms still stately trees up to 50 feet tall with 3 inch long cones a purplish-black, on the grassy slopes at the foot of cliffs, with *Rhododendron Przewalskii* Max., growing on the outskirts. The entire head of the valley, now narrowed to a ravine, is flanked on both sides by immense crags and cliffs which reminded very much of those of the Yü-lung Shan of Li-chiang, the flora is however less rich and birds are very scarce.

It is a beautiful sight to watch a snowstorm rage over the mighty grey limestone crags, and lovely Rhododendrons in full bloom growing in front of one's tent. The strata of the upper cliffs is vertical and the walls appear to be split into huge upright segments a few yards thick; in the intervening spaces grow *Abies Faxoniana* and *A. sutchuenensis*, which also form forests, small trees at this altitude of 12,000 feet, with dark, somber needles; the undergrowth is composed mainly of *Rhododendron rufum* Bat., some 15 feet tall. (see Plate **6**). *Primula chionantha* is very common here, but rare is the lovely *Meconopsis quintuplinervia* Reg., with pale lavender bells. Along mountain torrents which descend from the crags through side ravines grew the new willow *Salix pseudospissa* Görz, with rich yellow staminate catkins, this willow extended also on to the alpine meadows with a tiny purple Iris, Meconopsis etc. The soil is here black mud into which one sinks at every step. As one ascends higher the Rhododendrons decrease in size and only the two leaved species are to be seen, *Rh. Przewalskii* Max., forming dense masses. In the boggy meadows where the snow had recently melted grew the pale lavender poppy, the new deep yellow *Draba Rockii* O.E. Schulz, an undescribed variety of *Myosotis alpestris* Schmidt, a lovely blue forget-me-not, *Primula limbata* Balf. f. et Forr., with large deep ultramarine colored flowers and white peduncles, and sessile glabrous leaves; with it grew the new *Anemone Rockii* Ulbrich, a white-flowered species which frequents also the crevices of limestone crags, these with *Oxygraphis glacialis* Bge, a yellow composite and *Doronicum thibetanum* Cavill, only 3 inches tall, make up the swampy alpine flora.

Among the crags 500 feet higher than Kuang-k'e pass, at an elevation of 13,000 feet, subsist the yellow flowered *Arctous rubra* Nakai, a prostrate shrub with brick-red fruits, *Primula Purdomii* Craib, with large deep bluish purple flowers, its peduncle, pedicels and calyx a mealy white (farinose), and *Primula graminifolia* Pax & Hoffm., with wine-colored flowers. Still higher, at 13,500 feet in loose scree or limestone rubble grew the extraordinary composite *Saussurea medusa* Max., the inflorescence consisting of a large white cottony cone in which the purple flower-heads exsuding a fragrance of vanilla, and supported by purple bracts, rise from a rosette of very fragrant pinkish-green leaves.

The black muddy trail gives way to sharp limestone gravel rising steeply in zigzags to the pass called Kuang-k'e 光克, elevation 12,550 feet. On June 12th the stream Kha-gya (K'a-cha), which has its source here, was one long sheet of ice beneath which the water roared, the ice being covered with snow. From the steep limestone walls hung masses of ice, frozen waterfalls. The crags are bleak and fluted, honeycombed with holes and caves. The strata is vertical in most places.

A little distance beyond the pass a trail leads east up to a higher pass 13,150 feet whence a path descends into a ravine which brings one in front of the main Shih-men or Rock Gate. This part will be described separately. The main pass merges into a swampy plateau or meadow with limestone crags protruding everywhere (see Plate **7**). A little stream flows from east to west in a deepening ravine; all around high massive limestone mountains rear their bare crowns to a height of 15,000 feet. In a southeasterly direction there extends an immense limestone wall with tremendous black holes with patches of snow at each entrance.

On the limestone crags grew *Sorbus tapashana* Schn., a shrub 4-5 feet high, *Juniperus saltuaria* Rehd. & Wils., and on moss-covered boulders the tiny pale purple flowered *Primula pumilio* Max., while the previously mentioned Primula grew on the boggy meadow below the pass, with *Polygonum Hookeri* Meisn., *Rheum pumilum* Max., *Allium monadelphum* Turcz. var. *thibeticum* Reg., and another as yet undescribed *Primula* no 12407, with deep purplish blue flowers.

The southern wall of the ravine opposite Kuang-k'e pass, the Tibetan Gung-kha, forms the eastern sentinel of the tremendous rock gate leading into The-wu land (see Plate **8**), and through which a stream flows, a branch of the Pai-shui Chiang 白水江 or Pai-lung Chiang 白龍江. It flows south of the Min Shan, and in Ssu-ch'uan becomes the Chia-ling River 嘉陵江 which enters the Yangtze at Chung-ch'ing (Chungking).

Directly west the range is pierced by a pass, actually a dip in the range extending from north to south, filled with scree; over this a trail leads steeply up the Ch'e-pa pass and down into the long Ch'e-pa kou (valley) to the T'ao River west of K'a-cha Valley and Cho-ni. See the flora of Ch'e-pa kou. Due east, beyond the pass and up the ravine, a trail leads to the rear of the main Shih-men or Rock Gate, the chief landmark of the entire Min Shan. From the crest screes extend into the ravine. Along the foot of the limestone crest and immediate scree grow *Corydalis dasyptera* Max., with dull wine-colored blooms, yellow *Draba oreades* Schrenk proles *chinensis* O. E. Schulz, *Paraquilegia anemonioides* (Willd.) Ulbr., and *Meconopsis racemosa* Max., this species which also occurs in meadows lower down is very different from the Yün-nan *M. rudis* Prain. The latter has thick glaucous blue leaves with very stiff scarlet spines which contrast greatly with the leaves, they have not as Dr. Taylor[10] says only a purple spot at the base, one must see these plants growing to appreciate the difference. *M. racemosa* has green spines when young and pale yellow straw-colored ones when mature, the leaves are rich green, not at all glaucous, the spines are soft and the plant can be strocked toward the apices of the spines, while in *M. rudis* they are stiff and very thick at the base. I have never seen *M. rudis* elsewhere except in Yün-nan. The scree Primulas previously mentioned, and Saxifraga (not in flower in June) composed the scree plants.

On the alpine meadows below the pass and on the summit, the most common shrub was *Potentilla fruticosa* L. var. *parviflora* Wolf., a shrub 1 foot tall with very large, rich golden, yellow flowers. *Pleurospermum* spp.? not yet identified nos. 13094 and 13095, *Pedicularis armata* Max., with yellow flowers, the new *Pedicularis Merrilliana* Li, a deep purple flowered species and two Delphiniums as *D. Souliei* Fr., and *D. Maximowiczii* Fr., graced these alpine meadows in August with their large, rich, purplish-blue flowers.

Cushion-shaped saxifrages were then (June 12th) not yet in flower. The trail leads up to a pass 12,800 feet over marshy meadows to almost directly back of the main Shih-

---

[10] George Taylor: *An account of the genus Meconopsis*. With notes on the cultivation of the introduced species by E. H. Cox. With a foreword by Sir William Wright Smith. London: New Flora and Silva Ltd 1934. XIII, 130 pp., 29 pl.

men or Rock Gate. A similar ravine descends on the other side east. From the pass an excellent view can be had of the extreme eastern end of the range in which the Ta-ku-ma 大古麻 crag and rTsa-ri-khi-kha are situated.

As seen from here the range extends from east-northeast to west-southwest. Back of it are other lower limestone crags and bluffs which hide a complete view of the range. On the screes back of the main Rock gate *Saussurea medusa* Max. was fairly common, with most of the other plants found on Kuang-k'e. Directly south of the craggy midriff of which the much higher bluffs of the Rock Gate were a part, there is another much lower range extending from east to west, parallel to the Min Shan some 30 miles to the south of it, this is the mountain chain enclosing the Pai-lung Chiang; beyond this range is an unadministered part of Ssu-ch'uan which drives here a wedge between Ch'ing-hai and Kan-su provinces. Southeast some 80 or more miles distant there rose a pyramidal snow-capped mountain mass of which Yang-pu Shan 陽布山 is the highest point; it is the border between Cho-ni and Ssu-ch'uan.

Later in the summer during the month of August appeared a different flora in the alpine meadows of Kuang-k'e Shan between 12,000 and 12,500 feet, but less rich than the spring flora. *Morina betonicoides* Benth., displayed its cream-colored flowers, spiny bracts and pale green leaves in company with *Aster Vilmorini* Franch., an ornamental species with dark-lavender-purple flowerheads, the yellow flowered *Saxifraga montana* H. Smith var. *splendens* H. Sm., *Ligularia virgaurea* Max., and *Ligularia sagitta* Max.; two undescribed Saussureas nos. 14267 and 14168 bearing dark purple flowerheads and an equally undescribed *Swertia* sp. with greyish-blue flowers, as well as *Aconitum Szechenyianum* Gay and *A. rotundifolium* K. & K. var. *tanguticum* Max., the former with pale yellow, the latter with bluish-purple flowers, a yellow flowered onion *Allium chrysanthum* Reg., and *Trollius pumilus* Don, were inhabitants of the alpine meadows. Partial to the gravelly slopes were *Primula gemmifera* Batal., and *Delphinium coelestinum* Franch.

A prostrate shrub with large red fruits proved to be *Lonicera thibetica* Bur. & Fr., growing among boulders while a bluish-yellow gentian, *Gentiana tetraphylla* Kusnez, and *G. Przewalskii* Max., were here and there visible among *Poa bulbosa* L., and *Juncus exploratorum* Walker. An undetermined *Delphinium* sp. no 14640, hid its pale-bluish to purple flowers in bushes of *Potentilla fruticosa* var. *parviflora*, and was never seen in the open. Another *Delphinium, D. tanguticum* Huth., preferred the wet gravel at the summit of the pass, as did its congener *Delphinium Maximowiczii* Fisch., both with deep purple flowers. Nestling on mossy boulders, but a thousand feet below the summit we found a variety of *Saxifraga diversifolia* Wall.

During October there still flowered *Allium cyaneum* Reg., *Gentiana hexaphylla* Max. var. *caudata* Marq., *Saussurea polystichoides* Hk. f., with a rosette of lyrate leaves, white beneath, and a large single flowerhead in the center; associated with it was the *Saussurea katochaetoides* Hnd.-Mzt., a small plant with subsessile flowerheads, and leaves mealy white underneath and *Delphinium Forrestii* Diels, with short dense spikes and yellow flowers tinged blue. All the above were confined to the alpine meadows at the summit of the pass at 12,500 feet.

Below the pass at 11,000 feet fruited in October *Rubus idaeus* L. var. *strigosus* Max., with red, edible sweet berries, *Lonicera heteroloba* Bat. var. – a shrub 6-8 feet, with globose orange fruits on slender peduncles –, and at 12,000 feet *Ribes Meyeri* Max., bearing small black fruits on short racemes. *Lonicera thibetica* Bur. & Franch., a prostrate shrub with wooly leaves bearing orange fruits hugged the boulders on the summit pass and on the swampy alpine meadow flowered *Saussurea stella* Max., coloring its meadow red from the red rosette-forming leaves and red bracts supporting the centrally placed flowerheads. Immediately below the pass grew the new willow *Salix pseudospissa* Goerz, a shrub 4-5 feet tall.

From the summit pass a muddy trail leads to the streambed visible in the photograph[11] to an elevation of 11,500 feet where it enters the terrific gorge flanked to both sides by cream to pink colored limestone walls thousands of feet in height, terminating in pinnacles. Here a spring gushes forth at the foot of the northern wall, the waters of the spring joining the stream which rushes madly through the immense gorge.

At first we encounter only alpine meadows on which at 11,000 feet elevation *Pedicularis recurva* Max.*, Pedicularis labellata* Jacq., the former with pale red and the latter with large, rich, purplish-red flowers brighten the rich green sward with many other herbaceous plants found also on Kuang-k'e pass and previously enumerated. Along the foot of the walls grew masses of *Rhododendron Przewalskii* Max., *Lonicera hispida* Pall., with yellow to orange red fruits (July), and *Prunus stipulacea* Max.; on grassy slopes here and there *Delphinium tongolense* Franc., a deep blue-flowered species and a yellow *Ligularia* sp.? (no 13110) added to the color scheme. *Lonicera szechuanica* Batal., 4-5 feet tall with horizontal branches and globose red fruits and *Sorbus tapashana* Schneid., frame the meadows and with *Rhododendron Przewalskii* form the vanguard of *Picea purpurea* forests which rise from the foot of the cliffs, while the lichen *Cladonia gracilis* (L.) Willd. var. *elongata* (Jacq.) Floerke, and *Peltigera aphthosa* (L.) Willd., grow with the deep Hypnum moss in the shade of the Rhododendron. With the purple spruces grew *Abies Faxonica* Rehd. & Wils., here only 40 feet tall, at an elevation of 11,000 feet, while lower down its place is taken by the dark green *Picea asperata* Mast., which never ascends to the same altitude as *Picea purpurea* in this region, which made R. C. Ching surmise that it is not as hardy a species as *P. purpurea*. Just the reverse seems to be the case for *Picea purpurea* finds its northern limit at La-brang or Hsia-ho hsien, while *P. asperata* extends far north to the Nan Shan or Richthofen Range where it forms pure stands and where *P. purpurea* dares not venture.

As a whole autumn alpine flowers are scarce, they have no time to develop as the summers are very short and the latter merge soon into winter, snow can be expected as early as August in these regions and as late as June, leaving just 2 1/2 to 3 months for the alpine flora to emerge from its winter sleep.

The scenery in this limestone gorge which leads into the Upper The-wu country or T'ieh-pu ti-fang 鐵布地方 is by far the grandest I have ever seen anywhere in West China, and while there are no snow peaks or ranges with eternal snow as in Ch'ing-hai or Hsi-k'ang, yet the Kan-su Min Shan and the The-wu country in particular are unique and most remarkable in the whole of West China. The oftener one beholds this stupendous work of nature the more awe-inspiring it becomes.

---

[11] Photograph not identified.

It is at the same time so weird and amazing that one stands aghast and spellbound, unable to find words to adequately describe it. It must be seen to be realized.

Enormous talus slopes extend from the rockwalls into the streambed and with every step the scenery becomes more and more majestic. Through the gorge which widens considerably in its upper half, mountain ranges are visible in the distance which lie beyond the The-wu border in Ssu-ch'uan (Szechuan). To both sides are sharp limestone ridges densely forested with conifers mostly *Picea* and *Abies*. Larix is here wanting but common on the eastern end of the range, but never forming forests. The tall, spire-like spruces and firs, whose lofty crowns seem to out-reach the mighty pinnacles of the cliffs mitigate the overpowering impression these gorges make on one.

The cliffs in places are vertically split and divided like the leaves of a book stood upright, yet firs cling to them and birches fill the spaces between them. The elevation is here 11,000 feet and the walls rise from 3,000 to 4,000 feet vertically into the sky.

Rhododendron rufum and *Rh. Przewalskii* abound here, also *Rh. capitatum, Rh. thymifolium* and *Rh. cephalthoides* who with willows, *Salix pseudospissa* Görz, *Salix rehderiana* var. *brevisericea* Schn., *Salix myrtillacea, Salix hypoleuca* var. *kansuensis* Görz, *Salix plocotricha* Schneid., and *Salix alfredi* Görz, and rich green birches *Betula albo-sinensis* Burkill, with *Berberis kansuensis* Schn., Lonicera, etc., compose the ligneous plants.

Primulas abound on the moss-covered boulders as *Primula stenocalyx* Max., with large purplish-blue flowers, and the lady slippers *Cypripedium nutans* Schl., its flowers a reddish brown and striped, and *C. tibeticum* King, common along the stream. In crevices flourished the sturdy new cushion plant *Anemone Rockii* as also among the rocks in the streambed at 10,600 feet, it is a very showy plant with dark green foliage, white flowers and lobed petals. *Primula flava*, shunning dampness, grew in dry dusty caves and hollows with *Viola biflora* L. A lovely *Corydalis* its flowers dark red to reddish purple (not yet described no 12495) selected the very foot of the cliffs with the new *Pleurospermum pseudo-involucratum* Wolff. *Euonymus alatus* Reg. var. *apertus* Loes., a shrub with rambling branches and small reddish brown flowers adorned the rocky bank of the stream with *Sorbus Koehneana* Schneid., and the lovely shrub *Abelia brachystemon* (Diels) Herb. Edinb. with stems 10 feet high: the herbaceous *Aster heterochaeta* Benth., and the pale lavender *Arabis? alaschanica* Max., a crucifer first known from Mongolia found a foothold among the stones along the river as did the lovely lavender-purple *Cardamine macrophylla* ssp. *polyphylla* (Don) O. E. Schulz, while the caprifoliaceous *Triosteum pinnatifidum* Max., preferred shady places, and the *Saxifraga kansuensis* Mattf., a fleshy white-flowered herb, chose the crevices of the cliffs at 11,500 feet with *Sedum venustum* Praeg., with reddish flowers.

The trail leading down the gorge is gravelly and clean and the roaring waters of the stream as pure as crystal. The air is the purest which can be desired. The whole ensemble was purity and beauty itself. The vegetation is rich here if not in species, in numbers and in beauty, yet strange to say birds were scarce.

At an elevation of 9,700 feet the most extraordinary spot in the entire gorge is reached, it is a narrow defile through a limestone spur which completely blocks the valley. The stream has cut the narrow chasm through the solid wall and roars through it over huge boulders. Over this torrent a traject has been built of heavy timber, strong stout logs rest on each side in holes chiseled into the rock wall, and over these logs a trail as wide as the defile leads over the madly rushing waters.

Every step reveals new beauty in this over ten mile long gorge; here *Anemone Rockii* is plentiful as is the gorgeous blue poppy *Meconopsis racemosa* Max., with flowers of the riches blue. From this defile we emerge, after passing through another breach with limestone walls thousands of feet in height, into a long amphitheater surrounded on all sides by enormous mountains of limestone crowned with crags of the most fantastic shapes. The scenery is so extraordinary that it appears almost unreal and recalls an extravagant stage setting. This amphitheater, the heart of the Upper The-wu land or Shang T'ieh-pu, is called Brag-sgam-nag or the Black Stone Box pronounced Drag-gam-na.

Upper or Shang T'ieh-pu Land, Its Flora and Its People

Drag-gam-na
The jewel of the Upper The-wu country is undoubtedly Drag-gam-na or the place of the Stone Box, well named for it is indeed enclosed all around by enormous limestone crags of great scenic beauty. What has been said of the Pearl Mosque in the Fort of Delhi should be applied to Drag-gam-na «if there is Heaven, it is this». Only the inhabitants would belie such a statement for most of them are potential, if not de facto robbers. [Plate **9-11**] It is an amphitheater oblong in outline and I doubt if there is anywhere in the world a similar spot, or one that could in any way be compared to it. The Dolomites of Europe are certainly grand and massive, but they lack the gracefulness and the subtle charm with which these mighty crags have been arrayed by nature. Everywhere one looks there is elegance and yet immensity, tiers upon tiers of greyish pink rock are contrasted by the various shades of green provided by the stately conifers, birches and other deciduous trees. The whole is divided by a crystal stream which has cut its way to it, and where it is joined by another coming from the south, whence united they were confronted by an enormous crag, but which they finally conquered by splitting it, leaving two huge turrets like immense cathedral spires on either side.

This massive gate is the entrance to Yi-wa kou 亦哇溝, a valley that leads the mighty torrent south into the Pai-lung Chiang or White Dragon River.

Undoubtedly this huge basin was once a lake until the waters forced an outlet through the above rockgate. Now gentle slopes, formed by the soil etc. washed down from the crags and spurs extend towards the center where the stream flows. Nestled at the foot of the northern crags is the romantic lamasery Lha-bsam dgon-pa (Lha-sam Gom-pa) belonging to the Yellow Sect but peopled by wild The-wu monks. This lamasery, situated on top of the sloping hill overlooks the village of Tong-wa with its terraced fields on which the The-wu cultivate barley and wheat. West of sTong-wa are the rest of the villages in the order named Nye-ri, sTa-re, rTi-pa all The-wu names and inhabited by The-wu whom the Chinese designate as T'ieh-pu 鐵布. That they form one of the most interesting types of aborigines there can be no question.

Everywhere one looks are lovely forests interspersed with luscious alpine meadows; the average elevation within Drag-gam-na is 9,500 feet. Looking west Drag-gam-na is closed by a high mountain called Ben-chhu-ri, the eastern slopes of which belong to Cho-ni, the western ones to Ssu-ch'uan, beyond are grasslands inhabited by Drogpa or herders, robber nomad tribes of sTag-tshang-lha-mo and the Zhag-sdom.

The conifer forests of Drag-gam-na are composed of several species of Picea and Abies, no single species of either genus forms pure stands as does *Picea asperata* Mast. in the north. Nearest to it is *Picea purpurea* which ascends to a higher altitude than the other Picea in this region, and having no competitors, except a few intruders of Abies, occupies most of the available space.

Around the alpine meadows at 9,700 feet forests extend to the foot of the mighty limestone cliffs. The trees composing these forests are mainly conifers carpeted thickly with Hypnum moss. *Picea purpurea* Mast., reaches here 50 to 60 feet, with slender drooping branches and recalls *Picea likiangensis* of Yün-nan. *Abies sutchuenensis* Rehd. & Wils. (see Plate **12**), reaches over a hundred feet in height, has ascending branches and a greyish-brown flaky bark. This is the most common Abies; *Abies Faxoniana* Rehd. & Wils., is less common but reaches greater dimensions trees of 180 feet or even more are not uncommon with trunks three feet in diameter, the branches are short and descending, the bark is drab-colored and longitudinally furrowed. These are the main species of Abies in this forest; at a lower elevation occurs another species *Abies recurvata* Mast., in Yiwa kou (Valley) at 9,000 and below 9,000 feet, it reaches 60-80 feet, the needles are large, broad and sessile, the large cones are of a dull greyish brown to dull black. It had not been previously recorded from Kan-su, as its real home is in northwest Ssu-ch'uan where it forms extensive forests south and west of Sung-p'an. A fourth Abies but very rare, also not previously recorded from Kan-su, is the huge *Abies chensiensis* Van Tiegh., a tree more than 150 feet tall and a trunk with 4 feet and over in diameter. It occurred only in the extreme southeast of the Lower The-wu country. It was first found by Père David[12] in 1872 on the Ch'in-ling Shan 秦嶺山 in Shensi, but was however found by me also in northwest Yün-nan on the Mekong-Salwin divide (see: *The Ancient Na-khi Kingdom* etc. Vol. II Pl. 184). In the fir forest a number of birds disported themselves mostly in search of food and here we collected the large black woodpecker *Dryocopus martius khamensis* (Buturl.) besides a number of others.

There remain now *Picea Wilsonii* Mast., and *Picea asperata* Mast., I personally believe that there are more than these two species of Picea plus *P. purpurea*, and as I have previously remarked *Picea Meyeri* Rehd. & Wils., and *P. Schrenkiana* Rehd. & Wils., should not have been sunk and made synonyms of *P. asperata*. The latter may be a very variable species but there are Picea on the Min Shan of such different habit that I cannot believe that they are all *P. asperata*.

Several trees of entirely different branching habit, color of needles, etc., grew next to each other, one could at a glance differentiate these; they were all of one height and one age. I specially took photos showing these divergencies, but Wilson judged these different trees from herbar material only, ignoring their dissimilarities in habit etc.

Undoubtedly most of the Picea and Abies found on the Min Shan attain a height of more than 100 feet, especially in such protected areas as Drag-gam-na.

---

[12] Armand David, 1826-1900, Lazarist missionary and scientist, collected about 3,000 plant species in China. See Bretschneider: *History*, 837-870, with analysis of the three research trips through China.

Along the slopes and banks of brooks below the villages enumerated grew *Berberis dasystachya* Max., *Caragana Maximovicziana* Kom., and in the shade of willows along meadows *Saxifraga tangutica* Engl.; the banks of the main stream were lined with willows as *Salix hypoleuca* Seem. var. *kansuensis* Görz, *Salix Rehderiana* Schn. var. *brevisericea* Schn., *Salix myrtillacea* And., *Berberis Silva-Taroucana* Schn., not previously recorded from Kan-su and *Berberis kansuensis* Schn.

On the rich meadows of Drag-gam-na at 9,500 to 10,000 feet luxuriated *Primula tangutica* Duthie, not exactly a beauty with dark red, almost black flowers, *Pedicularis chinensis* Max., with deep golden flowers, *Pedicularis rudis* Max., one of the largest louseworths pale-yellow flowered venturing on grassy slopes and into spruce forest where it reaches 5 feet in height.

On the rocky banks of little streams descending from the massive limestone spurs blue bells as *Adenophora* sp? (no 13105) bloomed in July, with *Prunus stipulacea* Max., and a new variety of the araliaceous *Acanthopanax Giraldii* Harms var. *pilosulus* Rehd., a shrub 5 feet tall with red spines. In the forest of spruces and firs or their outskirts flourished the beautiful, white flowered, fragrant, new variety of *Philadelphus pekinensis* Rupr. var. *kansuensis* Rehd., the mock orange, the only representative of the genus in Kan-su, the type of which is no 12587. It is widely distributed in the province but mainly on limestone ranges as on the Min Shan and Lien-hua Shan, but extends to north and northwest of Lan-chou. With it grew the comely *Daphne tangutica* Max., 2 feet tall, with rich purple flowers, identical with *D. Wilsonii* Rehd., it extends to west Ssu-ch'uan and west Hu-pei. On drier, grassy slopes, we find *Delphinium tongolense* Franch., with deep blue flowers; it is rare at the lower (9,600 feet) levels but frequent at 11,000 feet on the grassy slopes which join the cliffs at that elevation. Partial to the semi-shade of the spruces is the composite shrub 8-10 feet high, with scandent branches and lavender flowers, *Pertya sinensis* Oliv., first known from Hu-pei (Hupeh), as is the large red fruited *Sorbus tapashana* Schn., *Juniperus squamata* var. *Fargesii* Rehd. & Wils., a tree 20 to 30 feet high with small oval, black fruits delights on the drier slopes at the foot of the limestone crags with *Caragana jubata* Poir., here spreading, and thick-stemmed, while on the crags the handsome new *Anemone Rockii* Ulbr., with flowers one inch across, white and deep blue beneath forms large cushions 2-3 feet in diameter, its leaves glossy and a rich dark green, seeds were introduced under no 13626. Associated with it is *Anemone narcissifolia* L., growing in crevices, or also forming cushions over boulders, its leaves glaucous bluish and flowers a delicate lavender blue; pendant on vertical cliffs, the most delicate of all, *Paraquilegia anemonioides* (Willd.) Ulbr. nodded from crevices with its mauve-lavender blooms ast the slightes breeze.

Rhododendron capitatum Max., 2-3 feet tall, its small leaves dark brown beneath, and flowers a purplish blue, covered the exposed steep mountain slopes like heather. *Betula albo-sinensis* Burk., occasionally ventured from recesses in the limestone walls out into the open, either to form copices or to grace with its bright green foliage and glossy, copper-colored bark, the outskirts of the somber conifer forests, or ascended among the mighty cliffs to an elevation of over 11,000 feet. The heavy, compact, globose shrubs of *Rhododendron Przewalskii* Max., mostly chose the protection of mighty stalwart *Abies sutchuenensis* in whose immediate vicinity they formed rank.

side. Back of the rock gate is a huge rampart which is contiguous with the rear part of the range, this rampart is circular in outline, and unscalable. Immense talus slopes extend from the foot of the rampart within the gate which is in itself a jumble of colossal boulders the size of a cottage. It is the home of herds of blue sheep, and should one climb into the rock gate as we did, one is exposed to being stoned by the rocks sent flying by the escaping animals.

The Approach to the Shih-men
Beyond the last village in K'a-cha kou (valley) Kha-rgya ya-ru there is a lovely meadow known as Chhu-chhui-dzong, elevation 9,450 feet yonder which a stream debouches into the K'a-cha stream on the right, coming from the south. This valley is called Tsha-lu Nang or Cha-lu kou 扎路溝 in Chinese. By following this valley up stream for about 8 miles one comes to a preliminary rock gate at an elevation of 10,950 feet, through which a torrent has cut its way.

The vegetation up Tsha-lu valley is very similar if not identical with that of K'a-cha kou, *Picea purpurea* is here very symmetrically cone-shaped recalling *Cupressus funebris* trees, *Rosa Biondii* Crep., and *R. omeiensis* Rolfe, with *Ribes glaciale* Wall., are found on the outskirts of the forests, also *Lonicera trichosantha* Burm. & Fr., and *L. deflexicalis* Batalin, while the yellow poppy *Meconopsis integrifolia* covered the grassy slopes up to 10,500 feet. At the head of Tsha-lu valley is a lovely meadow in front of a terrific Rock Gate, the preliminary one mentioned, its strata is vertical (see Plate **14**). Unlike the two high bluffs of the main rock gate, which are blunt, those of the former are sharply pointed. In front of the rock gate is a small stream or brook joining the stream on the left or west. Vertically rise the whitish grey limestone walls over a thousand feet above the meadow; on the grassy slopes which extend up to the cliff, are forests of *Picea purpurea* while in the meadow flourish the yellow *Ligularia sibirica* Cass. var. *speciosa* DC., the pink *Paeonia Veitchii* Lynch, the green-flowered *Codonopsis viridiflora* Max., the scrophulariaceous *Scrofella chinensis* Max., first described from Ssu-ch'uan, the labiate *Galeopsis tetrahit* L., a native also of Europe, and *Delphinium tongolense* Fr., *Primula gemmifera* Batal., and the lovely satiny blue *Meconopsis racemosa* Max., preferred the vicinity of the streambed. In the shade of willows hid *Aconitum laeve* Royle, its flowers flesh to pink-colored, as did the purple-flowered *Pedicularis labellata* Jacq.

Rhododendron capitatum occupied the open grassy slopes with *Potentilla fruticosa* L. var. *dahurica* Ser., and var. *parviflora* Wolf, the former white and the latter golden-yellow flowered. These were not confined to the head of the valley, but covered the open slopes the whole length of the valley, the yellow-flowered variety almost completely drowning out the white one.

From the meadow the trail leads immediately into a tremendous defile with overhanging limestone cliffs of vertical strata. A few log bridges assist the traveller across the turbulent stream; the canyon is only thirty feet wide in the widest part, the water cutting and undermining the rockwalls nearly 2,000 feet in height. The rock in the defile is of a yellowish-grey to reddish limestone. From this chilly canyon one emerges into an open valley which leads southwest, confronting one with the main Shih-men or

Rock Gate of the Min Shan (see Plate 15). In July the summit crags were entirely free of snow and ice except here and there on a rockshelf lay patches of snow. The broad meadow through which the stream flows is boggy and composed of peat in which thousands of the yellow-flowered *Cremanthodium plantaginium* Max., grew exuberantly. Shrub-covered slopes, and ridges extend steeply to the valley floor which narrows, till one again reaches a broad meadow framed by magnificent forest which extends to the foot of the cliffs (see Plate 16).

Willows, as *Salix denticulata* And., the new *Salix hypoleuca* Seem. var. *kansuensis* Görz., *Salix sibirica* Pall., and the new *Salix Alfredi* Görz, etc., covered the hillsides with the shrubby Rhododendrons previously mentioned. At the head of the valley, directly southwest, at the foot of the rock gate is a lovely alpine meadow at an elevation of 11,750 feet, a difference of 800 feet between the two rock gates and a distance of three miles, the conifer forest surrounding the alpine meadow consists mainly of *Abies sutchuenensis* Rehd. & Wils., and *Abies Faxoniana* Rehd. & Wils., stately somber trees, on the outskirts of which is on array of the white flowered *Rhododendron Przewalskii* Max., with its bright green foliage and yellow petioles. In the fir forest occurred a species of *Senecio* as yet undetermined (no 12997); *Cacalia Potaninii* C. Winkl., was limited to the gravelly streambed, with *Meconopsis racemosa, Primula gemmifera* Bat., and another species of *Cacalia* no 13028 probably new. Among boulders above the forest the yellow *Saxifraga confertifolia* Engl. & Irmsch., formed beautiful cushions. The birch *Betula albo-sinensis* Burk., with glossy copper-colored bark and *Sorbus tapashana* Schn., grew among boulders at the foot of the Shih-men, but the latter was more often found in company with the firs and Rhododendrons. *Arctuous ruber* Nakai, although not common, was found here and there among boulders.

At the foot of the cliffs of the Rock Gate near a small lake or pond an attractive rich blue flowered onion *Allium Henryi* Wright, an undescribed Cremanthodium with nodding yellow flowerheads (no 13049) grew in clusters with *Aconitum tanguticum* (Max.) Stapf., its flowers bluish purple with yellow tinge.

Saxifraga Giraldiana Engl., formed cushions in the fir forest, and among willows grew the new *Thalictrum Rockii* Boiv.; *Meconopsis punicea* Max., flaunted its brilliant scarlet flowers in the meadow, while the prostrate willow *Salix flabellaris* And., and a form of it, forma *spathulata* And., covered boulders and grassy slopes mat-like. The new *Astragalus longilobu*s Peter-Stib. grew on the gravelly slopes at 12,000 feet with the new *Anemone Rockii* Ulbr.; *Crepis paleacea* Diels aff., was restricted to grassy slopes, and in crevices among the rocks grew a species of *Hedysarum* (no 13063), with *Vicia unijuga* A. Br., var? and an undescribed *Oxytropis* (no 13065). One single *Primula limbata* Balf. f. & Forr., clung to the rockwall, while *Saussurea nigrescens* Max., with purple flower heads flourished in moist meadows in the shade of willows and firs.

Rosa Biondii Crep., *Spiraea longigemmis* Max., and *Ribes glaciale* Wall., the latter a shrub with long whip-like branches, were confined to the fir forest; with them in the shade grew also the arborescent *Rhododendron rufum*, often accompanied by *Rh. Przewalskii* on the outskirts of the forest.

At the very foot of the rockwall flocks of snow pigeon *Columba leuconota gradaria* Hart., had their home, they nested in crevices and holes in the cliff. This pidgeon is peculiar to high altitudes and was never observed below 11,000 feet.

Beyond the house-sized boulders which lie scattered in the Rock Gate one reaches a flat gravelly space whence enormous screes extend steeply to a circular rampart without outlet. Through the gate, looking northeast, a wonderful vista opens over the grasslands. We estimated the thickness of the walls forming the gate 1,200 feet, the width of the gate at the top 1,800 feet, and at the bottom 900 feet. In the extreme rear is a higher rock peak, part of the main crest which is however not visible from below. In the actual rock gate grew a huge *Rheum palmatum* var. *tanguticum* Max., compositae, aconites, etc., as described previously.

To either side of the rockwall forming the Min Shan, from in front of the Shih-men, a valley extends, the left or eastern one being shorter and narrower while the right or western one is broader and much longer, it ends in a pass 13,150 feet in height, whence one can descend to Kuang-k'e pass which leads north to K'a-cha valley, and south into Drag-gam-na. This valley which seems to have no name is rather rich in alpines. Towards the pass the western ravine widens considerably with meadows in the central part giving way to scree on the upper slopes towards a high ridge, the northern flank of the ravine, with limestone crags rising from the rubble. [plate **17-18.**]

In the alpine meadows at 12,500 feet, less than half way up the valley, clumps of *Saussurea poophylla* Diels, *Trollius pumilus* Don, growing singly, the deep orange-yellow *Corydalis Rheinbabiena* Fedde, the handsome very fragrant *Primula gemmifera* Bat., *Aster flaccidus* Max., with purplish-blue ray-florets, enlivened the green of the turf with their rich colors. On moist gravelly slopes at 13,000 feet a lovely rich-blue forget-me-not, a variety of *Myosotis alpestris* Schmidt, subsisted in the cold in these bleak uplands, with the cushion-forming, rich yellow flowered *Potentilla biflora* Willd., which however delighted more to dwell among huge limestone boulders than in the moist fine gravel. In the latter a *Hedysarum* (no 13010) with large flesh-colored to rich-pink flowers, and a deep tap-root most difficult to dislodge, found a foothold as did two probably new composites an Artemisia with yellow flowers and silvery leaves (no 13013) and a *Cremanthodium* sp.? also yellow flowered but with a purple involucre this latter extended to 13,500 feet elevation. At the same altitude *Primula graminifolia* Pax & Hoffm., hugged in the foot of the cliffs in the scree.

The new *Astragalus longilobus* Pet.-Stib. a pale-yellow species descended as low as 11,500 feet, while *Leontopodium linearifolium* Hand.-Mzt., formed clumps, as did the yellowish purple *Saxifraga Przewalskii* Engl., but at 13,000 feet, in meadows with *Gentiana Przewalskii* Max., a white flowered species with steel-blue markings.

In very coarse gravel at 13,500 feet, the intense ultramarine blue and white *Corydalis melanochlora* Max., its delicate stems contrasting strangely with the rough, sharp, limestone, appeared unharmed and delighted in its coarse environment; with it *Cremanthodium humile* Max., a really humble herb, with yellow flowerheads and silvery leaves, shared its milieu. Another companion proved to be the curious, rosette-forming *Crepis rosularis* Diels aff. with central flowerheads, its hairy leaves dull greyish-lavender, and anchored in the gravel by a long taproot. Of Ranunculaceae there remain to be recorded the *Delphinium tanguticum* Huth, and *Aconitum rotundifolium* K.

& K. var. *tanguticum* Max., both flaunting large, deep bluish, purple flowers, and forming large clumps in the scree, while the small, single flowered *Delphinium Pylzowi* Max., kept 500 feet above the two former in the same type of environ.

There seemed to be no end to the alpine plants which prevailed in these high, bleak regions among gravel and boulders, with here and there a bit of green sward. Among the scree plants between 13,000 and 13,500 feet, the very handsome large cushion-forming *Androsace yargongensis* Petitm., with pale pink flowers, *Arenaria Kansuensis* Max., producing dense cushions and white flowers, often prefering large boulders, the curious *Saussurea medusa* Max., already recorded from back of the Shih-men, are the most noteworthy.

Here and there, on patches of turf amidst the gravel, bloomed a *Gentianella* (no 13031), the purple *Pedicularis Davidi* Fr., *Leontopodium calocephalum* (Fr.) Bod., and *Saussurea epilobioides* Max. var. *cana* Hand.-Mzt., with lanceolate leaves white beneath, and purple flowerheads all around 13,500 feet elevation. At lower altitudes, at the 12,000 feet level we encountered a yellow *Allium* sp. (no 13037), the yellow *Pedicularis anas* Max. var., and the mat-forming *Oxytropis melanocalyx* Bge. *Meconopsis quintuplinervia* Reg., with pale, drooping, single flowers, was present at 13,000 feet, also in patches of turf, but not common.

From the pass down the steep ravine to Kuang-k‘e across which a trail leads south into the Upper The-wu country, and one north to the T‘ao River, at elevations of 13,500 – 12,500 feet occurred the cushion-forming *Arenaria Przewalskii* Max., *Fritillaria cirrhosa* D. Don var. *ecirrhosa* Fr., *Delphinium Maximowiczii* Franch., the crucifer *Megacarpaea delavayi* Fr., a tall herb with lavender-purple flowers, and the deep orange flowered cushion plant *Saxifraga pseudohirculus* Engl.

From Kuang-k‘e Pass to Ch‘e-pa kou (Valley)

Ch‘e-pa kou 扯巴溝 is one of the longest valleys which extends from the summit of the Min Shan to the T‘ao River, west of K‘a-cha kou. A trail leads from Kuang-k‘e pass, elevation 12,550 feet, west-southwest to Ch‘e-pa kou pass 12,500 feet over the western end of the Min Shan. The northern wall is a long narrow ridge of limestone boulders whence steeply sloping screes descend to the trail into the V-shaped valley. The ascent over the loose scree is very difficult as one sinks into the rocky debris at every step. The top of the spur is 13,840 feet and offers a marvellous view over the top of the Min Shan the highest part of which is in the southeastern end. Many of the plants found on the screes and patches of turf here, occur also back of the Shih-men, and on the screes leading to Kuang-k‘e (q. v.). At the highest point 13,800 feet in the scree flourished *Cremanthodium humile* Max., with single yellow flowerheads and leaves white-wooly beneath, a delicate pinkish-lavender-flowered *Geranium* sp? rare and as yet undetermined (no 13184), and *Delphinium Pylzowi* Max., with large, dark purplish-blue flowers; forming cushions over boulders was the lovely *Androsace tapete* Max., here white flowered *Meconopsis racemosa* grew at 12,000 feet and lower with *Aconitum anthora* L. var. *anthoroideum* (Rehb.) Reg., a yellow flowered aconite, and *Gentiana Piasezkii* Max., an erect plant with rich blue flowers.

From the pass the valley extends northwest; alpine meadows commence at 11,500 feet and there we find *Gentiana quinquinervia* Turr., a fleshy herb with erect spikes and white bluish flowers, its leaves forming rosettes; *Saussurea epilobioides* var. *cana* Hdl.-Mzt., its flowerheads a dark purple, the intense blue flowered *Delphinium grandiflorum* L., first known from Siberia and *Ligularia virgaurea* Max., while the meadows were mostly of *Poa sphondylodes* Trin.

The valley is rather narrow, its slopes covered with *Potentilla fruticosa* L. var. *parvifolia* Wolf, a deep yellow flowered variety, willows, Sorbus, *Rhododendron rufum* Batal., *Rhododendron Przewalskii* Max., and *Rhododendron capitatum* Max. Of trees *Picea purpurea* Mast., forms pure stands on the slopes and along the stream. *Sibiraea angustata* (Rehd.) Hao, with *Lonicera* occurs here as in other valleys, where they are common. At 11,200 feet there is a huge rock gate with lateral spurs extending thousands of feet into the sky. It is a very picturesque spot enhanced by the tall *Picea purpurea* trees which flank it to both sides. The stream has cut a defile through this barrier passable only at low water. Beyond the defile the stream is lined with willows and only a short distance beyond we encounter the first The-wu village called hGro-chhui-thang (Dro-chhui-thang), in Chinese To-chu 多朱, elevation 11,100 feet. Not far ahead a valley opens from the left called gTsang-dgah-yas-khu (Tsang-ga-ye-khu); it extends from south-southeast to north-northwest. here are lovely terraces or level meadows whence the mountains slope gently to the limestone crags, the slopes being forested with *Picea purpurea*, tall stately trees, in whose company grow also *Abies Faxoniana* Rehd. & Wils., and *Abies sutchuenensis* Rehd. & Wils. Higher up the valley at an elevation of 11,200 feet occured *Juniperus saltuaria* Rehd. & Wils., beyond which the valley is enclosed by mighty limestone walls crowned with pinnacles and spires. Tall erect, *Caragana jubata* Poir., covered the upper hillside like huge tall candles, as in the upper part of K'a-cha kou.

Beyond the mouth of Tsang-da-ye-khu the main valley narrows considerably and turns from northwest to north; forests extend all the way to near the mouth of the valley where it debouches into the T'ao River. The-wu men were picking the seeds of a wild *Polygonum* which they grind and use as tsamba. The The-wu of this valley are renowned for their hostility not only to strangers but they are also sworn enemies of the Cho-ni prince who did not dare penetrate into their fastnesses. It was here that The-wu barred my way with drawn sword. Between Dro-chui-thang and the village of rMe-ri-shol, Me-ri-shöl, the Chinese Mai-li-shih 買力什, are forests of *Picea asperata* Mast., and the usual shrubs along the streambed as in K'a-cha kou; a very fine-leaved shrub, *Caragana brevifolia* Kom., with slender pendant branches grew in the shade of spruces and junipers of which latter genus only a few trees occurred.

Me-ri-shöl is a large village of several hundred houses, situated at the mouth of a broad lateral valley or between the confluence of Ch'e-pa kou and the valley from the west. The village people were most unfriendly and scowled at us as we passed, a friendly greeting on our part being not only scorned but answered by men with drawing their swords from their scabbards. As a matter of fact some The-wu from Drag-gam-na had told the Cho-ni prince that we were intending to go from the upper The-wu country to the T'ao River via the Ch'e-pa kou (valley), whereupon he sent runners to try to stop us from going that way on account of the savage manners of the inhabitants of Ch'e-pa kou. As it was, the runners never caught up with us, and did not dare follow us into the Ch'e-pa Valley. In spite of their hostility the people seemed to be religious, for over every doorway was a little wooden windmill, wooden cross pieces, the ends of which were carved into large scooped out ladles, below each such windmill was a prayerwheel. Under the eaves of their roofs hung shoulderblades and other bones of yak inscribed with sacred prayers or formulas, which the people either shook themselves or let the wind do the swaying of them, and thus send the prayers written on them heavenward.

From Me-ri-shöl on the valley is very broad, barley fields and villages alternating. On the left bank of the stream beyond a cantilever bridge is a yellow lamasery called Chhos-bshad dgon-pa, Chhö-she Gom-pa, known in Chinese as Ch'e-pa kou Hsin-ssu 扯巴溝新寺 or the New Monastery of Ch'e-pa Valley. Here grew *Delphinium Henryi* Franch., in meadows with *Delphinium albo-coeruleum* Max., with bluish-white flowers, at 9,600 feet. After passing many lateral valleys we come to the last above the village of Hsieh-wu 斜無, also called Hsiao-wu 肖吾, the Tibetan Sheu, at 9,500 feet; the low hillsides opposite are here covered with spruces *Picea asperata* Mast., and willows. The stream is broad and meanders over the valley floor from south-southwest to northeast to the T'ao river. Among the The-wu of Sheu are dwarfs and many cretins, probably due to intermarriage, for every village is at feud with the other, the only way peasants can secure women from other villages is to kidnap them. The men are small of stature and appear degenerate and no comparison to the stalwart The-wu of Drag-gam-na or the still more virile and giant-like Drog-wa of the grasslands.

The broad valley appears lonely, the trail crosses the stream several times, part of it having been obliterated, it follows the path of the water; one trail, built over tressels, follows the hillside high above the valley floor in a zigzag manner; every bush in the valley was piled high with debris indicating that the stream can become at times a terrific torrent. It seems to cut deeper and deeper into the western valley wall, forsaking its old bed in the center of the valley. The mouth of the valley is very broad and joins the broad banks of the T'ao at an elevation of 9,080 feet. Here on the right bank is the village of Ma-ru, the Chinese Ma-lu 麻柔, and opposite on west bank is another called Yag-rug, Ya-ru, in Chinese Ya-lu 牙柔, and on a bluff overlooking the village the monastery of Yag-rug dgon-pa, Ya-ru Gom-pa. In Chinese the monastery is known as T'ieh-tang Ya-lu Ssu 迭當牙柔寺. The inhabitants of Ch'e-pa kou (valley) are known as the T'ieh-tang tribe 迭當族.

The T'ao River makes here a large curve issuing from northwest turning to south-southeast and passes in front of Ma-ru village where there is a ferry. Here on the banks of the T'ao grew a lark-spur which we had not encountered elsewhere, *Delphinium sparsiflorum* Max., a plant 4-5 feet tall with pale blue flowers.

The hills are bare and low. A large alluvial fan extends into the T'ao Valley built by the Ch'e-pa stream. The northern hillsides enclosing the T'ao River are rocky and bare, while the southern slopes are covered with scrub and forest common in other parts of the T'ao Valley.

The Smaller Valleys of the Northern Slopes of the Min Shan

The smaller valleys extending from the T'ao River south, but which do not reach to the backbone of the Min Shan are from west to east: La-li kou 拉力溝, Shao-ni kou 勺尼溝, Ma-erh kou 馬兒溝, and Po-yü kou 波峪溝, to mention the most important ones. The easternmost in the former Cho-ni prince's domain is Ta-yü kou 大峪溝, and extends as far as A-chüeh 阿角 where a valley joins it from the southwest called Ta kou 大溝 or Great Valley, this carries the main stream into Ta-yü kou; a smaller one joins it from the south called the Hsiao kou 小溝 or the Small Valley. This latter leads to a pass designated as Tsa-ri Khi-kha, on the top-level of the Min Shan, but not the highest part, nor does it bring one directly into the valley of the Pai-shui Chiang in Lower The-wu land.

Ma-erh kou (Valley)
At a narrowing of the T'ao, a short distance east of Cho-ni is a well-built bridge the only one over the T'ao River which one has to cross in order to reach any of the valleys on the south, there being only two primitive ferries at Ch'e-pa kou and K'a-che kou, respectively. Ma-erh kou is a little west of Cho-ni; its steep slopes, like those of K'a-cha kou are forested with spruces and firs, mainly *Picea Wilsonii*, *Abies sutchuenensis* and *Abies Faxoniana* which form forests at the head of the valley; the undergrowth being again *Rhododendron rufum* and *Rhododendron Przewalskii*. Lovely meadows alternate with forest on the valley floor, and willows abound with many other shrubs along the stream and on the grassy open slopes.

Picea asperata Mast., is more confined to the lower levels at 9,000 feet and covers even the steep slopes near the mouth of the valley overhanging the T'ao river. *Picea purpurea* is restricted again to the upper levels near the head of the valley. *Ribes Meyeri* Max., is common along streams and in spruce forest with *Aster heterochaeta* Bth. Of interest is here *Saxifraga fragrans* var. *platyphylla* H. Smith, an orange flowered species found only here and in Hsiao kou, east of Ma-erh kou but not west, it is apparently a rare species in Kan-su which found its way north from Ssu-ch'uan. The rose-red *Paeonia Veitchii* Lynch, is often met with along streams and on shady banks up to 10,000 feet elevation. *Iris ensata* Thunb., a short, thick compact plant here not more than a foot high grows on meadows and along trails and roadsides at 8,500 feet.

Thermopsis lanceolata R. Br., is usually found under willows in meadows, while *Prunus stipulacea* Max., flourishes with *Ribes glaciale* Wall., on margins of spruce forest at 9,000 feet, with *Daphne tangutica* Max. A beautiful farinose form of the pale cream-colored *Primula chionantha* much handsomer and sturdier than its representative of the alpine meadows of Chung-tien 中甸 in Yün-nan, and *Primula Purdomii* Craib, 1/2 to 1 foot tall, another handsome species with dark purple flowers, are found in alpine meadows at an elevation of 10,000 feet; the latter is spread over the northern slopes of the Min Shan but not on the southern. It does however extend to Ra-gya and the Yellow River gorges and to the Am-nye Ma-chhen in Ch'ing-hai. The species was described from a cultivated plant grown from seed sent by Purdom. On the grassy slopes of the valley at 10,000 feet we encounter *Daphne Giraldii* Nitsche, with *Rosa bella* Rehd. & Wils., also *Rubus amabilis* Focke, and the large flowered *Clematis macropetala* Ledeb., climbing over bushes, its flowers very ornamental and of a rich lavender purple. The woody climber *Clematis Fargesii* Franch., with large white flowers in long erect pedicels covers willow bushes at 10,000 feet; here and there in the spruce forest we meet the very fragrant orchid *Habenaria conopsea* Benth., and *Ligularia sibirica* Cass. var. *speciosa* DC., the common palmately lobed *Senecio acerifolius* C. Winkl., which finds its way from the shade to the open grassy slopes; it is also found in the Upper The-wu country and in central Kan-su. Its associate in Ma-erh kou is *Mimulus nepalensis* Benth., the only locality where we found it on the Min Shan or in Kan-su proper.

Among rocks at lower elevation grew *Polygonatum sibiricum* Ledeb., while *Poa sphondylodes* first known from Hu-pei (Hupeh) usurped the slopes and reached a height of 3 feet; it extended as high as 12,500 feet especially on the western part of the Min Shan. The woody climber *Rubus pileatus* Focke, was confined to the spruce forest, and on the rocky slopes of this valley we found the first maple on the Min Shan, *Acer caudatum* Wall. var. *multiserratum* Rehd. No Acer is found in the western Min Shan, but the genus is well represented south of it in the warmer valleys. None was encountered either in K'a-cha kou or Ch'e-pa kou. *Berberis parvifolia* Sprague, a 2 feet tall shrub with small spathulate leaves was confined to the rocky drier slopes, with *Vicia cracca* L. Near the head of the valley at 10,000 feet grew *Astragalus monadelphus* Bge., 2 feet tall with the willow *Salix plocotricha* Schn., its reddish catkins, contrasting with the dark green leaves whose petioles and young shoots were also red; this willow found also in K'a-cha kou, forms at high elevations dense flat-topped scrub over considerable areas. On open meadows *Morina chinensis* (Bat.) Diels, a spiny herb with greenish flowers was found only in this valley and not elsewhere on the Min Shan, but it does extend into Ch'ing-hai to the Yellow River gorges north of Ra-gya.

In willow scrub we met with *Euonymus Giraldii* var. *angustialatus* Loes., with *Viburnum erubescens* Wall. var. *gracilipes* Rehd., the latter a tree 25 feet tall, with edible fruits according to Cho-ni natives. *Clematis glauca* Willd. var. *abeloides* (Max.) Rehd. & Wils. f. *phaeantha* Rehd., with purplish-brown flowers covered willows and other bushes. The 25 feet tall *Sorbus Prattii* Koehne, with large leaves and white to pink fruits was found with the spruces and Abies at 10,500 feet elevation.

Meconopsis punicea and the homely *Primula tangutica* grew in meadows near the head of the valley and on the outskirts *Potentilla fruticosa* L. var. *dahurica* Ser., the white flowered variety formed bushy borders, but encroached also on to the open hillsides. *Senecio argunensis* Turcz., and *Aster trinervius* Roxb., were restricted to the lower elevations at 9,500 feet but not common, and so was *Lonicera Ferdinandi* Fr., which descended as low as 8,500 feet to the mouth of Ma-erh kou and the banks of the T'ao River.

Mention must finally be made of *Gentiana Farreri* Balf. f., with prostrate rosettes and deep sky-blue flowers in the meadows at 9,000 feet where it blooms in August and September; in its company grew *Allium kansuense* Reg., a purplish flowered onion also found south of the Min Shan; it does not grow singly here but forms dense tussocks.

This accounts for most of the plants found in Ma-erh kou a lovely valley but not very broad. It is so densely wooded at the head that it was impossible to ascend any of the spurs to a summit ridge to obtain a view of the intervening spurs between the head of Ma-erh kou and the backbone of the Min Shan, without cutting a clearing. With such a view however one was rewarded by climbing to the highest grass-covered spur at the head of La-li kou.

La-li kou (Valley) and the Intervening Spurs of the Min Shan

La-li kou 拉力溝 is one of the loveliest valleys in the Cho-ni district, and unlike the other smaller valleys connects with K'a-cha kou by a pass 12,000 feet above sea-level. This pass leads southwest to a lateral valley of K'a-cha kou called gSer-zhu-na (Ser-zhu-na), and in Chinese Se-shu-na 色樹那 by which name it is known to the people of Cho-ni. Se-shu-na valley has its source east and south of the summit ridge which encircles most of the valleys to about the length of La-li kou, namely about 12 miles from the T'ao River.

The vegetation between the mouth of La-li kou and the bridge at Cho-ni is the same as elsewhere in the T'ao River valley on the southern bank; three species of Picea cover the slopes, willows along the T'ao River, the white fruited Sorbus, Lonicera, several Berberis, Prunus, Betula and Hippophaë but none except those already mentioned elsewhere. The trail on the right bank of the river is broad enough to be termed a road for it is frequented by bullock carts. Yaks are employed dragging spruce logs along the road and over the bridge to Cho-ni where 1,000 logs 15-20 feet long and 10 inches to a foot in diameter sold for $120.00 silver, then worth about $80.00 U.S.

Above 9,000 feet *Picea purpurea* Mast., is the most prominent conifer and forms nearly pure stands. The valley narrows and the forests become denser. A lateral valley called Tso-do (kou) leads east over a ridge into Ma-erh kou and thence to Po-yü kou 波峪溝 and into Ta-yü kou 大峪溝. Trails also lead into K'a-cha kou via another lateral valley called in Tibetan gYon-lung (Yön-lung) or Left Valley. *Abies sutchuenensis* and *Abies Faxoniana* join *Picea purpurea* at 10,000 feet whose undergrowth was mainly moss. At 11,000 feet *Abies Faxoniana* mainly composes the forest with *Rhododendron rufum* Batal., *Rhod. Przewalskii* Max., *Rhod. capitatum* Max., and *Rhod. anthopogonoides* Max. Scattered here and there grew the erect *Caragana jubata* Poir., and on the slopes *Juniperus formosana* Hayata, with willows as undershrubs as the new *Salix cereifolia* Görz, *Salix Wallichiana* And., etc.

In this forest every branch and leave was encased in ice in the middle of October, which could be pulled off like a stocking. At an elevation of 11,500 feet, Abies gave way to dense thickets of *Rhododendron rufum* Bat., their cracked and twisted branches formed an impenetrable network, it was a natural lattice work. No other species was mixed with it and to ascend the ridge it was necessary to cut a trail through the tangle of light brown trunks and branches, the ground being thickly covered with snow up to one's knees. Higher up the willows and *Potentilla fruticosa* L., bushes became mixed with Rhododendron scrub till at 11,750 feet the top of the spur, here grass-covered, which formed the head of La-li kou, was reached after a four hours climb through Rhododendron thickets.

Before us lay a magnificent panorama, directly south was the Great Rock Gate or Shih men or the Min Shan which stretched from southeast to northwest. The terrific chasms between the enormous limestone crags stood out clearly, beautiful clouds swept over the steep precipices of mighty pinacles and peaks filling the clefts between them.

Looking north one could see T'ao-chou Old City with its few trees situated among bare loess hills. North-northwest loomed up the high range called Ta-mei Shan 大煤山 over which a pass leads to Hsia Ho or La-brang. The valley of the T'ao River could be followed plainly between a maze of bare mountains which in the distance merge into grasslands. North-northwest rose the bold limestone massive of Lien-hua Shan a more or less isolated mountain but which is in a line with an enormous limestone wall of equal height extending to south-wouthwest.

Beyond Ta-mei Shan there towered a bold mass of crags the Chinese T'ai-tzu Shan 太子山 or the Mountain of the Heir Apparent, and the Tibetan Am-nye Nyen-chhen (gNyan-chhen); its separates La-brang from Lin-hsia 臨夏 the former Ho-chou 河州. These observations were made from an elevation of 11,750 feet. From a pass to the west, elevation 12,000 we took photographs of the Min Shan whose extent ranged from E 25.60 to W 12.60 of the compass.

In the immediate foreground are shallow valleys partly wooded with junipers, extending south, these merge into a valley extending from east to west. Towards the main backbone of the range sharp deep valleys extend from south to north into the one seemingly parallel to the Min Shan. The ridges separating these deep V-shaped valleys are as sharp as a knife edge.

As seen from the summit pass La-li kou, the highest point of the Min Shan appears to be the western bluff of the main Rock Gate. However the highest point is in the southeastern end of the range (q.v.). The whole is a maze of valleys and sharp ridges to map which could only be done from a plane.

At the mouth of La-li kou overlooking the T'ao, on the loess bluffs grew *Berberis Silva-Taroucana* Schn., with oval scarlet fruits, a shrub 4 feet tall, also the crimson-carmine-fruited *Cotoneaster multiflora* Bge., *Berberis vernae* Schn., *Betula japonica* Sieb. var. *szechuanica* Schn., with a dark, silvery-grey bark, and many other shrubs already enumerated in the description of the T'ao River flora.

Shao-ni kou 勺尼溝 *(Valley)*
The valley is named after a little village situated on the south bank of the T'ao River near a large band from northeast to south. The valley extends from southeast to north-northwest and has three branches, being not longer than about 6 miles. Near its mouth are large groves of *Crataegus kansuensis* Wils., *Malus kansuensis* Schn., Pyrus, Prunus, especially numerous being *Malus toringoides* Hughes. The hill-sides near its junction with the T'ao are densely forested with Picea, mixed with *Rosa Sweginzowii* Koehne, *Betula japonica* Sieb. var. *szechuanica* Schn., 60 feet or more tall with a special predilection for loess slopes at 8,500 feet, also *Populus Simonii* Carr. On the flat spaces towards the T'ao river the new hybrid *Malus kansuensis* x *toringoides* Rehd., is quite common, it is a lovely tree about 25-30 feet tall with a large spreading crown, its leaves are deeply tri-lobed and tomentose beneath. Seeds of it were introduced by us under no

14925 and distributed by the Arnold Arboretum. On the slopes near the stream the lovely new *Philadelphus pekinensis* var. *kansuensis* Rehd., with large white fragrant flowers scented the air and stood out against its somber background of spruces and firs.

Juniperus formosana Hay., a shrub or small tree from 3-15 feet tall delighted in dry loose shale of the Shao-ni kou bluffs facing the T'ao River. Along the Shao-ni stream grew many *Berberis*, willows, and Cotoneasters. No Abies were seen in the lower part of Shao-ni valley but occurred near its head at 10,000 feet elevation, the same species as found in La-li kou; with them grew the 15 feet tall *Sorbus Prattii* Koehne, with large white fruits, larger than those found at lower elevation (see *Sorbus Koehneana* Schneid.) its branches are long, and straight ascending. There is only a short trail into the valley which soon leads into the streambed, owing to its narrowness and difficulty of ascending into it; it is covered with absolutely virgin forest untouched by man or beast. The most interesting feature of the arborescent flora is the many wild rosaceous trees as enumerated above which grow on the little plain in front of Shao-ni kou and along the T'ao River between it and La-li kou.

Po-yü kou 波峪溝 *(Valley)*

Po-yü Valley is smaller than Ma-erh kou and is densely forested like La-li kou. A village called Po-yü after which the valley is named or vice-versa is situated east of the mouth of the stream on the south bank of the T'ao River, and east of Cho-ni. Here the Cho-ni prince had his summer home, and it was in this valley where he met a violent death at the hands of his own people.

Back of the village is the somewhat shallow, small gulch or ravine of Po-yü; to east of it is another ravine, and across a spur, whence a small rivulet flows into the little stream, on the slopes of a hill, is a small monastery called Kha-dog dgon-pa, (Kha-do Gom-pa) the Chinese K'ang-to Ssu 康多寺, at an elevation of 9,100 feet. South of the lamasery are a lovely meadow and some fields which adjoin magnificent *Picea* forest. Here a trail leads over a spur in to a diminutive valley and to the great Ta-yü kou 大峪 溝 which leads directly south to the summit of the Min Shan. This is the last pass over the range in Cho-ni territory, save a farther eastern one which leads to the Min District or Min Hsien 岷縣 formerly known as Min Chou. It was in this latter area in the extreme southeast of Min Chou where Farrer and Purdom worked. This region was also visited by Berezowski[13] in 1886.

Near the mouth of Po-yü Valley is an extraordinary array of herbaceous plants many of which are not found to the east of it, while in the upper part of the valley the vegetation is the same as in Ma-erh kou and La-li kou. At the 8,500 foot level, on grassy slopes, and in the meadows near the T'ao River, as well as on the valley floor there blossomed in July the pale pink leguminous *Astragalus melilotoides* Pall., the everlasting *Anaphalis lactea* Max., a new labiate, the blue *Scutellaria scordiifolia* Fisch forma *pubescens* Diels, only here and not elsewhere, *Leontopodium Smithianum* Hnd.-Mzt., *Allium tanguticum* Reg., its flowers a silky, glossy purple, equally rare and found

---

[13] Mihail Mihajlovič Berezovskij, Russian biologist; see E. Bretschneider: *History*, 1023, 1033. Berezovskij accompanied Potanin on three research trips in Central Asia; he was particularly interested in zoology.

here only, as was the geraniaceous *Biebersteinia heterostemon* Max., yellow-flowered and apparently more common south of Lan-chou where it occurs along moist banks of cultivated fields, but is not known outside of Kan-su. The deep purple labiate *Nepeta macrantha* Fisch., is found everywhere on the Min Shan, and with it the lovely *Pedicularis torta* Max., its yellow flowers set off by a purple keel, the new *Pedicularis cristatella* Penn. & Li, with purplish red blossoms, and *Hypericum Przewalskii* Max., also found in the Yellow River gorges of Ch'ing-hai province. Very common in meadows was *Ligularia Przewalskii* Max., but only in the eastern part of the T'ao River, east of Cho-ni, yet it is of wide distribution having been first recorded from Mongolia; equally common was the yellow flowered *Artemisia Sieversiana* Willd., *Sedum aizoon* L., less so *Asparagus brachyphyllus* Turcz., first known from Northern China, with which grew two as yet unidentified *Potentilla* spp. nos: 12911 and 12912. *Pedicularis striata* Pall. var. *policalyx* Diels, a purplish-flowered species luxuriated in moist meadows at the edge of woods with *Geranium eriostemon* Fisch., and *Polemonium coeruleum* ssp. *vulgare* (Ledeb.) Brand, a deep bluish-purple flowered herb.

Among rocks thrived the new *Thymus serpyllum* L. ssp. *mongolicus* Ronn., known from here and the grasslands between Labrang and the Yellow River, also Tibet, and Altai Mountains, the yellow *Pedicularis cranolopha* Max. var. *longicornuta* Prain, also known from Hsi-k'ang in the south, and the campanulaceous *Adenophora Smithii* Nannf., the latter also in meadows but rare, as it it in the grasslands west of La-brang.

On the alpine meadows higher up the valley between 9,500 and 10,000 feet grew *Meconopsis punicea*, and in woods *Rubus amabilis* Focke, a shrub 3 feet and small orange berries, first described from Shensi; on the hillsides and outskirts of conifer forest the usual shrubs were encountered and elsewhere on the northern slopes of the Min Shan, except *Cotoneaster racemiflorus* K. Koch var. *soongaricus* Schn., which is confined to the eastern part of the Min Shan where it covers the drier open valley slopes and lends a distinct color to the landscape due to its grey foliage; it bears an abundance of fragrant, small white flowers, and small red fruits borne singly on short peduncles. It was first described from Sungaria as the varietal name indicates.

Ta-yü kou 大峪溝 *(Valley)*
The name of this valley actually means Great Ravine Valley, the word yü denotes also a mountain pool, but no pools occur in that long valley; a large stream, which issues from a deep valley west of the village A-chüeh 阿角 forms the main stream of Ta-yü kou which commences at the latter village. It is there joined by a valley coming directly from the south known as Hsiao kou 小溝 or Small Valley in contradiction to the one issuing from the west which carries the larger stream and is hence known as the Ta kou 大溝 or Great Valley, yet Hsiao kou is much longer. It was not possible to follow the Ta kou to its source as no trail leads up it any distance, and the stream is a great mountain torrent full of rocks and boulders which it is impossible to ascend.

Ta-yü Valley is a long one, about 25 miles to the village of A-chüeh, from its mouth, and from there it is about 8 miles to Cho-ni. The village is also called A-yi-na 阿亦那, a transcription of the Tibetan name which I was unable to ascertain. E. H. Walker gives a different Chinese character for the second syllable viz., chüan 絹, and calls it A-chüan. The name as it occurred on the Cho-ni prince's map was as written above, namely A-chüeh, the people of Cho-ni being used to speaking Tibetan pronounce the last syllable nasally, hence it sounded like chüan. Like most northern Chinese the Kan-su people add a guttural «r» to syllables ending in vowels, hence Ching's «Archuen». The Chinese are the worst offenders when it comes to romanizing Chinese names, each individual having his own romanisation; they have never learned to use the Wade-Giles system, which, while not altogether satisfactory, is the one universally adopted in English speaking countries, and is the standard Mandarin romanisation used on maps.

Ta-yü kou is a rather broad valley carrying a large stream whose waters are white which would indicate the source to be a glacier, but as no glaciers occur anywhere on the Min Shan, it must have its source in some hidden snow fields. The vegetation of the valley is rather uninteresting, being poor in comparison to K'a-cha kou. Much of it has been destroyed through cultivation for not less than eleven villages are situated in the valley.

Between 8,600 and 9,000 feet elevation the hillsides are open and covered with scrub of which *Cotoneaster racemiflorus* K. Koch var. *soongaricus* Schneid., is the most common and also most conspicuous as its greyish foliage lends a peculiar tone to the landscape. It occurs with Lonicera, Berberis, tall willows, Spiraea, Prunus and poplars.

On grassy banks *Galium verum* L., a rubiaceous herb vaunted its showy yellow flowers in company with two lilies, *Lilium tenuifolium* Fisch., its flowers a deep orange to firy red, and the orange flowered *Lilium Davidi* Duch., cultivated in the south for its bulbs which are boiled, stuffed with meat and eaten. It is very similar to the foregoing. Both also grow along streams and outskirts of forests. It is the only place where *L. Davidi* has been observed on the Min Shan, while *L. tenuifolium* grows also in other valleys but only on the northern slopes of the range.

Paeonia anomala L. occurs both on grassy slopes and in the shade of shrubs and trees, also in spruce forest at 9,000 feet elevation. On swampy meadows in the upper part of the valley we find Aster, Pedicularis, and the lavender to red flowered *Primula conspersa* Balf. f. & Purdom, its stem and calyx a mealy white. It was especially plentiful near the village Chan-chan-ni 占占尼, half way up the valley at 9,500 feet. It occurs elsewhere on the Min Shan as well as on Lien-hua Shan, its northern limit. Here also we encounter *Gentiana Farrerii* Balf. f., which opens its large sky blue flowers streaked with white in the middle of September, its pale green leaves form flat rosettes, in company with *Gentiana striata* Max., its flowers a dark lemon yellow, growing to a height of 2 feet.

At the very head of the valley near A-chüeh, *Picea purpurea* Mast., forms forests, and on the open rocky slopes *Juniperus distans* Florin, grows to a height of 35-40 feet with trunks over a foot in diameter. Along the streambed we find the usual shrubs as *Berberis diaphana* Max., *Berberis kansuensis* Schn., the latter with large suborbicular leaves on long petioles and oval fruits on long drooping racemes. Both species ascend also to the spruce forests with *Cotoneaster acutifolius* Turcz., a shrub 5 -6 feet, with oval pubescent leaves glaucous beneath and black fruits. These are the outstanding plants found in Ta-yü kou.

Although the valley is broad, the stream, half way up the valley beyond the village of Chan-chan-ni, flows for most of the distance in a narrow channel of reddish limestone with a vertical strata. On the very edge of the vertical rockwall overhanging the stream grow *Picea Wilsonii*, *Picea purpurea* and *Juniperus distans*. Near the village of Brag-las (Dra-le) the Chinese Cha-lieh 扎列, the stream described a broad curve, and is then shut in by the steep valley slopes, partly denuded and partly covered with the willows *Salix myrtillacea* And., *Salix sibirica* Pall., the new *Salix cereifolia* Goerz, and *Salix plocotricha* Schn.; here also occurs *Betula albo-sinensis* Burk., and here and there a spruce. The trail leads high above the stream till near the village of A-chüeh where there are several gently sloping terraces and lovely groves of spruces.

A-chüeh itself is situated at an elevation of 9,500 feet on the eastern slopes of a lateral valley called Ta kou 大溝 or the Great Valley which opens out from the southwest and carries the main stream. The valley is trailless and hence impossible to ascend as previously stated. High limestone crags are visible in the distance. A-chüeh is a small affair composed of a few barn-like houses with flat roofs of tamped earth on which grass grew luxuriantly.

The Ta kou stream receives an affluent from the west and unites with it southwest of A-chüeh. This valley is called Changolo (spelling unknown), or Ts'e-tz'u kou 測次溝. At A-chüeh proper at 9,000 feet elevation among bushes on the edge of meadows bloomed in July *Polygonatum sibiricum* Led., a rather rare plant only found besides here in Ma-erh kou (valley); it was first described from Siberia. At 9,000 feet *Salix plocotricha* Schneid., lined the streambed, while *Sorbus Koehneana* grew with birches on the slopes and edge of conifer forest at 10,000 feet above the sea.

Eastern Min Shan

The Mountains West of A-chüeh
Immediately back of the village of A-chüeh is a valley called Changolo in Tibetan and Ts'e-tz'u kou 測次溝 in Chinese; there is no other way of ascending the valley except by following in the streambed, yet it is easier to ascend than the larger valley called Ta kou where the streambed is steep and so full of boulders that any attempt to advance into the densely forested valley becomes impossible. Ta kou is absolutely virgin territory, untrodden by any man, in a primeval state, and a sample of what the Min Shan and its valleys were before the advent of man.

The streambed of Ts'e-tz'u kou was lined with the usual willows, Lonicera, *Potentilla fruticosa* L. var. *dahurica* Ser., with white flowers and the variety *Purdomi* Rehd. with lemon yellow flowers, the latter not previously recorded from Kan-su, the description having been first drawn from plants raised from seed collected by Purdom in northern China. Spiraea and Sibiraea were also common. The underbrush became so dense that any further progress became impossible. The only way to get out of the valley was by climbing zigzag the steep valley slopes to the top of the grass-covered ridge. From this spur a grand view could be had up the Ta kou or Great Valley. On the ridge itself a trail leads to the rocky range beyond over a spur which separates the Ta kou and Ch'i-pu kou 其卜溝 (valley). Apparently a very rough trail crosses the latter valley and the spur beyond whence one reaches the backbone of the Min Shan. This is a short cut to the Hsia T'ieh-pu 下鐵布 Land, but is not feasable for pack animals except perhaps yak.

On the ridge grew *Picea purpurea, Sorbus tapashana* Schn., *Rhododendron rufum* Bat., and *Rhod. Przewalskii* Max., at an elevation of 10,500 to 11,000 feet. Here were outcroppings of yellowish red limestone covered with a scrub vegetation which consisted of *Berberis diaphana* Max., *Salix oritrepha* Schn., and *Potentilla fruticosa* L. var. *dahurica* Ser. The meadows were a mass of flowers, especially common was the large deep wine to purple flowered *Primula Woodwardii* Balf. f., which ascended from 10,000 feet to 12,500 feet elevation; the Tibetan lady slipper *Cypripedium tibeticum* King, Pedicularis, etc. From the very top of the ridge one could look down into Ch'i-pu kou (valley) from an elevation of 11,550 feet. The valley extends from north to southeast, whether it cuts through the main range or flows into the Hsiao kou was impossible to see or to learn, at any rate it is a strange direction for a stream to flow when the backbone of the range is only about 6 miles distant. We saw no lateral valley in the Hsiao kou which carried a stream big enough to indicate that the torrent of Ch'i-pu kou was having its outlets into Hsiao kou.

Looking south and southwest from the top of the ridge we were confronted by limestone crags covered with bushes mostly willows, the new *Salix Ernesti* Schneid., *Salix oritrepha* Schn., *Rhododendron anthopogonoides* Max., *Caragana jubata* Poir., here a branching shrub 2-3 feet tall, also the purple flowered *Rhododendron capitatum* Max., which extends to swampy meadows. At the foot of the crags *Juniperus saltuaria* Rehd. & Wils., forms forests at 12,000 feet, also *Abies sutchuenensis*, the former reaches here a height of 40 feet, while the latter for an Abies was of small stature. The trail follows down the slopes of the ridge facing Ch'i-pu kou and thence up a ravine. The meadows here are swampy; *Caragana jubata* Poir., is common everywhere and the steep crags are forested with *Juniperus saltuaria* Rehd. & Wils. Among the Rhododendron bushes were hollow-stemmed yellow *Crepis Hookeriana* C. B. Cl., with a globose infloresence, the lavender *Meconopsis quintuplinervia* Reg., *Meconopsis punicea*, but most abundant of all was *Primula Woodwardii* Balf. f.; at 12,700 feet in alpine meadows the orange flowered *Pedicularis oederi* var. *heteroglossa* Prain, the pale red new *Pedicularis calosantha* Li, and the pale blue, new forget-me-not *Microula Rockii* Johnst., and *Trollius pumilus* painted the meadows in all the colors of the rainbow. On limestone boulders the dainty white-flowered *Androsace tapete* Max., formed cushions, from the crevices of the cliffs hung delicate bunches of *Paraquilegia anemonioides* and the yellow *Astragalus yünnanensis* f. *elongatus* Simps., and on mossy slopes along the limestone crags the pale yellow *Corydalis dasyptera* Max., was at home.

Other plants but not so common here and there in the alpine meadows were *Polygonum Hookeri* Meisn., the new rich purple *Pedicularis chenocephale* Diels, and the yellow composite *Cremanthodium bupleurifolium* W. W. Sm.

On meadows in forest clearings at 10,600 feet occurred the yellow *Ligularia altaica* D.C., first described from the Altai mountains, and the beautiful *Ajuga calantha* Diels, a perfect nosegay, two dark green leaves closely pressed to the ground and a bunch of dark blue flowers rising in the centre, all not higher than about 2 inches; and lastly the fleshy-leaved, white-flowered *Pinguicula alpina* L., of the family Lentibulariaceae, which grew embedded in moss on limestone rocks.

The trail continued along the ridge higher and higher between the crags and cliffs with a vertical strata, here and there a gap in the cliff permitted a marvellous view down the steep valleys towards the main backbone of the Min Shan (see Plate **19**).

Hsiao kou 小溝 *or Small Valley*

Hsiao kou or Small Valley is, as it carries a much smaller stream than the Ta kou or Large Valley, an affluent of the latter, yet actually it is an affluent of Ta-yü kou for it joins the latter below the confluence of the two main streams. According to Western usage the name Ta kou or Great Valley should be applied to the whole length of Ta-yü Valley but Chinese and Tibetans give individual names to all the various branches of streams. Thus Ta-yü valley or rather its stream has no source and begins at the confluence of the two upper branches.

Hsiao kou is a fairly long valley, about 12 miles or more long, and has its headwaters in the summit of the eastern end of the Min Shan called rTsa-ri khi-kha (Tsa-ri Khi-kha), a grassy flat 11,700 feet above the sea. The actual pass is further south and is 11,250 feet elevation. It has two branches the eastern one comes from Tsa-ri Khi-kha and the southern one from the pass north of Chha-tshad-thig (Chha-tshe-thi). All the waters from these passes or the Tsa-ri Khi-kha flat drain north into the Yellow River via the T'ao, but it is not the actual Yellow River – Yangtze watershed. From the 11,250 foot pass a trail leads almost vertically into the Chha-tshe-thi valley whose floor is 9,700 feet or 1,550 feet below the pass. This valley is joined by another small stream from the south which has its source in a pass 10,900 feet high, united they form the Sir-li-hdra, and flow east, and then probably north into the next valley east of Ta-yü kou. This last pass 10,900 feet high and known as gYen-chhen-run-sgo (Yen-chhen-rün-go), is the actual Yellow River – Yangtze divide.

Hsiao kou is intersected by four narrow limestone spurs at different intervals, from east to west, through which the stream cut narrow defiles forming Rock Gates or Shih-men, which during the rainy season in the summer become impassable. One is often obliged to wait for days, as we experienced, before it becomes possible to negotiate them.

From the village of A-chüeh the trail crosses the main stream of Ta-yü kou over a wooden bridge and then a small bridge over the stream which descends from Hsiao kou a wooded valley. There are no habitations in this valley which is very wild indeed. The valley opens into Ta-yü kou from southeast, but its source is south in the high plateau called Tsa-ri Khi-kha; the word Khi-kha as remarked elsewhere is a Tangut one and means range. It is up this valley and across Tsa-ri Khi-kha that a path leads into the extreme southeastern Hsia or Lower T'ieh-pu Land.

In the distance loom up high crags and luscious, green alpine meadows. Dense conifer forests exist here composed of *Abies sutchuenensis* Rehd. & Wils., *Picea asperata*, *Picea Wilsonii* and *Picea purpurea*. It is also here that the first larches *Larix Potanini* Bat., are encountered on the northern slopes of the Min Shan. The tree is entirely absent in the western end, both south and north of the range. It is common however, to the south of the range but only in the eastern and southeastern end of the Min Shan.

The valley is not rich in plants, more so than Ta-yü kou, but less than K'a-cha kou in the western part. Yet a number of species are only found in this part of the Min Shan, some of the Ssu-ch'uan (Szechuan) species have found here their northern limit but have not yet been able to extend westward.

The larches occur only from the first rock gate or Shih-men on, where they grow on the rocky slopes and on top of cliffs. Along the streambed and immediate slopes to both sides of the stream we find *Lonicera nervosa* Max., a shrub 3-4 feet with pink flowers, various *Salix*, *Spiraea alpina* Pall., with spreading branches and creamy-white flowers, *Sibiraea angustata* (Rehd.) Hao, one of the most common plants scattered also over meadows, *Lonicera trichosantha* Burm. & Franch., its branches spreading horizontally and bearing sulphur-yellow flowers, and a pubescent form of *Spiraea longigemma* Max.; with the above grew *Betula albo-sinensis* Burk., here a shrub or small tree 15 feet tall, its bark a rich copper color and its foliage of the lightest and softest green. On the meadows along the bank and valley floor the commonest plants were a purple aster, *Aster tongolensis* Franch., the new *Pedicularis paiana* Li, its bright yellow, large flowers contrasting beautifully with the purple asters, and the yellow-flowered *Pedicularis lasiophrys* var. *sinica* Max., *Paraquilegia anemonioides* (Willd.) Ulbr., and *Anemone demissa* H. f. & Th., festooned the limestone cliffs, below which, on boulders in the streambed grew the orchid *Orchis chusua*, displaying rich purple flowers.

At 9,450 feet the valley narrows, a limestone spur cutting clear across it, and here we meet with our first obstacle in the shape of a very narrow defile, most difficult to negotiate owing to the fierce torrent which rushes through it, washing both walls. This is the first of four rock gates one must pass in order to reach the head of the valley and the top of the range. We were held here for four days and finally built a trail close to the wall and huge boulders in the streambed by cutting big larches, laying them across. and filling the spaces with rocks and masses of willows. Had one fallen into the stream he would have been crushed to pieces among the boulders. The mules had to be led carefully across the improvised trail, and the loads carried sideways on their saddle frames. The rocks are here limestone and of a reddish-yellow color.

The steep hills are clothed with the loveliest forest of birches and *Larix Potanini* Bat., their branches are long and slender and droop like weeping willows. The Yün-nan Larix is to my mind a different species, of an entirely different habit of *Larix Potanini*; the Kan-su species has a pale, soft, grayish green foliage. Beyond the defile grew roses and in the meadow the lovely *Ajuga calantha* Diels, described previously. Everywhere the valley is hemmed in by high crags and forests interspersed with flower-studded meadows. In the dense spruce and larch forest in moss *Allium victorialis* L., a native of the European alps found here a new home with *Polemonium coeruleum* L. ssp. *vulgare* Brand, while on the outskirts of the forest among rocks thrived the new, fleshy, *Sedum progressum* Diels, *Cotoneaster adpressus* Bois., a shrub 1 foot tall covering boulders, *Saxifraga fragrans* var. *platyphylla* H. Smith, a yellow-flowered species, *Sedum henryi* Diels forma *gracilis* (♂), and the satiny blue-flowered *Meconopsis racemosa* Max., with pale yellow spines, at the foot of limestone cliffs; with it but on moss-covered boulders grew the orchid *Amitostigma monanthum* (Finet) Schlecht., a white flowered species spotted purplish brown. It was the only place where we collected this orchid not only on the Min Shan, but in the whole of Kan-su; that it was very rare is indicated by the fact that we secured only one single specimen.

The new *Saxifraga kansuensis* Mattf., was partial to the crevices of limestone cliffs and like the former very rare, this being the second locality where we encountered it on the Min Shan. *Aruncus sylvester* Kostel grew in meadows on the margins of forests of larch and spruce with *Salix Rehderiana* var. *brevisericea* Schn., and in the shade of the forest, *Sorbus tapashana* Schn., here a tree 18 foot high. Under the latter on mossy slope flourished a pale yellow *Corydalis* forming bushes 1 – 1 ½ feet tall, probably new (no. 12834). Its companions were a single *Primula gemmifera* Bat., *Primula Woodwardii* Balf. f. and the rare *Primula alsophila* Balf. f. & Farrer, the only place where we encountered it.

On the open grassy slopes *Meconopsis punicea* commenced at 10,000 feet to color the landscape with its brilliant scarlet flowers. On meadows surrounding larches at 10,500 feet *Rheum palmatum* L. floribus rubris, 4-5 feet tall, crimson flowers and very large leaves, its otherwise fleshy stem woody at the base, occupied considerable space. Retreated to the shade of the larches we met the araliaceous *Acanthopanax Giraldii* Harms, a spiny shrub 3-4 feet high and greenish flowers, also *Euonymus Przewalskii* Max., 4-8 feet tall, with scandent quadrangular branches and dark reddish-black flowers. Beyond appeared the first *Rhododendron rufum* Bat., in company with Abie*s* and Betula.

Beyond the fist rock gate the stream had to be crossed eighteen times, and two other rocky defiles had to be negotiated, the trail being the streambed. The fourth limestone defile was at 9,900 feet and from here the geological formation changes completely, the limestone has now given way to conglomerate.

The gorge we now enter is of the above formation and reddish in color. The massive blocks of conglomerate rise thousands of feet into the air to both sides. They appear rounded and as if cut into layers by the torrential rains which occur here nearly daily throughout the summer. Larix trees grew on the top and on ledges of the massive walls, also an occasional spruce and firs. *Juniperus saltuaria* Rehd. & Wils., became common and formed practically the sole tree growth on the summit plateau of Tsa-ri Khi-kha. Willows continue to be the principle deciduous trees and shrubs with the shrubs enumerated.

Near the head of the valley, 10,500 feet elevation, now reduced to a ravine, the conglomerate cliffs rise vertically thousands of feet, the trail leading steeply between crags. The grassy slopes were marshy and for pack animals most difficult to ascend. Finally the trail led along the edge of the crags, once on actual ledge of conglomerate wall, and to the summit of a grassy plateau intersected by a deep chasm from southeast to northwest, one branch of the Hsiao kou which has here its source at 11,700 feet. The snow-covered conglomerate walls rise here straight several thousand feet from the grassy flat or small plateau. A trail leads over the swampy meadow to Min-chou, the present day Min Hsien.

Tsa-ri Khi-kha is the highest pass on the eastern part of the Min Shan and its southern and drops abruptly into a valley called Chha-tshe-thi, the pass into the valley being 11,250 feet, the ground intersected by the chasm mentioned rises thus 450 feet to the foot of the northern crags. From the 11,250 feet pass above Chha-tshe-thi, small ravines extend north or northnortheast and in these is the source of the western branch of Hsiao kou. These ravines merge into larger ones which separate the crags to the north from the main plateau. The trail down Hsiao kou follows first down in the narrow ravine, better called ditch so narrow that a small donkey without any load can just pass through.

Tsa-ri Khi-kha, the Last Pass Over the Eastern End of the Min Shan in Cho-ni Territory

The flora of Tsa-ri Khi-kha is a disappointing one. The scenery however is magnificent beyond words, it would make a wonderful stage setting for «The dawn of the gods». No one who crosses this plateau into Lower T'ieh-pu Land lingers here for it is the happy hunting ground of the T'ieh-pu robbers. It was here that we encountered bandits with whom we exchanged shots resulting in one of our men only wounded and one T'ieh-pu killed by my Cho-ni escort furnished by the Cho-ni prince.

Among bushes along the little watercourses which criss-cross the swampy plateau, *Meconopsis quintuplinervia* Reg., and *Primula tangutica* Max., were common; the bushes represented mostly the prostrate *Lonicera thibetica* Burm. & Fr., *Potentilla fruticosa* var. *dahurica* Ser., and var. *Purdomi* Rehd., the latter found only on the eastern end of the Min Shan, *Salix* etc.

The flora of the alpine meadows was very poor in comparison to that of Kuang-k'e Shan, but certain plants occurred here which were peculiar to the eastern end of the range, as *Meconopsis psilonomma* Farrer, a very distinct species which could never be mistaken for any other of the lavender or purple species of Meconopsis. To begin with the flowering stalks are terete and *hollow*, spiny-haired, the spines dark red, the plant is 1-2 feet tall, the flowers are quite large semi-drooping and deep purplish blue, they are suspended from the very tip of the stalk in such a way as to appear artificially attached and not contiguous with the stem. It only occurs on Tsa-ri Khi-kha in thick turf, most difficult to uproot, at 12,500 feet elevation and nowhere else. Farrer's Ardjeri is A-chüeh, and it is precisely there where it grows. At Tsa-ri Khi-kha it meets *Meconopsis quintuplinervia* Reg., a much less robust species, very easily uprooted, with solid, not hollow stems, smaller flowers and much paler in color, and the red *Meconopsis punicea*. The former is usually found in Rhododendron scrub as *Rhod. anthopogonoides* Max., and the latter either in open meadows or on the edge of forests.

Abies Faxoniana Rehd. & Wils., clings to the foot of the cliffs or grows in groves near the cliffs with *Abies sutchuenensis* Rehd. & Wils., *Picea Wilsonii* and *Picea asperata*. *Rubus pileatus* Focke, belongs to a lower level with Larix in the shade of which it grows.

The conglomerate cliffs had a flora of their own as the umbelliferous *Pleurospermum Franchettianum* Hemsley, which grew in crevices, *Anemone demissa* Hook. f. & Th., and in the autumn *Gentiana Szechenyii* Kan., with deep bluish-purple flowers, it did not ascend to the summit but remained in more protected areas on the conglomerate cliffs, where it was associated with *Saxifraga lumpuensis* Engl., this latter plant also grew on detached boulders to near the summit. This is the only instance that it has been observed on the Min Shan, it does however occur also on the northern limestone mountain called Lien-hua Shan q.v.

In the alpine meadow at the summit, between 12,500 and 12,700 feet, thrived *Primula optata* Balf. f. & Farrer, a very handsome plant with bluish purple flowers, stems 1-2 feet tall, and linear oblong glabrous leaves; *Pedicularius szechuanica* Max., typica Li, its flowers a rich crimson to purplish elected only the highest parts of the meadow at 12,750 feet. *Cremanthodium Limprichtii Diels*, grew scattered over the summit its nodding yellow flowerheards being quite conspicuous in the rich green swampy meadow. The lovely *Androsace Mariae* var. *tibetica* (Max.) Hnd.-Mzt., its white flowers sessile, favored huge boulders which it covered, forming a cushion, it was usually accompanied by *Lloydia tibetica* Bat. var. *lutescens* Franch., a liliaceous herb with yellow flowers striped a darker yellow.

The new crucifer *Megacarpaea Delavayi* Franch. var. *grandiflora* O. E. Schulz, over 2 feet tall, its flowers a rich punkish lavender, preferred the grassy slopes near Rhododendron and Juniper forest.

Somewhat below the summit in alpine meadows grew *Pedicularis Davidi* Franch., *Morina betonicoides* Benth., and a species of *Allium* (no. 12616).

The northern end of this grassy plateau where it drops vertically into Chha-tshe-thi valley is flanked by enormous red conglomerate cliffs at the foot which grow Abies, Picea, and *Juniperus saltuaria* Rehd. & Wils., the alpine meadows sloping gently to the naked cliffs, in gentle undulations with little hills and rock outcroppings here and there. These little hills were covered with firs and Rhododendrons as *Rhod. rufum* and *Rhod. Przewalskii*, but mainly with the somber *Juniperus saltuaria*.

The Valley of the Pai-lung Chiang 白龍江 and the Southern Slopes of the Kan-su Min Shan

The whole area south of the Min Shan in the Cho-ni territory is divided into two tracts, the western one from Drag-gam-na to, and including Pe-zhu (dPal-gzhu), on the south bank of the Pai-lung Chiang or the White Dragon River, is reckoned to the Upper The-wu or Upper T'ieh-pu or Shang T'ieh-pu 上鐵布 Land, and the eastern tract from the bridge immediately beyond Pe-zhu to and inclusive the new monastery of Wang-tsang men-chhe Dom-pa (dBang-gtsang man-chhe dgon) is designated as the Lower The-wu or Lower T'ieh-pu or Hsia T'ieh-pu 下鐵布, pronounced in Kan-su Ha T'ieh-pu. To the latter area belong however all the valleys north and south of the Pai-lung Chiang, east of Pe-zhu, and the last eastern valley called Ma-ya kou 麻牙溝 extending north to the pass Lha-mo gün-gün (Lha-mo-gun-gun), and from the pass north to Tsa-ri Khi-kha (rTsa-ri Khi-kha) the eastern summit of the Min Shan. Still belonging to the Lower T'ieh-pu Land are the long valleys extending south of the New Wang-tsang monastery namely A-hsia kou 阿夏溝 and To-erh kou 多兒溝, the latter is the longest and extends to a high mountain range called Yang-pu Shan 陽布山, the Tibetan Ta-ge La (rTa-rgas La) which is not only the southernmost border of the Cho-ni prince's territory, but it is also the border between the provinces of Kan-su and Ssu-ch'uan.

The Upper T'ieh-pu Country

The Upper The-wu or Shang T'ieh-pu country includes the region immediately below Kuang-k'e pass and the scenic amphitheater called Drag-gam-na already described. From Drag-gam-na the streams which have united from the various directions enter a Shih-men or rock gate directly south into a valley known as Yi-wa kou. This valley is about 15 miles long to its confluence with the Tshong-ri Nang which is the main branch of the Pai-lung Chiang, also called Pai-shui Chiang 白水江 or White Water River; it has two affluents called Tso-lu kou 作路溝 also called Bum-pa Nang (hBum-pa Nang). This valley sends an affluent into the Za-ri Khog (Zwa-ri Khog or Nettles Mountain valley; Bum-pa Nang has its source in a mountain called Am-nye La-gu (Am-nye La-dgu). Za-ri Khog is not the main branch of Pai-lung Chiang but an affluent of the much longer Tshong-ri nang which flows between two lamaseries viz.: Dang-zhin-gi-ser-thri Gom-pa (Dang-zhin-gi-gser-khri-dgon) on the north bank, and Kir-di Gom-pa (Kir-rdi-dgon) of Tag-tshang Lha-mo (sTag-tshang-lha-mo). The source of the stream, the main branch of the Pai-lung Chiang or Pai-shui Chiang is east of the Je Khi pass, hence the valley and stream between the source and the lamaseries are called the Je Khog (rJe-khog). All these streams are however in Ssu-ch'uan, in that unadministered wedge between Ch'ing-hai and Kan-su provinces. The source of the Ch'a-lu kou is not known; it arises in the west and is a considerable stream where it enters the Pai-lung Chiang; in fact in Cho-ni, i.e. T'ieh-pu country, that stretch of the river which joins the stream from Yi-wa kou 亦哇溝 is known as Ch'a-lu kou 茶路溝 which is a Chinese transcription of its Tibetan name Tsha-ru Nang or Tsha-ru valley, the real Chinese name of the river is Hsiang-chih Ho from its source.

The confluence is called Chhe-khu-kha (Chhe-khu-kha), and is at an elevation of 7,850 feet. The Yi-wa kou (valley) widens here considerably and forms a triangular grassy plain, the stream hugging the western foot of the mountains which are forested. The confluence is the border between Cho-ni – Kan-su province and Ssu-ch'uan, and is no-man's land, inhabited by notorious Tibetan robber tribes.

A short distance from the confluence (south), on the north bank is situated the hamlet Wa-chi-k'o, opposite the mouth of an affluent from the west called Tsha-de-go-zhi (Tsha-sde-sgo-bzhi). The stretch from the mouth of Tsha-ru or Ch'a-lu kou to the long valley called Wa-pa kou q.v., which has its source southwest of Drag-gam-na and debouches into the Pai-lung Chiang west of Pag-shi gong-ma Gom-pa (sPag-shis-gong-ma-dgon) monastery, is known as Dro-tshu (hGro-tshug).

The Valley of Yi-wa – Dro-tshu

As already remarked the Drag-gam-na River after entering the massive Shih-men or rock gate south of, and facing Tong-wa village, becomes the Yi-wa Ho 亦哇河 and the long valley the Yi-wa kou. At the entrance to Yi-wa kou are two immense boulders which just permit one rider to pass between them, but for loaded animals it is necessary that the loads be removed and carried sideways on their racks. The elevation within the defile is 9,000 feet and the heights of the cliffs or sentinels to each side about 1,000 feet or more; the western part of the portal connects with a spur which rises to high wooded limestone crags, and extends west the whole length, south and southwest of Drag-gam-na to the Ssu-ch'uan border. [Plate **20**] Only a short distance beyond, less than two miles one is confronted with a second rock gate much higher than the first. [Plate **21**] The stream makes a sharp curve around the massive eastern cliff, and the trail leads into a semi-circular cove in which are situated two villages one east, Ga-khu (dGah-khu) and one west called Na-chia 那加, elevation 8,800 feet.

On the gravelly banks of the stream between the two limestone Rock Gates we encounter *Saussurea parviflora* (Poir.) DC., which carries its pale-pinkish small flowerheads in large corymbs; *Aster Fordii* Hemsl., with pale blue ray-florets; *Nepeta macrantha* Fisch., a blue labiate first found in the Altai Mountains; the umbellifer *Tongoloa elata* Wolff., gregarious along the streambed found also in Ssu-ch'uan besides here on the Min Shan, and the tamarix-like shrub *Myricaria alopecuroides* Schrenk. On the steep slopes of the valley *Picea wilsonii* Mast., a spruce reaching gere 100-150 feet in height, with trunks 2-3 feet in diameter forms dense forests, between the last Rock Gate and another a little over a mile or so beyond. *Aralia chinensis* L. var. *nuda* Nakai, a small spiny tree 10 feet tall with spreading branches and yellow flowers is common at 9,000 feet and lower among bushes of willows *Lonicera*, etc.

The gorge becomes constricted by encroaching limestone spurs, and remains so for a considerable distance till the hamlet of Pe-thung is reached. *Picea wilsonii* forms magnificent groves. Here in this forest we shot the small, brown, parrot-billed *Suthora conspicillata* David. In gravel and on slate flourished a species of Senecio, 3-4 feet tall with small yellow flowerheads, as yet undescribed (no 14573); along the stream *Clematis brevicaudata* DC a wood vine which covered bushes of Berberis, etc. occurs also in the far northern Nan Shan, and was first recorded from between Peking and Jehol; with the latter grew *Viburnum glomeratum* Max., a shrub 5-6 feet tall, with black fruits, and in the shade of the spruces the fern *Athyrium filix-femina* (L.) Roth, with *Phlomis umbrosa* Turcz.

At the lamasery of Chho-og Gom-pa (Phyogs-og-dgon-pa) the Chinese Cha-ha Ssu, and the village of Go-dzü-na (sGo-dzul-nag) the former situated on a terrace above the west bank, elevation 8,180 feet, are still forests of spruces, but here we found at 9,500 elevation above the forests of *Picea wilsonii*, a silver fir not before recorded from Kan-su, it is the only place where this fir, *Abies recurvata* Mast., has been found so far, outside of the mountains, west of Sung-p'an 松潘, in the extreme northwest of Ssu-ch'uan where it forms extensive forests. It is here, on the Min Shan, a tree 50-80 feet tall and easily distinguished by its greyish-brown, dull cones, and its bluish white needles.

Beyond the village of Go-dzü-na the rock formation changes completely, the hard grey limestone gives way to mica-slate, shale and schist. Tall conifers still form forests on the western valley walls of Yi-wa kou, the crest of which is the border of Cho-ni and Ssu-ch'uan. With the change of the rock formation the scenery also changes, the only beautiful aspect of the whole is the magnificent groves of *Picea Wilsonii*, the only spruce found in this valley. At 8,000 feet elevation we meet the first pine trees *Pinus tabulaeformis* Carr., near the village of Drü-tsho-na (hBrul-htsho-nag), the vegetation takes on now a xerophytic character, the valley slopes are now covered with scrub vegetation of *Cotoneaster multiflorus* Bge., *Crataegus kansuensis* Koidz., and *Pyrus pashia* Ham., with brilliant red fruits one inch across (no. 15090) certainly different from the *Pyrus pashia* found in the south of Yün-nan and Burma where its fruits are never larger than a small marble and black.

The oak is here a small tree 10-15 feet high and forms the main growth on the hillsides to near the mouth of Tsha-ru valley, the usual Berberis, Hippophaë etc., occur along the stream, while *Pinus tabulaeformis* covers the hillsides in part. The stretch from the confluence of the Tsha-ru valley to the mouth of Wa-pa kou is known as Dro-tshu (hGro-tshug) and harbors the above dry xerophytic type of vegetation, it is a continuation of Yi-wa kou. The mouth of Tsha-ru valley is at an elevation of 7,850 feet its slopes are grass-covered and partly terraced, a village being situated not far from its confluence with the Yi-wa kou stream.

Opposite the village of Wa-chi-k'o a valley debouches from the west called Tsha-de-go-zhi, here the hills are low, composed of conglomerate, gravel, schist and loess. We find here the small cushion-forming yellow-flowered *Caragana tibetica*, a real sign of arid conditions.

Directly opposite the mouth of Wa-pa kou is a high peak with limestone cliffs and crags, called Tsha-ri-ma-mön (Tshwa-ri-ma-smon) about 13,500 feet in height (see Plate **22**).

From Wa-pa kou to Pe-zhu on the Pai-lung Chiang
The Wa-pa kou River flows in a very deep gorge which must be forded and the high valley wall ascended to attain the large grassy terrace on which the monasteries Pa-shi gong-ma Gom-pa and Gong-ma-nang Gom-pa (dGong-ma-nang dgon) are situated, the latter being lower than the former. In Chinese they are called Tien-ha shang-ssu 殿哈上寺 and Tien-ha hsia-ssu 殿哈下寺, respectively. The elevation of the terrace is 7,900 feet.

The Pai-lung Chiang received several affluents from the southern slopes of the Min Shan, the first east of Wa-pa kou is called Ra-na Nang (Ra-sna-nan) after the village of Ra-na, the Chinese La-na 拉那 situated on the west bank. [Plate **23**] Within the confluence of this and another small stream, before it reaches the Pai-lung Chiang is the lamasery of Za-kö Gom-pa (gZah-bskos) situated on a bluff. Opposite on the south bank of the river are ruins, the remnants of square walls and moats. This is the ancient site of the city of T'ieh-chou 疊州 over which a viceroy ruled under the T'ang emperor T'ai-tsung about 627 A. D. The city of Tieh-chou was first established during the northern Chou dynasty between 557 and 581 of our era. During the Sui dynasty 589 to 617 A.D. it was abolished but then reestablished at the beginning of the T'ang as stated above. It fell later into the hands of the wild Tibetan tribes whose name is derived from the ancient city. In the records of Cho-ni it is stated that during the Han dynasty B.C. 206 to 25 A. D., the name of the ancient land of the Shang T'ieh-pu was T'ieh-chou. Undoubtedly excavations among these ruins would bring interesting objects to light. However, during the occupation of the land by the wild and war-like T'ieh-pu tribe this would have been impossible without a well armed troop of soldiers as protection. East of the ruins are the villages of Kon-re and Sa-nge and west of them Shih-tzu 什子 and La-lu 拉路. Beyond is the valley of La-lung-pa, and a village of the same name at 7,400 feet elevation; opposite is the Bön Lamasery of Sa-rang gom-pa, the Chinese Sa-

lang ssu 撒浪寺. On the same side of the river, a short distance beyond is the hamlet of
Bo-tsha-kha (sBo-tsha-kha).

At the village of Ngo-ngo, elevation 7,350 feet, a lovely valley opens out from the
south called Do-ro-phu (rDo-ro-phu) this is just half the distance between Wa-pu kou
and Pe-zhu (dPal-gzhu). The Valley is densely forested and we explored it thoroughly.
Opposite Ngo-ngo on the north bank of the river is the lamasery of Tra-shi Gom-pa
(bKra-shis-dgon-pa) the Chinese Cha-shih ssu 扎什寺. There are several villages still
on both sides of the river and one more Bön or Black Sect lamasery situated on a bluff
on the north bank called Chhi-chhe Gom-pa (Phyis-phye-dgon-pa), the Chinese She-she
ssu.

The last village in the upper T'ieh-pu land is that of Pe-zhu on the south bank. The
entire stretch of the river from Wa-pa kou to Pe-zhu is called Dön-dra-le Chhu (gDon-
hgra-sle-chhu) by the T'ieh-pu, they delight in giving various parts of a river a different
name thereby causing much confusion.

The Vegetation Between Wa-pa kou and Pe-zhu
The flora of this stretch of the river which flows in a rather broad valley is of a
xerophytic nature. Above the north bank are high grassy terraces, while the south bank
is level with the river. The entire valley walls back of the villages on the south bank are
densely forested with the robust pine, *Pinus tabulaeformis* and other conifers, while the
north bank is gravelly and dry and covered with a scrub vegetation composed of
Cotoneaster, Berberis, *Prinsepia uniflora* Bat., which the The-wu call Ti-ti.

Near and at the hamlet of Bo-tsha-kha are large groves of *Populus cathayana* Rehd.,
first described from Ssu-ch'uan, forests of pines, spruces, pears and willows. The rock
formation is loose shale, slate covered with loess, hence the valley is broad and open.

Back of the village of Ngo-ngo, elevation 7,350 feet, a valley extends south from the
south bank of the river. This valley proved of special interest as it was densely forested,
and for this reason we explored it thoroughly.

The Forests of Ngo-ngo
The main conifer in this rather narrow gorge proved to be *Picea Wilsonii* Mast., which
grew to great height from 100-150 feet with trunks up to 4 feet in diameter, it covered
the valley slopes to the exclusion of nearly everything else. On its outskirts tall birches,
Betula albo-sinensis Burk. var. *septentrionalis* Schn., reaching 60-80 feet and 2 feet in
diameter with red bark and large leaves, lined the margins of the forest as if to protect
the spruces, at 9,500 to 10,000 feet.

Along the stream, above the entrance to the valley, the 30 foot oak *Quercus
liaotungensis* Koidz., with oblong sinuate leaves and large acorns lined its bed with
Cornus macrophylla Wall., a tree 25 feet tall, and the tortuosely branching *Caragana
tangutica* Max., a shrub 5-8 feet tall. This latter shrub extended also into the Picea
forest. Climbing over trees, the huge liana *Clematoclethra lasioclada* Max., bearing
paniculate flowers and black fruits with a reddish tinge, swung from tree to tree like a
huge giant snake, its woody stems 40-60 feet long. The new *Viburnum betulifolium*

Bat., forma *aurantiacum* Rehd., a shrub attaining 15 feet, with orange colored fruits borne in drooping panicles preferred the streambed and edge of forests as did the yellow-flowered *Caragana densa* Kom., *Hydrangea Bretschneideri* Dipp., and the black-fruited *Lonicera nervosa* Max. Scattered among the spruces grew *Euonymus alatus* Reg. var. *apertus* Loes., its branches round, leaves elliptical, and bearing dehiscent capsules with four carpels divided to the base; the seeds are large black with an orange arillus. With it was associated *Rubus pileatus* Focke, and the 6-10 feet tall *Cotoneaster lucidus* Schlecht., with oval to oblong black fruits borne in pairs at the apices of the small twigs. At 8,000 feet open, treeless slopes were occupied mainly by *Cotoneaster nitens* Rehd. & Wils., a shrub 6-10 feet with slender rambling branches, thickly tomentose leaves and carmine fruits, and *Spiraea wilsonii* Duthie, which also extended to the edge of forests. Seeking the shade of spruces was the araliaceous *Acanthopanax Giraldii* Harms, a shrub reaching 15 feet in height, with 3-5 foliolate leaves, densely spiny from the base to the tips of the branchlets. Of other deciduous trees *Acer tetramerum* Pax var. *betulifolium* Rehd., attained a height of 35 feet, distinguished by its ovate, acuminate, deeply serrate leaves and large fruits in simple racemes; it was confined to the 9,000-9,500 feet level where it was often smothered by another giant liana of the same genus as the previous one, but representing another species, viz. *Clematoclethra integrifolia* Max., its entangling woody runners reaching a length of more than 60 feet and several inches in diameter. Its leaves are small and sharply dentate, and its black fruits are single, axillary and borne on a slender peduncle. Wherever Clemathocletra occurs the forests are dense and mixed, i.e., composed of conifers and deciduous trees.

Abies Faxoniana Rehd. & Wils., formed stands at the very head of the valley with *Sorbus Koehneana* Schn., above or among which grew *Juniperus squamata* var. Fargesii Rehd. & Wils., extending to the otherwise bare hillsides at the ehad of the valley.

From Ngo-ngo to Pe-zhu
In front of almost every village there is a wooden bridge over the Pai-lung Chiang, spurs extend here and there from the Min Shan into the valley forcing the stream to describe curves. On the south bank there are no spurs, the valley wall rising steeply from the valley floor, cultivated by the The-wu of the various villages. Wheat and barley and buckwheat are the main crops. Villages are situated on the valley floor, usually near the mouth of lateral valleys, while the lamaseries are on bluffs or terraces above the river, but only on the north bank for there are no terraces on the south bank.

Chhi-chhe Gom-pa or She-she ssu is situated on a terrace above the river; the valley is here quite broad and flat grassy terraces extend east, while the stream flows close to the pine forested steep valley walls. The water of the stream is a milky white hence its name Pai-lung Chiang = White Dragon River or Pai-shui Chiang = White Water River. The trail down stream always follows on the terrace on the north bank but near Ngon-gon, the Chinese Kung-ku 工古, it descends to the stream bed. The scenery along the whole valley here is very beautiful; the banks of the stream are lined with lovely poplars, *Populus cathayana* Rehd., while pine forests extends all along the southern slopes of the valley. At Ngon-gon, elevation 7,180 feet, there is a cantilever bridge, the village being situated on the south bank. At the bridge we were confronted by a singular scarecrow (see Plate **24**); in a pile of rocks there rose a tall straw-man, a tall pole with a long crosspiece for outstretched arms, hung over with long straw, and a wreath of straw where the head was intended to be, and both ends of the crosspiece tied with straw, allowing loose straw to descend over the ends to represent the hands. His left hand carried a long stick. Evidently it was to scare off some evil influence or demon. On enquiry we were told that some cattle epidemic was then in progress and this straw-man was to prevent the evil spirit who spread it from bringing it to the village. The tops of the end-posts of the bridge were primitively carved to represent human faces, wearing round caps, while opposite the end of the bridge, in front of the village entrance, were four posts of unequal height, to represent human beings, their faces were long, their mouths wide, some had mustaches and goatees. They were embedded in a pile of rocks, and were to guard the bridge, and to prevent demons or other evil influences from crossing the bridge and entering the village. These primitively carved figures reminded of African fetishes and had certainly nothing to do with Buddhism; however most of the The-wu of this region are adherents of the Bön religion, the pre-Buddhistic religion of Tibet. [Plate **25**] Today they are known as the Black Lama Sect, considered enemies of Buddhism.

Not far from Ngon-gon is the village of Pe-zhu, the last in Upper T'ieh-pu Land. Pe-zhu, written dPal-gzhu, is a beautiful spot, especially the land opposite on the north bank. It resembles open park land and promised botanically to be of interest. The forest we encountered here was of an open type, like a savannah forest and harbored quite a number of trees and plants not encountered elsewhere. The most prominent tree was *Juniperus chinensis* L., which Cho-ni people called T'an-hsiang-mu 檀香木 or Sandalwood tree on account of its fragrance. It was here a tree 40-50 feet tall with a round crown, white or pruinose fruits, and formed groves. It is one of the most widely distributed trees in China, but was the only place on the Min Shan where we encountered it. It is reported to grow also in the Ta-ra valley south of the Pai-lung Chiang (See Plate **26**)

It is associated here with two species of oaks, one mentioned previously *Quercus liaotungensis* Koidz., and *Quercus Baroni* Skan., the latter a tree 25-30 feet tall with a spreading crown. The undershrub was composed of *Sageretia theezans* Brogn., a shrub with long whip-like branches occuring also in Li-chiang and Yün-nan in general, *Berberis vernae* Schn., widely distributed in Kan-su, *Pyrus pashia* Ham., with spherical greenish-yellow fruits resembling apples, and *Rhamnus leptophylla* Schn., a stiff shrub with horizontal branches and globose black fruits.

Along the banks of the river in gravelly soil grew *Prunus salicina* Lindl., *Clematis brevicaudata* DC., and *Clematis fruticosa* Turcz., an erect shrub 2-3 feet tall, with erect branches and dark yellow flowers; *Picea asperata* and *Pinus tabulaeformis* descended to within a short distance of the river bank at 7,300 feet; near Pe-zhu in the forests of Chi-ni-no on the north bank occurred *Sorbus hupehensis* Schneid. var. *aperta* (Koehne) Schn., a large tree 40-50 feet tall with trunks 2 feet in diameter, large leaves, and white fruits borne in large and ample panicles. *Crataegus kansuensis* Wils., with red fruits, like in Cho-ni on the T'ao River, thrives here on the banks of the White Dragon River with *Acer Maximowiczii* Pax, *Rosa Wilmottiae* Hemsl., *Ribes moupinense* Fr. var. *tripartitum* Jancz., a shrub 5 feet tall with deeply tri-lobed leaves, and black fruits borne in long racemes, and *Euonymus nanoides* Loes. & Rehd., the latter extending up grassy slopes in the valley where the soil is mostly loess; here also occurs *Prinsepia uniflora* Batal., and most of the shrubs found higher up in the valley.

In the Juniper grove at Pe-zhu we shot a new bird *Fulvetta cinereiceps fessa* B. & F. Beyond Pe-zhu a wooded ridge extended to the stream and the valley had the appearance of changing its character.

The Lower T'ieh-pu Land or Hsia T'ieh-pu 下鐵布

From Pe-zhu to Wang-tsang

The Pai-lung Chiang from the border of the Upper T'ieh-pu to the mouth of the valley of Ts'ao-shih 草什溝, a distance of about 25 miles is called by the The-wu Chhu-lung Kar-po (Chhu-klung dkar-po) or the White Nâga River, the word klu pronounced lu is a serpent spirit, the Indian Nâga and equivalent to the Chinese dragon. Its two most important affluents are first the Ta-ra Nang (rTa-ra Nang) the Chinese Ta-la kou 達拉溝 and further east the Wang-ts'ang kou 旺藏溝. The head of the latter valley meets the Ta-ra Nang, being separated by a spur. The river drops from Pe-zhu where it is 7,100 feet to 6,100 feet at Ts'ao-shih kou, a drop of 1,000 feet in 25 miles. Both above mentioned valleys debouch from the south into the Chhu-lu-kar-po or White Dragon River.

Its affluents from the north, that is from the Min Shan, number three, but are not as long as those joining the river from the south, as their sources are not far from the crags of the Min Shan. They debouch at right angles to the river, while those from the south flow parallel to the river, in long valleys, before entering the latter.

The small district east of Pe-zhu is called K'a-pa-lu 卡巴絫 and is dissected by two streams, between these are situated two lamaseries, one above the other. The upper one is called K'a-pa shang ssu 卡巴上寺 meaning upper K'a-pa temple, its Tibetan equivalent is Kha-pa-lu gong-ma Gom-pa or the Kha-pa Nâga upper lamasery, and the lower K'a-pa hsia ssu 卡巴下寺 or the lower K'a-pa temple, in Tibetan Kha-pa-lu dgon-ma Gom-pa. The upper being the better one belongs to the Yellow Sect while the lower is a Bön or Black Sect lamasery. On the opposite bank is the village of K'a-pa-lu. The most important and oldest one in the region is the Yellow Sect Lamasery of Wang-tsang, south of the river, at an elevation of 6,490 feet. Its Tibetan name is Wang-tsang-Yön-chhe Gom-pa (dBang-gtsang yon-chhe dgon) or the Monastery of Great Knowledge.

Beyond Pe-zhu the river makes a sharp curve around a spur forested with lovely pines which hang over the torrent stream. From here on the valley changes into a gorge or in fact a canyon.

The trail is exasperatingly narrow. To both sides of the river are huge forests of pine (*Pinus tabulaeformis* Carr.) associated with the oaks recorded from Pe-zhu, *Pyrus pashia* Ham., the new *Crataegus kansuensis* forma *aurantiaca* Wils., and *Pistacia sinensis* Bge. The scenery is magnificent and enhanced by tall trees of *Populus cathayana* Rehd., willows etc. The village of Ni-shih-k'a 你什卡, the first one in the lower T'ieh-pu country is situated east of the gorge, till we come to a lateral gorge emptying from the south, the notorious Ta-ra Valley, notorious on account of the Ta-ra The-wu tribe who inhabits it. It is a long valley and extends for several miles behind, and parallel to the southern valley wall of the Pai-lung Chiang. They recognize no authority, are a law unto themselves, pay taxes to no one and are the terror of all the other The-wu south of the Min Shan. They dress in scarlet from head to foot and are the most warlike people next to the Go-log of the Am-nye Ma-chhen. In the winter when the rivers are frozen the Ta-ra The-wu emerge from their fastnesses and rob and plunder. Woe to the traveler who encounters them for he is taken as a slave never to emerge again from their stronghold.

Beyond the village of Ni-o 你峨 the Tibetan Nyi-ngön (Nyi-sngon) and where the Ma-ni kou 麻尼溝 empties into the river from the south, the river makes a right angle turn north at 6,800 feet elevation and flows into a limestone canyon, extremely narrow and filled with enormous boulders, the white foam dashing from rock to rock. The gorge continues till the lamasery of Wang-tsang is reached. All the scenic beauty has been left behind, the people are poor and eke out a precarious existence.

Here we encounter a more or less xerophytic vegetation. *Koelreuteria paniculata* Laxm., a sapindaceous tree here 15 to 18 feet tall with bladdery fruits, and spreading branches, occupied the gravelly banks, often overhanging the river, with *Pistacia chinensis* Bge., 25-30 feet tall bearing large clusters of red fruits (September), with *Rhododendron micranthum* Turcz., a 4 foot shrub with small linear leaves, and small white flowers arranged in dense racemes, *Wikstroemia chamaedaphne* Meisn., a bush 4 feet with rich yellow flowers, extending up the rocky slopes with the foregoing, the composite shrubs *Microglossa salicifolia* Diels, 4 feet tall, and *Aster incisus* Fisch., 3 feet tall and bearing white flowers, formed often large clumps. The new *Indigofera Bungeana* Walp. forma *spinescens* Kob., formed spiny cushions with pink flowers on slaty slopes near Wang-tsang, but as the species itself occurred in identical situations it cannot be an ecological form as believed by Kobuski.   Other associates in the limestone canyon delighting in the dry gravelly, rocky banks and slopes were *Plectranthus discolor* Dunn, the white *Aster incisus* Fisch., *Clematis aethusifolia* Turcz., *Spiraea uratensis* Fr., *Cotoneaster multiflorus* Bge., the lilac *Syringa oblata* Lindl. var. *Giraldii* Rehd., its flowers pink or violet, first described from Shensi, a very ornamental shrub and worthy of cultivation, then *Aster albescens* (C. B. Cl.) H.-M. which forms large clumps bearing white flowers, the new *Spiraea Blumei* G. Don. var. *microphylla* Rehd., only found here in this canyon, the rosaceous *Macleya microcarpa* (Max.) Fedde, and the Tree of Heaven *Ailanthus altissima* Swingle, here a stunted tree overhanging the river; on grassy slopes outside the gorge it reaches a larger size but does not attain the large size known from colder regions of China. Others to be mentioned occurring on the dry valley slopes are *Clematis Gouriana* Roxb. var. *Finetii* Rehd. & Wils., climbing over Berberis, not previously recorded from Kan-su, and *Clematis fruticosa* Turcz., only found here on the gravelly slopes of the river bank and not elsewhere on the Min Shan.

Of poplars in these arid gorges the only one observed near the Wang-tsang monastery at 6,490 feet, was the Lombardy poplar *Populus nigra* L. var. *italica* Duroi, it did however not have the tall columnar crown of the Italian tree.

Wang-tsang Valley and Its Forests
The valley of Wang-tsang has its source more than 60 miles west-southwest, back of, and parallel to the spur which hems in the Pai-lung Chiang; a spur separates it from Ta-ra Valley which carries a large stream almost as large as that of the Pai-lung Chiang itself. It debouches into the latter river through a narrow gorge east of the village of Wang-tsang back of which are fairly large trees of *Malus baccata* Borkh., bearing round yellow apples. [Plate **27**]

The mouth of the valley near the village is bare, but further up where the stream issues through a narrow part of the valley the forest commences. It is a very deceptive valley, and one could continue for more than three days ere reaching the end. There is no real trail except to where the forests commence after which it is necessary to either follow up the streambed or cut one's way.

The initial forest is a deciduous one followed by a mixed forest and pure conifer forest near the head which gives way to Rhododendron and willow scrub and the latter to the bare grassy crests of the spurs.

It is one of the few valleys left harboring primeval forests, undisturbed by man or domestic animal. It is a joy to explore such a region free of cosmopolitan weeds and foreign plant intruders. There one feels that nature has been left uninterfered with since the beginning of time.

The flora of this botanically rich valley may be divided into three or four distinct zones, the first ranging from 7,000 feet to 8,000 feet, from 8,000 to 9,000 feet, from 9,000 to 10,000 feet and 10,000 to 11,000 feet. To the first belong both ligneous and herbaceous plants which have found their way from the more arid valley of the Pai-lung Chiang, for the first trees we meet are *Quercus liaotungensis* Koidz., which line the streambed at an elevation of 7,200 feet, it forms groves and ascends the hillsides, together with *Quercus Baronii* Skan. [Plate **28**]

On their outskirts grow *Ligularia yesoensis* Fr. var. *sutchuensis* Fr., the legume *Campylotropis macrocarpa* (Bge.) Schindl., a shrub 10 to 12 feet tall with purplish-pink flowers, the yellow *Jasminum humile* L., not previously recorded from this province, and *Viburnum glomeratum* Max. var. *Rockii* Rehd., a new variety occurring only in this valley with *Macleya microphylla* (Max.) Fedde.

To the 7,000 foot level and extending to above 7,500 feet towards the next zone belongs deciduous forest composed of *Acer tetramerum* Pax. *betulifolium* Rehd., *Tilia chinensis* Max., *Viburnum betulifolium* Batal. forma *aurantiaca* Rehd., *Acer davidii* Fr., *Zanthoxylum setosum* Hemsl., with bright red fruits and black shining seeds, *Cornus macrophylla* Wall., *Malus kansuensis* Schn., *Acer pictum* Thbg. var. *parviflorum* Schn., *Meliosma cuneifolia* Schn., also found in Yün-nan, and the composite shrub *Pertya sinensis* Oliv. The many maples with their brilliant red petioles and drooping racemes or panicles of winged fruits, the clusters of bright red berries of the Viburnums, and the canopy of green of many shades produced a beautiful pageant.

In the shades of the maples delighted the araliaceous tree *Acanthopanax leucorrhizus* Harms, the var. *glabrescens* Rehd., of *Hydrangea Bretschneideri* Dipp., *Sorbus hupehensis* Schn. var. *aperta* (Koehne) Schn., *Aralia chinensis* L. var. *nuda* Nakai, and *Lonicera chrysantha* Turcz. var. *longipes* Max.

On the bank of the stream fighting for light grew *Budleia albiflora* Hemsl., *Berchemia pycnantha* Schn., and Rhamnus, while over them trailed the new *Smilax rubriflora* Rehd. In meadows we encounter *Aconitum anthora* L., and among ferns and *Rodgersia pinnata*, *Saussurea stricta* Fr., most of them belonging to southern climes. *Picea Wilsonii* belonged here to the lower levels of 7,500 feet where it reached 150 feet in height, and shaded *Betula japonica* var. *szechuanica* Schn., a tree 40 feet, white bark with black rings, *Euonymus Giraldii* var. *angustialata* Loesn., *Rosa sertata* Rolfe, the latter nearly spineless, and growing with ferns and canebrake in the shade up to 8,000 feet.

Epiphytic on old tree trunks were the ferns *Polypodium ciliophyllum* Diels, *Adiantum latedeltoideum* (Christ) Chr., and *Cyclophorus sticticus* (Kze) C. Chr.

The very fragrant *Benzoin umbellatum* (Thbg.) Rehd., joined the birches and *Acer maximowiczii* Pax, and *Acer caudatum* var. *multiserratum* Rehd., on the margin of the tall Picea forests. Over the maples trailed the huge liana *Clematoclethra lasioclada* Max., and extended up to 8,000 feet. In dense shade grew the erect shrubby 4-5 feet tall *Smilax trachypoda* Norton, while here and there, but not common, we found in mixed forest the ash *Fraxinus platypoda* Oliver. The tallest and largest of all conifers encountered here was *Abies chensiensis* Van Tiegh, called Lao-li by the The-wu, a tree 150 feet in height and trunks of over 4 feet in diameter (see Plate **29**). This rare silver fir had not been recorded previously from Kan-su nor from Yün-nan where I found it in the Si-la valley west of the Me-kong on the Salwin-Mekong divide. In open grassy areas seeking light flourished *Cotoneaster obscurus* var. *cornifolius* Rehd. & Wils., a 10-15 feet high shrub with whip-like branches and black fruits known to the The-wu as Tsa-p'u. To moss covered trunks adhered the lichen *Sticta Henryana* Muell.-Arg., and at their foot revelling in the shade thrived the fern *Cystopteris moupinensis* Fr. From the plants cited here it can be seen that they represent a more southern element, which found its way across the border into the deep sheltered valleys of southwest Kan-su.

The last to belong to the deciduous forests of the 7,500 foot zone is the new shrub *Acanthopanax stenophyllus* Harms forma *angustissimus* Rehd. To the plants enumerated here must be added others already recorded and more commonly encountered in similar valleys of the Min Shan.

Among the plants belonging to the 8,000 to 8,500 feet belt, may be mentioned *Picea Wilsonii* which extends from the lower level up here where it prefers the steep slopes of the valley; on grassy slopes occurs the mock orange, *Philadelphus pekinensis* Rupr. var. *kansuensis* Rehd., and in the dense forest *Hydrangea longipes* Franch., *Ribes Vilmorini* Jancz., and *Rosa serrata* Rolfe; the orchid *Habenaria cucullata* (L.) Hoefft., thrives in mossy ground but in open spaces surrounded by forest.

Clematis lasiandra Max., with tri-foliolate leaves and purple flowers covers shrubs and trees with its vines, and *Athyrium filix femina* (L.) Roth var. *cyclosorum* Rupr., thrives on shady banks along the stream, while the species proper grows among shrubs. The almost spineless *Caragana tangutica* Max., prefers shady banks among rocks with

Rubus pileatus Focke, *Ribes moupinense* Fr. var. *tripartitum* Jancz., *Cotoneaster acutifolius* Turcz. var. *villosulus* Rehd. & Wils., and *Berberis kansuensis* Schn.

Picea asperata makes here also its appearance where it invades the forests of *Picea Wilsonii*. Among boulders and mossy banks in the dense birch and spruce forest the luxuriate ferns *Polystichum Braunii* (Spenn.) Fée, and *Polystichum molliculum* Christ, with *Dryopteris filix mas* (L.) var. *Khasiana* Clarke, and *Athyrium acrostichoides* (Sw.) Diels, while the cosmopolitan *Polypodium lineare* Thunbg., adheres to mossy tree trunks.

In clearings of birch and spruce *Senecio nemorensis* L., flaunts its yellow blooms, and *Aconitum volubile* with trusses of purple flowers festoons the surrounding bushes. Into the 9,000 foot zone ascend from the previous ones *Betula japonica* var. *szechuanica* Schn., and its companion *Picea Wilsonii*, while the fir *Abies Faxoniana* Rehd. & Wils., makes here its first appearance. Of shrubs *Lonicera szechuanica* Batal., and *Acanthopanax Giraldii* Harms, seek its shade as does the liliaceous *Clintonia udensis* Trautv. & Mey. These do however ascend with the Abies into the 10,000 foot belt.

A red-barked birch takes now the place of the white barked one, namely *Betula albo-sinensis* var. *septentrionalis* Schneid. This tree reaches a height of 80 feet and trunks 2 feet in diameter, unlike its relative of the lower levels, which grows on the outskirts of the conifers, this birch forms forests of its own with spruces and Abies up to 9,600 feet elevation.

The forests become denser on the valley floor and here we meet for the first time a new poplar with straight boles 100 feet of more, it is *Populus szechuanica* Schn. var. *Rockii* Rehd.; a magnificent tree with drab to greyish brown bark longitudinally furrowed; it is associated with Picea, and a species of Arundinaria (canebrake) as undergrowth. Its bole is often 80 feet or more tall before the first branch appears. (See Plate **30**). On the mossy trunks of this poplar we often find the ranunculaceous herb *Actaea spicata* L. var. *erythrocarpa* Ledeb., with purplish black fruits. It is the only place where it was observed in Kan-su. In this dense forest we also find a wild cherry *Prunus pubigera* var. *Prattii* Koehne, a tree 30 feet tall, its yellowish fruits borne on long racemes and the *Hydrangea Bretschneideri* var. *glabrescens* Rehd., with panicles borne on long erect peduncles.

On the hillsides for the first time we encounter *Pinus Armandii* Franch., on limestone cliffs, the only place where we observed it south of the Min Shan in Kan-su, and north on the limestone mountains of Lien-hua Shan, its northern limit.

We come now to the last tier of forest at 10,000 feet to the 11,000 foot belt.

Abies Faxoniana joins here *Abies sutchuenensis*, and *Picea purpurea* reaching a height of 120 feet always the last of spruces, not happy except at high altitudes, but humid atmosphere, unlike *Picea asperata* which prefers here the lower levels on the Min Shan but drier climate; therefore it can exist on the drier ranges in the north as far as the Nan Shan facing the Gobi Desert. *Abies sutchuenensis* reaches heights of 100 feet with trunks 3 feet in diameter and forms pure stands, with its appears *Rhododendron rufum* Bat.

There were still a number of trees to be found in this valley which were neither in flower nor in fruit hence unidentifiable. In the upper part of the valley progress was most difficult as it became narrower, and fallen logs and thickets obstructed one's way. Everywhere the ground is covered with moss and canebrake grows scattered just as in the Abies forests of western Yün-nan. Here in these forests we shot many birds, one new one among the birches and willows, *Fulvetta cinereiceps fessa* B. & P., also the water crow or Shui-lao-wang *Caeruleus immansuetus* B. & P.

The valley seems endless, higher up the forests are more uniform and composed of fewer species, the trees become smaller and the Abies give place to scrub Rhododendrons and willows, the usual species found at these heights. Eternal twilight reigns and the rays of the sun rarely penetrate into the central part of the valley.

From Wang-tsang to Ma-ya kou 麻牙溝

From the valley of Wang-tsang to the mouth of the Ts'ao-shih kou (valley), the Pai-lung Chiang flows in an east-southeasterly direction and almost straight a distance of about 6 miles, the trail leading on the right or south bank of the river.

At Wang-tsang village there is a fairly good bridge across the Pai-lung Chiang, it spans it at a very narrow place among bare hills or covered with xerophytic shrubs. Above the right bank consisting of steep bluffs are terraces cultivated by the people of Wang-tsang village.

The mouth of Wang-tsang valley is very narrow where it debouches into the Pai-lung Chiang similar to the tributaries of the Yellow River in the grasslands. As we ascend, the gorge of the river becomes more and more arid; here and there grows an oak *Quercus liaotungensis* Koidz., and on the shale and schist *Lespedeza floribunda* Bge. finds a foothold with the cushion-forming *Indigofera bungeana* forma *spinescens* Kob., *Pistacia chinensis* Bge., and the sapindaceous *Koelreuteria paniculata* Laxm. *Pinus tabulaeformis* Carr., covers the upper slopes of the valley while *Juniperus chinensis* L. grows scattered on the lower; confined to the very banks of the stream overhanging the water we find *Juniperus chinensis* var. *pendula* Fr., a fairly large tree 40-60 feet tall with long slender branches, but rather rare here, with *Microglossa salicifolia* Diels, and *Aster incisus* Fisch. *Cotoneaster multiflorus* Bge, *Wikstroemia chamaedaphne* Meisn., and *Rhododendron micranthum* Turcz., are confined to the gravelly slopes. On the cliffs grow *Selaginella involvens* a rosette forming plant also common on limestone cliffs in Yün-nan, and a pink Allium.

At the mouth of Ni-pa kou 你巴溝, elevation 6,100 feet, on the left bank of the river, the gorge of Pai-lung Chiang narrows considerably and the geological formation which up to here consisted of slate and shale changes again to old limestone. The trail leads 50-100 feet above the stream. On the right opens a valley called Ts'ao-shih kou which is densely wooded in its upper part.

The xerophytic character of the vegetation continues and becomes even more pronounced. Most of the shrubs are armed with spines, and some form cushions as the Indigofera previously mentioned. The river flows in a terrific limestone chasm which it has carved for itself through a spur which stretches across the valley from the Min Shan. The gorge is not wider than 60 yards, the river roaring deep below with a deafening noise in its limestone prison.

The limestone cliffs support here and there a pine tree and rise to a height of 1000 feet on the right bank. In the center of the gorge is a cantilever bridge, the loosely placed central portion which in time of danger from invading Tibetans can be pulled away, sways fearfully as one crosses, especially with loaded animals (see Plate **31**). For a short distance the trail leads above the left bank of the torrent, only to cross again to the right over another cantilever bridge. Ere crossing the first bridge the trail descends in short, steep zigzags to the foot of the reddish-gray limestone cliff, not on terra firma, but on 2 feet wide tressels, the vertical pieces resting on rocks below, the crosspieces stuck, one end into the wall of rock, the top is covered with rocks and gravel, but large holes permit the roaring waters to be seen below. In places the boards are no more joined to the wall and a long breach extends the length of the trail. It is a ribband of a trail swaying as one crosses, with abysses of 100 feet to both sides, of course without railing.

From the second bridge on the trail is even worse, it spans vertical chasms and chutes in the limestone cliff, or is bridged-over, zigzag fashion at a steep angle with a gulf to both sides. It is a most hazardous undertaking to lead a loaded caravan across such a chicken ladder of a trail. At the end of the gorge which is known as the Ma-ya Chha-lung, and is about one mile long we come again to mica slate, shining silvery, also shale and here and there a block of limestone embedded in the shale.

Beyond the gorge the valley widens somewhat, with a small pine covered amphitheater to the right. Diagonally across debouches the Ma-ya Valley into the Pai-lung Chiang. Up this valley leads a trail to Tsa-ri Khi-kha, one of the most scenic stretches to the summit of the eastern end of the Min Shan. Near the village of Ma-ya, the valley widens and a bridge, elevation 6,000 feet, leads across to the village and valley. The trail continues down stream where the river makes a sharp bend towards east-northeast. At the bend of Pai-lung Chiang towards Hsi-ku Hsien 西固縣 there debouches a long valley called A-hsia kou, but actually it is the To-erh kou which descends from Yang-pu Shan 陽布山, the border of the Cho-ni prince's domain and Ssu-ch'uan. These two valleys are described in the next chapter.

Around a bend of the Pai-lung Chiang, against dry, scrub-covered cliffs or slopes nestles the new Wang-tsang lamasery or Ma-ya Hsin ssu 麻牙新寺 or New Ma-ya Lamasery. [Plate **32**] In Tibetan it is called Wang-tsang men-chhe Gom-pa (dBang-gtsang-man-chhe dgon), situated at an elevation of 6,290 feet. the xerophytic vegetation is the same as mentioned previously.

The Valleys of Ts'ao-shih, A-hsia and To-erh
Ts'ao-shih kou 草什溝 is the valley next to Wang-tsang and has a most peculiar configuration. We explored it thoroughly. A-hsia kou 阿夏溝, a transcription of the Tibetan A-ja Nang (Â-bya-nang), is a long valley but apparently not so long as the one which it joins about 6 miles before it empties into the Pai-lung Chiang. That valley is called To-erh kou 多兒溝 which is also a Chinese transcription of the Tibetan name Do-ro Nang (rDo-ro nang). The first valley has its source apparently southwest of the spur which shuts in the valleys of Ts'ao-shih kou and Pai-lung Chiang, but where it is, and how long the valley is I could not learn. In the valley some distance up, a lamasery

is situated called Nang-go Gom-pa (Nang-sgo-dgon) in Tibetan, and Na-kao Ssu 納高
寺 in Chinese.

Unfortunately time did not permit to explore the two valleys A-hsia and To-erh,
although I traversed part of A-hsia and the whole length of To-erh in the late winter
when we were obliged to leave Kan-su, for the political conditions of the country were
such that further delay would have been disastrous. The country had then already gone
communist and the Rainbow flag of China had disappeared and a red flag had taken its
place. All our collections had been sent to the coast to be shipped to America, and my
twelve Na-khi assistants and myself made our way to Sung-p'an in northwest Ssu-
ch'uan, a journey which took us 19 days through wild and lawless country. No
collecting could be done, and would have been impossible as it was late in the winter
when all plants were dormant or covered with snow.

We did not know of the existence of Yang-pu Shan, a mountain over 13,000 feet in
height in the south and still in the Cho-ni prince's territory, for otherwise I would have
made all efforts to explore it. There was however so much territory to cover, as far
northwest as to the borders of the southwest Gobi, the Am-nye Ma-chhen, to say
nothing of the Min Shan etc. that no time could be spared. Had political conditions
permitted I would have remained another year to explore the region thoroughly.

The Valley of Ts'ao-shih (kou) 草什溝 and Its Flora

Ts'ao-shih valley has a most peculiar configuration. The stream has two branches, both
issuing opposite each other, a spur separating them, one flowing northwest and one
northeast, the latter being the longer, both join other branches, one coming from west
and flowing northeast, the other has its source near the village of Pe-khar and flows
northwest, both meet and united flow for about 4 miles directly north, at right angles
into the Pai-lung Chiang. The four lateral valleys enclose thus a square of mountains
from the sides of which descend three small rivulets.

The plants found in Ts'ao-shih kou are the same as those of Wang-tsang Valley; at
its head are immense groves of the majestic *Abies chensiensis* Van Tiegh, which next to
Picea wilsonii is the most common conifer. Maples abound, Sorbus, Quercus, Tilia, and
the giant lianas Clematoclethra, and the rosaceous *Exochorda Giraldii* Hesse, a tree 15-
25 feet with a bark resembling that of *Lagerstroemia indica*. It inhabits the drier areas
of Ts'ao-shih kou, but we did not find it in the Wang-tsang valley although it may occur
there.

The main branch of the Ts'ao-shih stream issues from a narrow defile, but beyond
the valley widens, the slopes are mainly covered with *Pinus tabulaeformis* Carr., their
rich green constrasting from the dry arid slopes of the valley which below the pines are
covered with Berberis and *Rhododendron micranthum*, the latter very common and
reaching a height of 15 feet or more, it was introduced to cultivation by me under no
15004, distributed by the Arnold Arboretum of Harvard University.

A short distance beyond the defile is the village of Tsho-ru-zhi (mTsho-ru-bzhi) built
on stilts, the rear part on terra firma, as it is perched on the terraced steep hillside. The
eastern valley wall is terraced and cultivated, the western is covered with pines. The
valley branches above the village as explained in the introductory chapter, the western

arm extending deep into the mountain, it is this branch which is densely forested.

The fields are a mass of limestone rocks and above them are pines. At 7,600 feet were a few isolated huts; the trail passes between fields to the head of the valley, mainly a rock pile barren in the extreme. On the very top of the spur at an elevation of 8,750 is situated the village of Pe-khar (dPal-khar). The eastern branch of Ts'ao-shih kou does not end here but makes a sharp bend deeply into the mountains. Tall spruces covered the upper valley slopes and floor which lower down was filled with oaks such as encountered at Pe-zhu and Wang-tsang.

The spur on which Pe-khar (here pronounced Pa-kar) is situated separates To-erh kou from Ts'ao-shih kou, and extends down into A-hsia kou, the main stream of A-hsia entering To-erh kou at the foot of Pe-khar, while a small affluent with its source south of Pe-khar flows south into A-hsia kou, west of the spur extending from Pe-khar.

A-hsia kou and To-erh kou

From Pe-khar the trail descends very steeply 2,350 feet in zigzags the precipitous ridge; from here one could look up both valleys, the streams are clear and flow in trench-like gorges of limestone, the valley walls rising thousands of feet sheer from the streambeds and culminate into fantastic, snow-covered limestone crags. Beyond, Yang-pu Shan 陽布山 a formidable mountain mass, then a pearless white, rises some 30 miles southeast, at the head of To-erh kou. Other limestone mountains, also snow-covered, fill the whole triangle between A-hsia kou and To-erh kou, extending their snow-crowned summits some 7,000 feet or more above the streams which here flow at an elevation of 6,400 feet.

The trail crosses the narrow valley which extends from Pe-khar into A-hsia follows down stream and then up to A-hsia kou, crosses the latter over a frail, shaky bridge and then climbs 700 feet steeply over a terrible rocky trail to the summit of the ridge dividing A-hsia kou from To-erh kou, at 7,100 feet elevation, only to descend again to the To-erh kou stream which flows here at the same hight as the A-hsia stream, namely 6,400 feet above sealevel.

Crossing the To-erh kou stream over a narrow flimsy, railing-less bridge, the trail leads now up the To-erh kou on the right bank as far as the village of T'ai-ni-o 台你峨, marked on the C. G. S. L. S. W 10-n 2, as T'ai-li-ao 台里敖, here pronounced by the village people as T'ai-ling-ngo. The spelling and characters used (T'ai-ni-o) are those used in the records of the Cho-ni prince's territory, and occurred so on his map (an ancient one painted on the wall of his Ya-men, but since destroyed by Moslems). T'ai-ni-o is situated at 6,600 feet.

In A-hsia kou are situated two lamaseries, a Bön or Black sect one called Nang-go Gom-pa (Nang-sgo dgon) the Chinese Na-kao Ssu 納高寺) and a Yellow Sect one called A-ja Ya-nub Gom-pa (Â-bya-ya-nub-dgon).

The plants encountered up to T'ai-ni-o are entirely xerophytic ones composed of oaks, *Rhododendron micranthum* Turcz., *Prunus tangutica*, *Syringa oblata* Lindl. var. *Giraldii* Rehd., and those found in Ni-pa kou and lower Ma-ya kou q.v., only that pines predominate above the deciduous shrubs and trees.

At T'ai-ni-o the stream is crossed to the left bank, and the trail, a fairly good one, leads on that side of the river for 15 miles to another bridge at 7,675 feet elevation, a rise of 1,075 feet, the only bridge across the To-erh River on that stretch. The valley is arid and remains so until near the village of Yang-pu 陽布 near the head of the valley, at 9,300 feet elevation.

The trees encountered in the arid stretch are the above mentioned plus Rosa, Juniperus, Cotoneaster, Spiraea and *Pinus tabulaeformis*.

A distance of 6 ½ miles the trail brings the traveler to a Yellow sect lamasery called Ra-zid Gom-pa (Rwa-gziḍ-dgon) the Chinese La-tzu Ssu 拉子寺; it is not large but it is beautifully situated beyond a lateral ravine which descends from a snow-covered limestone peak crowned by crags.

Two and three quarter miles beyond is another lamasery belonging to the Sa-skya Sect, called Pe-ku Gom-pa (dPe-sku-dgon), the Chinese Pai-ku ssu 白古寺 (see Plate **33**).

Pe-ku monastery is situated at 7,400 feet elevation on the left bank of the Pai-lung Chiang. Its walls are broadly striped red and white as is the custom in all Sa-skya lamaseries.

The trail continues up stream through plowed fields and oak scrub vegetation as one encounters between Pe-zhu and Wang-tsang, past the village of Po-ku or Pai-ku, and crosses the river to the right bank which it follows to the village of Yang-pu, the last in the Cho-ni prince's territory, at an elevation of 9,300 feet.

The trail, some distance beyond the bridge, leaves the main stream which seems to reach deep into the heart of the limestone mountains, part of Yang-pu Shan, and follows up a narrow lateral valley. Farther ahead a valley opens in the branch which leads to Yang-pu village, up which a trail leads to Sung-p'an in northwest Ssu-ch'uan, it is shorter by a few days but impassable for pack animals.

In the upper part of the valley to Yang-pu, we meet now with both white and red-barked birches, poplars, large oaks, *Picea purpurea*, large Berberis, and Hydrangea bushes. Ere reaching Yang-pu the valley seems to divide into two horizontal branches, but actually it receives an affluent from the south where the stream describes a right angle curve. A short distance beyond is the hamlet of Yang-pu also called Shang T'a-yü 上塔峪, in Tibetan Ta-yü gong-ma (rTag-yul gong-ma), elevation 9,300 feet. South-southeast looms up the mighty snow-covered crest of Yang-pu Shan over which a pass leads called Ta-ge La (rTa-rgas-la) elevation 12,500 feet. The village of Yang-pu harbors 70 families of The-wu, then subjects of the Cho-ni prince. Opposite the village is spruce forest of *Picea asperata* which higher up gives way to *Picea purpurea*, *Abies sutchuenensis* and *Juniperus saltuaria*. There are several lateral valleys, the last one from the southeast leading to the pass. Here among the firs were several species of Rhododendron as are not met with elsewhere on the Min Shan, but it being still winter, there were neither flowers nor fruits, only a few old capsules remaining of which such seed was gathered as could be found. The whole mountain was deeply in snow and in order to cross it necessitated cutting a trail through the five or more feet deep snow under which were buried masses of *Rhododendron Przewalskii* (see Plate **34**), the last plant on Yang-pu Shan.

The valley which leads from Yang-pu village to the summit pass Ta-ge La or Yang-pu Shan Shan-k'ou 陽布山山口, is called Tu-mu-lö by the The-wu of Yang-pu the last village. The upper part of Tu-mu-lö is grass-covered and red sandstone is prominent at the mouth of it. Higher up it narrows and the slopes are densely forested with firs and Rhododendron at an elevation of 11,400 feet, after which the mountain was buried in snow for a depth of more than five feet. No rocks could be observed. South of the pass was more or less free of snow, the valley into which the trail led being called Chhu-nyi-drö (Chhu-gnyis-hgros) and in Chinese Ta-shen kou 大深溝 or the Great Deep Valley. The rocks on the south side of the pass are slate and schist, no limestone being visible. The timberline on Yang-pu Shan is at 11,700 feet and consists mainly of Abies with Rhododendron undergrowth mixed with the branching type of *Caragana jubata* Poir.

The valley floor south of the pass in Ssu-ch'uan was 10,200 feet, a drop of 2,100 feet. Here the trees are mainly *Picea purpurea* with a species of Arundinaria (canebrake) as undergrowth. Red birches are numerous but Rhododendrons were entirely absent, while north of the Yang-pu pass were at least ten species if not more, many of them with very large leaves not known from the Min Shan. On the south side Abies is absent, the purple spruce, *Picea purpurea* being the most common tree. As a whole the vegetation on the Ssu-ch'uan side although wild in the extreme, and uninhabited till the first village in Ssu-ch'uan is reached, called Ts'ao-pa 草垻, elevation 7,230 feet, is much less rich in species than on the north side in Kan-su.

From Wang-tsang to San[g]-pa kou 桑巴溝

The region here described leads from the Pai-lung Chiang north first through arid country up the valley of Ni-pa (kou) 你巴溝, thence over a pass 8,000 feet on a spur separating Ni-pa Valley from the much longer valley of Ma-ya (kou), which debouches west of the village of Ma-ya 麻牙 into the Pai-shui Chiang or Pai-lung Chiang. Ma-ya valley or its stream has its source south of a pass called Lha-mo-gün-gün (Lha-mo-gun-gun), elevation 11,250 feet.

To the west of this pass is a huge limestone mountain which stands out detached from the limestone range and visible from afar. It has the shape of an oblong, somewhat tapering, truncate block not unlike a sky-scraper. It is called Hsiao Ku-ma 小古麻, and is visible in Plate ?[14]. The Kan-su natives, particularly those of Cho-ni pronounce hsiao «small» – ga so the name becomes Ga-gu-ma, in contradistinction to another much higher one known as Ta Ku-ma or the Great Ku-ma. The meaning of these two names has already been explained.

Ta-ku-ma is between He-ra village in San-pa kou and an alpine meadow called Yor-wu Thang, west of the stream. This is the highest prominence of the Min Shan, but is not visible from the north of the T'ao River as it is too far south-southeast and back of the main range, on a southeasterly extension of it.

The subdistrict of San-pa, the Tibetan Sam-pa (bSam-pa), is along a stream or in the valley of the same name viz. San-pa kou or Sam-pa Khog which extends southeast below the Yellow River – Yangtze divide in the Lower T'ieh-pu Land, and is next to Drag-gam-na in the Upper T'ieh-pu country one of the most beautiful regions of the

---

[14] Photograph not found.

Min Shan. Its forests are magnificent and its scenery wonderful.

Like elsewhere so also here, the The-wu have given the stream which is known as San-pa or Sam-pa more than one name; from its source below a pass 10,900 feet elevation, which is the actual Yellow River – Yangtze divide, it is known as Do-ya-ya, where it enters the tremendous defiles of the Min Shan it becomes the Yor-wu-drag-kar (Yor-bu-brag-dkar), and from the village of He-ra it becomes the San-pa or Sam-pa river. It flows southeast into the Pai-lung Chiang.

Ni-pa kou 你巴溝 and Ma-ya kou 麻牙溝

The valley of Ni-pa which is west of Ma-ya, enters the Pai-lung Chiang at an elevation of 6,100 feet. It is an arid valley and apparently the stream has its source in or south of a limestone peak called Ma-ya Shan 麻牙山 which is quite inaccessible.

The slopes of Ni-pa Valley are mostly gravel, and hence covered with a xerophytic vegetation as is found in the Pai-lung Chiang valley. We meet *Ailanthus altissima* Swingle, here a tree up to 30 feet tall, at 7,500 feet elevation, growing isolated over the arid hillside; oak forest composed of two species *Quercus liaotungensis* Koidz., and *Qu. Baroni* Skan., harbors also *Deutzia albida* Bat., a shrub 10-12 feet, the only representative of the genus in Kan-su, and found only in this region on the Min Shan. Its leaves are small and whitish gray; and is of a typical xerophytic aspect. Scattered along the outskirts of the oak forests on the dry grassy spurs at 7,500 feet, was *Iris dichotoma* Pall., and lower down at 6,600 feet on the banks of the Ni-pa stream on shale grew the pink-flowered *Silene Fortunei* Vis. With it and scattered over the dry arid slopes at 6,800 feet grew *Jasminum humile* L., a shrub 4-5 feet tall with yellow flowers, frequenting also the oak forests; it is new to Kan-su.

Berberis kansuensis Schn., belongs to the 7,000 feet level and is found also in Ma-ya valley. With the above mentioned plants are also associated many of the plants found in the arid valley of the Pai-lung chiang, but mainly those of its upper slopes.

The valley of Ma-ya is a much longer and also a more interesting one than Ni-pa kou. As already remarked it has its source south of a pass called Lha-mo-gün-gün at an elevation of 11,250 feet and drops to 6,000 feet where it debouches into the Pai-lung Chiang, a difference of 5,250 feet in a distance of about 12 miles, this illustrates the steepness of the southern slopes of the Min Shan. Ma-ya Valley has several small affluents, three on the west side and one on the east side. In the valley west of the stream are situated a lamasery called Do-lo gom-pa (Dog-logs-dgon) and two villages. Ni-pa 你巴 is situated below the lamasery which is on that account called Ni-pa Ssu 你巴寺 by the Chinese. The second is called Zhi-ga (gZhi-dgah) and is at an elevation of 7,900 feet. The village of Ma-ya whence the valley derives its name is near and east of its mouth.

The lower part of Ma-ya kou, in Tibetan Ma-yag Nang, also Ma-ja Nang, possesses like Ni-pa kou a xerophytic flora. At 7,500 feet elevation the valley slopes are shale and schist and here thrive the prostrate, stiff shrub, *Euonymus nanoides* Loes. & Rehd., the oaks already mentioned, and attached to their branches the European mistletoe *Loranthus europaeus* Jacqu., *Indigofera bungeana* Walp., a shrub 2 feet tall, the lilac *Syringa pekinensis* Rupr., a shrub 10-15 feet, originally known from Peking here wild, with creamy white flowers, while *Syringa oblata* Lindl. var. *Giraldii* Rehd., occurs in the moister 8,000 foot belt, and of similar size.

Along the stream climbing over Berberis and Lonicera are the widely distributed *Dioscorea nipponica* Mak. (*Dioscorea quinqueloba* Thunb.), and *Humulus lupulus* L., often with stems 30 and more feet long. In the oak forest, in the dry rocky soil, flourish *Lespedeza formosa* Koehne, displaying purple flowers, on slate and shale slopes *Syringa Potanini* Schn., with long flexible, rambling branches, *Lonicera Ferdinandi* Franch., with pale brown, shaggy bark and gracefully drooping branches, *Rhamnus leptophylla* Schn., and *Koelreuteria paniculata* Laxm., which here at 7,500 near the village of Ni-pa, on an affluent of the Ma-ya stream, reaches its largest size, trees of 40 feet not being rare. *Aster Limprichtii* Diels, with white flowers is here a shrub 2-3 feet high among the oaks, and finally the beautiful *Philadelphus pekinensis* Rupr. var. *kansuensis* Rehd., and the ubiquitous *Hippophaë rhamnoides* which has a special predilection for the banks of the streams where it often forms impenetrable thickets.

As we ascend to the 8,000-9,000 foot zone (see Plate **35**), we enter deciduous and mixed forest; of conifers *Picea Wilsonii* and *P. asperata* predominate with *Juniperus distans* Flor., and *Juniperus squamata* var. *Fargesii* Rehd. & Wils., at 9,600 feet, associated with the rare *Betula delavayi* Franch., a small tree 15-20 feet, leaves evenly green on both sides and horizontal branches. This species was found only here on the Min Shan, and occasionally reached a height of 40 feet. The new rambling shrub *Smilax rubriflora* Rehd., red flowers, black fruits and pale papery leaves, was, with the two varieties of junipers the only rarity in this valley. No other plants were found here that did not occur in Wang-tsang valley, but not all of the Wang-tsang plants in Ma-ya kou. A Polypodium with simple pinnate fronds grew under *Exochorda Giraldii* Hesse, with *Caragana densa* Kom., and *C. tangutica* Max.; Acer, *Clematoclethra integrifolia* Max., and the new *Juniperus squamata* forma *Wilsonii* Rehd., a small tree 20 feet, but only found here.

At 11,000 to 11,500 feet *Abies Faxoniana* took the place of *Picea*, and *Juniperus saltuaria* Rehd. & Wils., took the place of *J. distans* Flor., *Salix Rehderiana* var. *brevisericea* Schn., formed much of the undergrowth.

As the valley is a thoroughfare as one might say, from the Pai-lung Chiang to the T'ao River, the forests have been much disturbed by grazing animals, hence there is no comparison to the forests of Wang-tsang valley which is the gem of the Lower T'ieh-pu country. Furthermore the Ma-ya valley is much shorter and wedge-shaped and rises too steeply considering the distance for a richer flora to develop.

At or near the summit pass Lha-mo-gün-gün, the above mentioned *Abies* is associated with *Picea purpurea* which descends however to 10,000 feet. Here also dwells *Betula albo-sinensis* Burk., the bark a grayish-black, no white at all, Rosa, Spiraea, and willows. Below the grassy summit-spur, lining the fir trees, was *Rhododendron Przewalskii* Max., and willows, but no *Rhododendron rufum* Bat., (see Plate **36**).

From Lha-mo-gün-gün the head of Ma-ya kou, a grassy pass 11,250 feet, the trail descends through birch and fir forest entirely composed of *Abies Faxoniana* Rehd. & Wils. (see Plate **37**), carpeted with beautiful light green moss. Here *Rhododendron* *rufum* Bat., was again common with Ribes, *Prunus stipulacea* Max., and *Sorbus tapashana*, etc. In the moss grew also the fern *Dryopteris Robertiana* (Hoffm.) C. Chr.

As we descend the Abies trees become taller, the forest is virgin and undisturbed, the trees associated with *Abies faxoniana* at 10,100 feet overlooking the Sampa Valley (San-pa kou), 2,000 feet below (reckoned from the pass) are Rosa, Ribes, Prunus, Sorbus, *Hydrangea longipes* Fr., Acer, a black birch *Betula* sp? (but identified as *B. albo-sinensis* Burk.), festooned with the long streamers of *Usnea longissima*, and *Salix Rehderiana* var. *brevisericea* Schn.

The trail leads through this forest which clothes a small amphitheater near the stream which finds its way through an immense shih-men or rock gate, the most prominent landmark of San-pa Valley.

The northern wall of this valley is composed of schist, shale and slate and probably sandstone superimposed with loess, but preliminary spurs south of it show limestone outcroppings. The river flows deep in a trench some 3,000 feet below the summit. The southern valley slopes, densely forested with the above described species, fall steeply into the river bed, while from the foot of the range which encloses the long valley, a gentle sloping, broad, terrace-like plateau extends two thirds of the width of the valley. The long terrace is however much broken up by ravines with streams descending from the steeply eroded hillsides. The latter are absolutely bare, with only here and there, in the steeply furrowed slopes, tree growth.

On the terrace, north of the stream are situated two The-wu villages the first is called Wu-ho or U-ho, the second Pen-dza (Pan-rdza). near the river, southeast of U-ho is situated a lamasery called Ser-thang Gom-pa (gSer-thang dgon-pa) called in Chinese Shai-tang ssu 曬當寺.

Beyond Pen-dza the northern valley wall is much eroded and otherwise broken up into deep chasms the steep slopes of which are forested; very steep canyons extend deep into the range which here culminates into truncate peaks.

The head of the valley is directly south of Lha-mo-gün-gün, two parallel streams have their source north of the pass, one western on the north side of Hsiao Ku-ma, and the eastern directly north of the pass on its way to the main San-pa kou, it flows through the huge rock gate visible in the photo (see Plate ?[15])

As already remarked the real source of the San-pa stream is south of Yen-chhen-rün-go (gYen-chhen-run-sgo) elevation 10,900 feet which is the Yellow River – Yangtze divide. Within the confluence of the small western branch and the main branch which descends from the north is situated the hamlet of He-ra, at an elevation of 9,320 Feet. This is the village northeast of U-ho, a distance of about 10 miles as the crow flies.

At 9,500 feet elevation *Betula albo-sinensis* var. *septentrionalis* Sch., makes its appearance with Picea, etc. Here a narrow spur divides the western branch of San-pa kou from another affluent which has its source in the crags to the northwest of which the Ta Ku-ma 大古麻 is a part, but stands detached, the highest crag of the Min Shan which I estimated to be between 17,000 and 18,000 feet in height.

Back (west) of He-ra are immense limestone crags thousands of feet in height, and everywhere one looks are forests of spruces and firs, which ascend the steep walls of the

---

[15] Not identified.

grayish-yellow, to reddish limestone. The scenery is majestic, but yet does not compare to that of Drag-gam-na. In one sense it is wilder and more romantic, or terrifying, as one is completely shut in by these towering giants of limestone bluffs which rise vertically from the streambed.

The limestone walls which hem in the upper part of San-pa kou, which I estimated at 16,000 feet and absolutely bare, have only their lower buttresses adorned with conifers.

From San-pa kou to Tsa-ri Khi-kha

This last stretch which closes the circle around the entire Min Shan in Cho-ni territory, is one of the most beautiful of the entire range. It has an individuality of its own. Primeval forests alternate with alpine meadows, mountains of shale and slate and schist alternate with limestone, and the latter again with conglomerate.

From He-ra the trail descends steeply to the small stream which has its source in the crags of Ta-ku-ma; leaving a limestone defile to the right (east) ascends the western mountain side crosses a circular meadow, and after passing over broad slopes at 9,500 feet elevation, with bushes of Salix, Rosa, Berberis and scattered conifers (see Plate **38**), enters magnificent primeval forest of *Betula albo-sinensis* var. *septentrionalis* Schn., *Abies Faxoniana, Picea Wilsonii*, Acer, Sorbus, Ribes, Hydrangea, Lonicera, *Juniperus squamata* var. *Fargesii* Rehd. & Wils., etc., at an elevation of 9,000 feet. In this forest we shot the golden black grosbeak *Perissospiza icteroides affinis* Blyth, who were feeding on the seeds of conifers, while pheasants were disporting themselves in this somber, virgin forest. mammals were not encountered, but bears were said to be common.

Moss covered the ground thickly, and of ferns *Dryopteris Robertiana* (Hoffm.) C. Chr., *Woodsia macrospora* C. Chr. & Maxon, *Notholaena Delavayi* (Baker) C. Chr. and others grew in the dense shade of the above mentioned trees; where light was more abundant *Aruncus sylvester* Kost., *Senecio nemorensis* L., *Ligularia yesoensis* var. *sutchuensis* Fr., *Aconitum volubile* Pall., and other herbaceous plants thrived. Along brooks bloomed a *Swertia* (no 14777), and *Cotoneaster adpressus* Bois., occupied the gravelly banks, while the fern *Athyrium spinulosum* Milde, grew along watercourses in general at 9,000 to 9,500 feet elevation.

From the above mentioned forest the trail led to a large meadow called Yor-wu-thang (Yor-bu-thang) at the entrance of the enormous gorge with vertical walls of gray, yellow and reddish limestone called Yor-wu-drag-kar (Yor-bu-brag-dkar). A lovely stream, the main branch of San-pa, of the purest crystal clear water, thundered and roared out of the rocky prison gate to flow gently through the meadow, its banks lined with tall Picea, some a hundred feet in height. The elevation of the rock gate up which the trail leads is 8,750 feet, while Yor-wu-thang (meadow) is 9,000 feet above the sea.

The Yor-wu-drag-kar Gorge

Yor-wu-drag-kar gorge is about 2 miles long in which distance the stream drops nearly 1,000 feet. Words fail to describe the beauty of the scene and yet here is nature in constant commotion. Thousands of tons of rock avalanches descend from the heights of massive limestone walls. The streambed is piled up with blocks of limestone the size of a cottage, one upon the other, the stream roaring deafeningly and invisible beneath. Here are boulders which came down from the dizzy heights with huge firs wedded to them, their trunks shattered to thousand fragments, from others moss-covered trunks grew horizontally across the streambed. Only here and there one obtains a glimpse of the foaming waters. Trees have found a foothold among these masses of rocks and in their interstices filled with soil, since ages past. Mighty monarchs of larches *Larix Potanini* Bat., Rhododendrons, and *Abies* have knitted them together with their roots.

The trail leads up this mighty canyon on the right bank of the stream, past the highest of all crags the Ta-ku-ma which rises seven thousand feet above the stream, its head lost in the clouds, its walls, as those forming the deep canyon, honeycombed with weird caves which penetrate deeply into the cliffs. Here and there the little trail has been obliterated or buried by new avalanches of rock making detours necessary. The trail hugs the base of the vertical and often overhanging cliffs thousands of feet in height with their turrets and battlements, each in itself, hundreds of feet high.

In the center of the canyon rises a huge pyramidal mountain which divides the former into two gorges. The trail descends here to the streambed crosses it over logs and ascends the one issuing on the right, north, close to the foot of the wall and through magnificent *Abies* forest which had established itself on long ago, fallen limestone masses, now covered with moss and disintegrated into debris. Arundaria, a slender bamboo or canebrake forms the main undergrowth, birches and willows abound with *Lonicera* and other shrubs while in the moss thrive the ferns *Woodsia lanosa* Hook., *Polypodium clathratum* G. B. Cl., and *Notholaena Delavayi* (Baker) C. Chr. Over rocks and boulders the trail emerges into a canyon where the streambed is perfectly flat and bordered by white sand. This is the beginning of Do-ya-ya (rDo-yag-yag) gorge.

Do-ya-ya Gorge
Although the Do-ya-ya gorge is only a little over a mile long, the stream drops 400 feet between Yor-wu-drag-kar and To-ti-pa-na where a lateral stream joins it from the west and immediately south of the Yellow River – Yangtze divide. Do-ya-ya is narrower than Yor-wu-drag-kar, the canyon walls rise vertical from the streambed which is here almost level and at an elevation of 9,900 feet (see Plate **39**) and without any boulder obstruction.

In this magnificent canyon lovely forests exist of Abies probably *Abies Faxoniana* Rehd. & Wils., also some spruces, but unfortunately none of the conifers was in fruit and hence could not be identified. The deciduous trees were mostly *Betula japonica* Sieb. var. *szechuanica* Schn., *Acer Maximowiczii* Pax, *Acer caudatum* Wall. var. *multiserratum* Rehd., *Acanthopanax Giraldii* Harms, the previously mentioned ferns, *Lonicera nervosa* Max., *Lonicera saccata* Rehd., and *Hydrangea longipes* Franch.

Higher up at 11,000 feet appeared *Juniperus saltuaria* Rehd. & Wils., at the foot of limestone crags where it is a tree 25-30 feet tall.

Where Do-ya-ya merges into Yor-wu-drag-kar from the north, another still narrower and impassable canyon opens from the southwest called Do-lo (rDo-lo), where they meet, the floor of the valley is level and at an elevation of 9,900 feet. This latter canyon is the hide-out of The-wu bandits who waylay travelers coming from A-chüeh q.v., on their way to Wang-tsang. They can overlook the junction and the valley in all directions. The Do-lo gorge is very narrow and blocked by immense rock avalanches so that it is impassable, its walls rise to terrific heights, vertically from the streambed.

Here we shot a low-flying bird *Nannus troglodytes idius* (Richm.), which lives in holes among the rocks along the streambed; otherwise birds were scarce in this canyon.

Higher up in the Do-ya-ya gorge *Larix Potanini* Batal., became more numerous, Abies formed still beautiful groves and the white flowered *Potentilla fruticosa* L. var. *dahurica* Ser., made its appearance along the streambed. The gorge narrows as we ascend, the trail rocky in the extreme leading up and down forcing us to cross the stream many times. From the west a deep ravine opens densely forested and carpeted with moss. Huge lichens, two feet or more across, covered the boulders, spec. no. 14867, *Lobaria pulmonaria* (L.) Hoff. var. *hypomelaena* (Del.) Crombre.

Rhododendrons abound here both *Rhod. rufum* and *Rhod. Przewalskii*. The crags are now less high and the slopes grass covered. This spot which is 10,300 feet elevation is called To-ti-pa-na. *Picea purpurea* with Abies and *Larix Potanini* surround here a lovely little meadow. Back of the ravine is visible a high red rocky peak composed of red conglomerate. *Juniperus saltuaria* forms now pure stands with here and there still an *Abies*; willows and the dark gray-barked birch, *Betula albo-sinensis* Burk., restricted to the higher levels.

At 19,700 feet Do-ya-ya merges now into an entirely different canyon. The rock changes here abruptly into red conglomerate, massive walls, smooth in appearance and with overhanging slabs which have the resemblance of dough of immense thickness squeezed out of the crevices. It is the same type of rock as found back of the Ra-gya Gom-pa on the Yellow River q.v. Where the conglomerate first makes its appearance it superimposes limestone which is visible beneath. These conglomerate walls have rounded tops and stand like massive sky-scrapers. The gorge is littered with enormous blocks of it, and is not easy to negotiate.

At last we emerge into an amphitheater, the Do-ya-ya stream, actually the San-pa kou, has its source to the east at the end the bare valley. By following the stream east one is on the trail to Min Hsien. The crags which crown the northern valley wall rise 1,000 – 1,500 feet above it and are limestone. The whole amphitheater is one large alpine meadow, dark scree descending from the crags on to the green turf.

Yen-chhen-rün-go (gYen-chhen-run-sgo), the Yellow River – Yangtze Divide
The grassy trail leads up a grassy slope west, the rounded meadow is the actual Yellow River – Yangtze divide, its altitude is 10,930 feet, limestone crags are again to both sides. The gentle rolling slopes merge into a peculiar triangular basin with an outlet north, through another rock gate or shih-men. The stream is called Sir-li-dra (Sir-li-hdra) and flows northeast and then north into a valley which empties into the T'ao River. The basin is filled with bushes of *Rhododendron capitatum* and *Juniperus saltuaria*, with here and there an Abies, and is framed by huge walls of conglomerate which are part of the great bluffs above Tsa-ri Khi-kha q.v.

The trail enters the defile and follows along the stream, the ground is thickly covered with moss in which Sorbus, Salix, Abies and Rhododendrons grow. The gorge is again typical limestone, and conglomerate has been left behind, the rock is tilted vertical and twisted. At 9,600 feet is another rock gate, the trail is built on logs and rocks and is a terror for man and beast. Here the stream receives a small affluent from the west which has its source in a peak called Tsa-ri-sri-mo (rTsa-ri-srin-mo). A short distance below the second rock gate the trail turns up a lateral valley called Chha-tshe-thi (Chha-tshad-thig) whose floor is at 9,700 feet elevation. It is filled with mossy *Abies* forest and *Athyrium acrostichoides* (Sw.) Diels, as undergrowth. The head of the valley is blocked by enormous conglomerate cliffs which form the western rampart around Tsa-ri Khi-kha.

From Chha-tshe-thi a trail ascends abruptly in zigzags up the valley wall. At 11,000 feet elevation is a small pass which leads steeply up between huge columns of conglomerate that stood erect like a row of giant posts; haze had filled the valley and each pillar was as if wrapped in the finest gauze, enshrowded by the mist like phantom ghosts. Above them was the summit pass, elevation 11,250 feet, and the plateau called Tsa-ri Khi-kha, q.v.

This concludes the entire circuit of the Min Shan.

Lien-hua Shan 蓮花山 or the Lotus Mountain

The region between Cho-ni and Lien-hua Shan is a high grassy plateau, interacted by valleys and ravines which harbor a rather scanty vegetation.

The only town encountered is the walled city of Lin-t'an, formerly called T'ao Chou New City spread out over an undulating area oblong in outline and situated at an elevation of 9,500 feet, ten miles from Cho-ni.

On the grassy slopes around the town we found *Scutellaria amoena* Wright, *Saussurea amara* DC., both purple flowered, while on the loess slopes, *Berberis Mouillacana* Schneid., a shrub 4 feet in height with glaucous reddish fruits, and *Cotoneaster adpressus* Bois., a prostrate shrub with red fruits found a foothold at an elevation of 10,000 feet.

Along ravines further northeast, here and there groves of *Picea asperata* Mast., made their appearance while *Pinus tabulaeformis* Carr., crowned the summit crests of limestone spurs which rise from a bed of schist and shale covered as usually with a deposit of loess.

These limestone walls form the side of a narrow ravine near the village of Ta-ts'ao-t'an 大草潭 whose stream debouches into the T'ao River. Among the rocks at the head of the Kan-kou Ho 甘溝河 or the Sweet Valley River grew the rosaceous *Sanguisorba canadensis* L., a long way from home, which *Aconitum volubile* Pall., embraced with its long coils; its flowers are here a pale, pinkish lavender, while in Yün-nan it displays deep purplish blue floral racemes in great profusion. On the grassy slopes flourished *Lonicera trichosantha* Bur. & Fr., which extends into Tibet and Ssu-ch'uan, a handsome shrub especially attractive on account of its bright red fruits. On loess banks occurred the blue-flowered aromatic verbenaceous shrub *Caryopteris tangutica* Max., two to three feet tall, the huge climbers *Polygonum aubertii* Henry, and *Humulus lupulus* L., the common hop, both with cream colored flowers, the former first described from Ssu-ch'uan, the latter from Europe whose dried ripe cones are used in the making of bitter beer.

In meadows and fields grew the Chinese pink, *Dianthus sinensis* L., with crimson flowers, and along streams the spiny shrub *Lycium chinense* Mill., with bluish-lavender flowers with the often, five feet tall, *Anemone vitifolia* Ham. var. *tomentosa* Max., which, with its pink flowers makes it a very desirable ornamental, worthy to be cultivated in a rock garden; it ranges from the Himalayas, Sikkim, to West China.

On the summit of the pass and on the upper slopes *Picea asperata* Mast., and *Picea purpurea* Mast., formed groves; the top of the pass is actually an alpine meadow. Wooded ridges, descend into a depression whence towers Lien-hua Shan 蓮花山 or Lotus Mountain whose peak is crowned by a temple; the approach is up a precipitous cliff by means of iron chains fastened into the rock wall.

Lien-hua Shan is a massive limestone mountain situated between Lin-t'an 臨潭 and the T'ao River 洮河. As it is north of the knee of the T'ao Ho and as there are limestone outcroppings before Lien-hua Shan is reached, it must be considered a northern extensive of the Min Shan 岷山, just as the Hsi-ch'ing Shan 西傾山 is a western extension of the latter. It rises from about 9,000 feet, this being the level of the surrounding country to 11,600 feet, and is composed entirely of old grey limestone. Like the Min Shan it is rich in plant species, much richer than any range to the north or northwest, and this includes the Am-nye Ma-chhen Range. Many of the plants occurring on the Min Shan also occur on Lien-hua Shan, especially is this true of the ligneous plants, as Rhododendrons, Abies, Crataegus and *Pinus Armandii*, all of which reach here their northern limit, except Rhododendron and of this genus, *Rhod. rufum* Bat., is not found beyond Lien-hua Shan. The flora is thus more related to that of the Min Shan than to that of the Nan Shan and its parallel ranges, and the non calcareous ranges to the west. A surprising number of new species have been found on that isolated mountain. The surrounding country is composed mainly of red sandstone covered with loess; it thus rises out of a bed of that formation. To the north of it, towards Lan-chou, desert conditions prevail.

Like most limestone mountains, Lien-hua Shan is botanically rich and deserves intrinsic study at all seasons of the year, except winter when it is covered with snow. A thorough botanical survey will undoubtedly bring to light a greater number of plants and new species as are here recorded.

The summit is a very steep limestone crag on which several small temples have been built. The mountain is a maze of depressions, valleys and ridges, one large, one main ravine extending into the T'ao River from its eastern slopes carrying the Lien-hua Ho or Lien-hua Stream. Ere the mountain itself is approached one meets with an array of woody plants. The trail to Ti-tao 狄道 now called Lin-t'ao 臨洮 skirts the mountain over ridges, spurs, and passes whence a view is obtained of the T'ao valley. On the eastern slopes is situated the village called Shan-shen-miao 山神廟 or the Temple of the spirit of the mountain. Deep rocky ravines extend anywhere from the mountain composed of limestone and conglomerate, as the eastern end of the Min Shan. One of these ravines extends into the Lien-hua stream which debouches into the T'ao River, the trail following it to its confluence where the T'ao River flows north between high, bare hills, covered with loess and grass. From the valley Lien-hua Shan appears as a deep, dark blue-green mass, its northern flank falling steeply to 7,150 feet and thus the mountain looks here much higher than from the southwestern side. The northern slopes are also much richer in species.

Around the base of the mountain the soil is a yellow loam, very slippery when wet. Here the only tall trees are willows mostly *Salix paraplesia* C. Schneid. Along the lower slopes in scrub forest we meet with *Lonicera hispida* Pall., a shrub 2-3 feet with yellow flowers and fruits, *Rosa Biondii* Crepin, 4 feet high, the flowers cream-colored or white; this rose extends also into the spruce forest up to 10,000 feet. On exposed banks we find *Potentilla fruticosa* var. *parvifolia* Wolf, not taller than 3 feet, and *Malus kansuensis* Schneid., a tree 20 feet or less with white flowers. The spruce forests extend from the 9,500 feet level to above 10,500 feet, and are composed mostly of *Picea asperata* Mast., while higher up its place is taken by *Picea purpurea* Mast., a stately tree with deep purplish-black cones. Both species occur also together at the 9,500 feet level with *Pinus Armandi* Franch., which here finds its northern limit. In the shade of the spruce forest thrives *Cortusa Matthioli* L. with purplish-red flowers resembling a Primula, and the new umbellifer *Ligusticum Weberbauerianum* Fedde & Wolff nov. spec., while *Ligusticum Pilgerianum* Wolff, prefers the more open alpine meadows at 10,000 feet. On open slopes and clearings in the spruce forest grows the hardy, but very slow growing *Rhododendron rufum* Batal. with lovely pinkish-white or rose-colored flowers, and thick leaves with a rufous indumentum, hence its name; it is here either a shrub of 10 feet or a small tree, its wood is hard and bark smooth and brown. This is its northernmost station while its confrere from the Min Shan, *Rhododendron Przewalskii* extends to the summit of the mountain, but is found much further north as in the Potanin Range, and on the Ta-pan Shan, flanking the Ko-ko Nor in the northeast. It is a much hardier species, but *Rhod. rufum* itself can endure temperatures of minus 20° Fahr., but then its foot is under a deep blanket of snow. I have however seen *Rhododendron Przewalskii* with its short trunk encased in ice up to the branches.

The deciduous forest is composed of *Tilia chinensis* Max., which extends to western Yün-nan, *Malus baccata* Borkh., a tree 15-20 feet with small red fruits which also frequents open slopes, especially on the northern faces of the mountain. *Viburnum Sargentii* Koehne var. *calvescens* Rehd., a small tree 10-12 feet with lobed leaves and bearing bright red fruits in late August, is also partial to a northern exposure at about 9,000 feet elevation. *Corylus Sieboldiana* var. *mandschurica* Schneider, a shrub 10-15 feet, with yellow hirsute fruits occupies the outskirts of the forests and open scrub on the northern slopes; higher up we encounter *Viburnum betulifolium* Batal., a shrub or small tree, with glabrous leaves and red   berries and with it *Aralia chinensis* var. *nuda* Nakai, with a few spreading branches and yellowish flowers; this shrub occurs also in company with the previously mentioned *Corylus*. Confined to the northern slopes of the mountain is *Quercus liaotungensis* Koidz., a small tree 20 feet, with spreading branches which is besides here only found in the Min Shan in the Lower T'ieh-pu country and not elsewhere in the west, but as its name implies was first described from the Liaotung peninsula in the northeast of China. The new *Crataegus kansuensis* Wils. extends from the T'ao River valley west of Cho-ni to the northern slopes of this mountain; it is a striking tree of 15 feet with red spines and red petioles, and equally red, young branches. Of lianas *Clematoclethra integrifolia* with purplish black fruits often covers *Malus baccata* and *Pyrus pashia* on the northern slopes, and like the oak occurs also on the northeastern end of the Min Shan, in dense forest among Acer, Betula and spruces.

Berberis diaphana Max., becomes a shrub of 5-6 feet while *Sorbus hupehensis* Schn. var. *aperta* (Koehne) Schn., attains the size of a fairly large tree of 40 feet, with a trunk of one foot in diameter, possesses large leaves and sharply serrated leaflets; its fruits are white and small, This is the only place where we encountered this variety, but it may be identical with a plant found on the Min Shan by R. C. Ching (no 920); the species itself was first described from Hupeh. It ascends into the spruce forest in company of *Sorbus Koehneana* Schn., a rather small tree also with white fruits, this latter species descends however to lower elevations.

Above the village of Shan-shen-miao the forest is drier and here we find associated with the oak and linden two species of maple, *Acer Maximowiczii* Pax, with tri-lobed leaves and flowers arranged in drooping racemes; on this mountain it reached however only a height of fifteen feet. This is its lower station whence it extends into the spruce forest with *Betula albo-sinensis* Burk., and *Acer tetramerum* Pax var. *betulifolium* Rehd., a shrub about 10 feet tall, and oval, sinuate, serrate leaves, and large fruits arranged in drooping racemes.

Other ligneous plants found around Shan-shen-miao are *Clematoclethra lasioclada* Max., a common woody climber over Acer, with long oval leaves, sharply dentate, and black fruits arranged in umbels; *Cornus macrophylla* Wall., a tree 20-25 feet with large, oval leaves, glaucous beneath, with its small purple fruits arranged in large cymes, and *Cotoneaster multiflorus* Bunge var. *calocarpus* Rehd. & Wils., a shrub 5-8 feet with oval acute leaves and red midribs, and dark red, large globose fruits, borne singly. *Acanthopanax Giraldii* Hams, with tri-foliolate serrate leaves, a spiny trunk 6-8 feet tall, and large black fruits borne in small umbels, *Sorbus Prattii* Koehne 10-15 feet tall with large white fruits (October), and *Rosa Sweginzowii* Koehne, a 4-5 feet high shrub pubescent throughout including the long pyriform fruit, all grew at an elevation of 9,500 feet forming the lower scrub forest. This gradually extends into the conifer forests from 10,000 feet composed of *Picea asperata* Mast. to 11,000 feet where *Picea purpurea* Mast. and *Abies sutchuenensis* Rehd. & Wils. take its place. This is the northern limit of *Abies* in the West of China.

To the 9,500 foot level belong also *Pyrus ussuriensis* Max. var. *ovoidea* Rehd., a tree 40 feet tall with large oval leaves, and dark green, globose fruits 1 ½ inches in diameter, and the large, 80 feet tall *Celtis Bungeana* with trunks 3 feet in diameter, leaves elliptical entire in the lower half, and black pea-sized fruits. Of shrubs or small trees mention must still be made of the lilac *Syringa pekinensis* Rupr., with white flowers and reaching 15 feet in height, the spiny *Ribes Giraldii* 4-5 feet tall, trilobed small leaves and small red globose fruits. *Prinsepia uniflora* Batal., another spiny shrub with globose red, edible fruits belonged however to the drier, exposed slopes of 9,000 feet, it grew also along brooks and streams, like its congener *Prinsepia utilis* of Yün-nan which is most common along streambeds and ditches. So much for the lower forest zone. To this zone belong also *Buddleia alternifolia* Maxim., a shrub 4-5 feet with linear lanceolate leaves, white beneath, *Clematis brevicaudata* DC., a woody climber with large, yellow fruiting heads 3 inches in diamater, and its confrere *Clematis tangutica* var.

obtusiuscula Rehd. & Wils.; the latter two are always found smothering bushes on the outskirts, also *Smilax Oldhami* Mig., with thin papery leaves and purplish black fruits, while *Humulus lupulus* L. climbs over trees to a height of twenty feet.

The upper zone from 10,500 to 11,000 feet is composed of the conifers already mentioned; as undershrubs occur *Rhododendron anthopogonoides* Max., *Salix plocotricha* Schn., a shrub 10 feet high which, while growing in the spruce forests, loves also light, and is more often found on the outskirts. Here also thrive *Malus kansuensis* Schn., 15-20 feet high with *Betula albo-sinensis, Euonymus Giraldii* Loes. var. *angustialatus* Loes., 8-10 feet tall, *Philadelphus pekinensis* Rupr. var. *kansuensis* Rehd., with fragrant white flowers, the herbaceous *Phlomis umbrosa* Turcz., with pale pink flowers, and *Lonicera heteroloba* Batal. Often found in moist situations under *Abies sutchuenensis* is *Tiarella polyphylla* Don., with delicate whitish pink flowers, while in the mossy (Mnium) ground under *Abies* and *Picea* the orchid *Orchis spathulata* Reich., with purple flowers, *Orchis chusua* D. Don, with deep rich purple flowers, the whitish green flowered *Herminium tanguticum* Rolfe, and the lovely *Primula aerinantha* Balf. f. & Purd. are at home. While all grew in the shade of *Abies* they often emerge into open alpine meadows; the Primula recalls the Yün-nan *Pr. pinnatifida,* with its lavender blue flowers but dull green leaves. *Valeriana tangutica* Batal. also loves moist shady places where in mid-July, it displays its pinkish flowers, frequently venturing into open alpine meadows in the conifer forests, or above them, at 11,000-11,500 feet.

In the open alpine meadows above the conifer forest we find quite an assortment of plants as *Saussurea Giraldii* Diels, with purple flowers and leaves whitish beneath, *Juncus leucomelas* Royle, *Adenophora* sp? (no 12719), *Senecio acerifolius* C. Winkl., with yellow flowers and palmately-lobed leaves, with the reddish-purple flowered *Pleurospermum Candollei* C. B. Clarke, which often is also found among rocks at 11,500 feet elevation, then *Pedicularis tristis* L. var. *macrantha* Max., flaunting sulphur yellow flowers, the grass *Beckmannia erucaeformis* (L.) Host., *Equisetum* sp?, *Primula conspersa* Balf. f. et Purd., with lavender purple flowers, *Aster Vilmorinii* Franch., with dark purple ray, and deep orange disc florets, the silvery leaved *Salix sibirica* Pall., *Dianthus superbus* L., *Adenophora* aff. *marsupiiflora* Fish, with tubular, blue-purplish tinged flowers, the white flowered *Scrofella chinensis* Max., which descends to 9,500 feet, and occurs also in spruce forest, while *Pedicularis chinensis* Max., with its long tubular, sulphur yellow corollas delights in drier situations on grassy slopes. Here also dwells *Trollius pumilus* Don, but more partial to wet meadows. Another *Pedicularis* with yellow flowers, *Pedicularis semitorta* Max., prefers wet meadows, while on the better drained grassy slopes a species of *Gentianella* (sect. Crossopetalum) no 12691, with *Oxytropis Giraldii* Ulbr. is at home.

On exposed grassy slopes the lovely *Lilium Duchartrei* Fr. var. *Farreri* Krause, with white flowers and longitudinal purple spots grew at lower levels. *Meconopsis quintuplinervia* Reg., here with deep lavender flowers and small hairy leaves adheres to the 11,500 foot level in open alpine meadows. *Pirola rotundifolia* L., keeps to the 10,000 foot level, as does the new orchid *Oreorchis Rockii* Schweinf. n. sp., and *Saxifraga lumpuensis* Engl., with red flowers. On the margins of the wet alpine meadows *Rhododendron capitatum* Max., with lavender-purple flowers forms uniform, dense thickets. In damp meadows at 10,000 feet elevation occur *Codonopsis viridiflora* Max., with greenish-purple flowers, and an unidentified *Saussurea* no 12740, as well as the pale, blue flowered *Aconitum laeve* Royl., and a *Ruellia*, no 12786. On drier meadows grew here the rare, white-flowered *Ajuga calantha* Diels f. *albiflora* with its leaves firmly appressed to the ground, also the scrophulariaceous *Euphrasia tatarica* Fisch., first described from Siberia; it usually loves shady banks, but here it grew in the open alpine meadow and had white flowers; a deep rich, purple-flowered *Allium* sp? no 12789 kept it company. *Parnassia Delavayi* Franch., was partial to watercourses in the alpine meadows at 10,000 feet elevation.

Above the alpine meadows among the limestone crags near the summit grew the following: *Primula stenocalyx* Max., a farinose form, with leaves white mealy beneath, and the flowers a pale blue or pale lavender; the new *Heracleum millefolium* Diels var. *longilobum* Norm., the type of which is 12734, with carmine red flowers, the entire plant being covered with white wool. On limestone boulders the white flowered *Androsace tapete* Max., formed cushions. *Saussurea* aff. *prophyllae* Diels, with purple flowerheads, a fleshy species of *Cacalia* with yellow flowerheads, and the new willow *Salix pseudospissa* Görz n. sp., a shrub 2-3 feet were rooted in rock crevices, while the scrub *Rhododendron anthopogonoides* Max., and *Salix oritrepha* Schn., grew scattered among large boulders. The flat spreading, red-fruited *Cotoneaster horizontalis* Decne, and the rich purplish-blue flowered *Meconopsis quintuplinervia* Reg., covered with yellow hair throughout, did not ascend to above 11,500 feet. On the very top of Lien-hua Shan 11,600 feet elevation, on the limestone crags and among crevices grew the bluish-purple *Aconitum tanguticum* (Max.) Stapf, the new, yellow flowered composite shrub *Tanacetum salicifolium* Mattf., and a host of others as *Primula aerinantha* Balf. f. et Purd., with lavender blue flowers; a straggling legume *Hedysarum* sp? no 12675 its flowers a rich pale purple; the 3 feet tall *Rheum acuminatum* Hook. f. & Thoms., with red flowers and the rosette-forming *Heracleum millefolium* Diels var. *longilobum* Norm. The yellow *Sedum Purdomii* W. W. Sm., and *Astragalus* with bluish-purple flowers no 12704, *Astragalus Moellendorffii* var. *kansuensis* with purplish-red flowers, *Leontopodium Jacotinianum* Bvd., and *Pedicularis* affinis *P. plicatae* Max., no 12784 with sulphur-yellow flowers formed the plant covering near and at the summit of Lien-hua Shan. Certain areas near and at the summit of the mountain were taken possession of by the robust *Rhododendron Przewalskii* Max., and *Rhododendron capitatum* Max., the first white, and the second purplish-blue flowered, two of the hardiest of all Rhododendrons.

Vegetation Along Water Courses and Especially in the Hai kou or Hai Valley on the Southwest Slopes of Lien-hua Shan

Hai kou is a deep valley on the southwestern slopes of Lien-hua Shan and is forested in its lower part mainly with *Pinus tabulaeformis* Carr., which reaches here a height of 70 to 80 feet and trunks of from 2 to 3 feet in diameter. Associated with it are *Malus baccata, Celtis bungeana,* Crataegus etc. The upper part of the valley is grass-covered and here we find a varied herbaceous vegetation through which are scattered shrubs such as *Ribes Giraldii* Jancz., *Lonicera heteroloba* Batal., and *Sorbus Koehneana* Schneider. At an elevation of 9,000 feet we find the thistle *Cirsium setosum* M. B., with lavender purple flowers along the stream, the bushy *Caryopteris tangutica* Max., endemic to Kan-su and Koko Nor, with a height of 1-2 feet, lavender flowers and greyish-wooly leaves, as well as *Anemone japonica* (Thunb.) S. & Z. var. *tomentosa* Max., with rich pale lavender flowers and large grey leaves of a silky texture. Among the rocks along the stream we encountered *Dracocephalum heterophyllum* Benth., known also from the Himalayas, and its congener *Dracocephalum imberbe* Bunge, first described from the Altai mountains with deep purplish blue flowers and leaves reddish purple beneath, *Bupleurum longeradiatum* Turcz., with small blackish-purple flowers and pale green leaves, and on meadows along the stream the orange-yellow flowered day lily *Hemerocallis Dumortieri* Morren, all at the 9,000 foot level. Higher up, but still along the stream bloomed *Hypericum* sp. (no 12772), *Achillea ptarmica* L., a European plant, *Anaphalis margeritacea* (L.) Benth. & Hook., *Erigeron acris* L., the rosaceous *Sanguisorbia canadensis* L., and *Rubus xanthocarpus* Bur. & Fr., bearing yellowish-red, edible berries, first described from Ssu-ch'uan. Of shrubs mention must be made of the yellow *Caragana frutex* K. Koch, 5-6 feet tall and two species of *Clematis, C. glauca* var. *abeloides* f. *phaeantha* Rehd., a climber over shrubs and *Clematis aethusifolia* Turcz., the former with purplish brown flowers.

This practically concludes the list of plants found on Lien-hua Shan; had time permitted, an intensified search of all the slopes and valleys at other seasons of the year would undoubtedly have brought forth many more species, but the mountain is also not high enough to develop a real alpine flora as is the case on the Min Shan. There is an absence of junipers; *Larix* is also wanting, although it occurs on the south and northeastern end of the Min Shan.

Lien-hua Shan with its varied flora is the last outpost in Kan-su, a rich oasis in a desert with which the meager flora of the Koko Nor region cannot be compared.

The approach to Lien-hua Shan from the south and again from the north is through arid loess country with a correspondingly poor flora, one would have thought that a mountain so close to the capital of a province, would have been botanically explored, yet its flora was less known than that of the remote Richthofen Range or Nan Shan, in the far northwest. As a whole, even taking into consideration the Kan-su min Shan, the flora of both Kan-su and Koko Nor cannot be mentioned in one breath with that of Hsi-k'ang and still less with that of Northwest Yün-nan, the richest of all China.

The Province of Ch'ing-hai

Present Extent
The province of Ch'ing-hai or Blue Sea, falls roughly between 90° and 103° E. Long., and 32° and 40° N. Lat. It is irregular in outline, much indented in the south and west, with a great part of it still unexplored. It is a high plateau whose mountains rise to over 20,000 feet, but in the northwest between 91° E. L., and 36-39 N. L., there is a great depression composed mostly of desert steppes and salt swamps known as the Tshai-dam (Tshwai-hdam), meaning salt swamp.

Three of the greatest rivers of Asia, as the Yangtze, Yellow River, and the Mekong have their sources in this province. The Salwin however originates in Tibet proper.

Prior to 1928 the province was smaller for part of its present, eastern territory, belonged then to Kan-su, the adjoining province, while a good part of its western region was reckoned to Tibet. With the exception of the changes in boundary to the east, those in the west are ill defined and marked more or less arbitrarily, for the Chinese did not control the territory.

The name by which the province is best known to the Western world is the Mongol one, Koko Nor, meaning Blue Lake of which the Chinese name is a translation; its Tibetan name is mTsho-sngon (Tsho-ngön = Lake blue), the adjective qualifying the noun follows it in the Tibetan syntax.

Historical Sketch

In ancient times the territory was known as the land of the Hsi Ch'iang 西羌 or western Ch'iang a large tribe related to the Tibetans, and later called the Ch'iang province. The Ch'iang were a nomadic people as the Chinese character for them testifies, it is composed of the radicals for sheep and man.

In the time of the Great Yü or Yü-kung 禹貢 period (tribute of Yü) B. C. 2205-2198 it was the land of the Hsi-jung 西戎 or western wild tribes. Prior to the Han dynasty B. C. 206-24 A. D., it became again the territory of the Hsi Ch'iang. After the Eastern Chin 東晉 317 A.D. it belonged to the T'u-yü-hun 吐谷渾, and in the early part of the T'ang dynasty, about 750 A. D., both the T'u-yü-hun and the Tang-hsiang 党項 or Tangut occupied the land, the former around the lake Koko (nor), and the latter the knee of the Yellow River. Later in the T'ang dynasty the T'u-fan 吐蕃, analogous to the Tangut, solely occupied the territory; they were all related to the present day Tibetans. The Tangut held sway till the advent of the Ming dynasty 1268-1644 A. D. after which the Mongol tribes, especially the Torgut Mongols, occupied the land up to 1616; at times (1512 A.D.) it was also under the sway of the Mongols of Ordos Tümed, who ruled more of the northern part, while the Tanguts were confined to the Yellow River area, from its source to within the knees, and east of it. In the beginning of the Ch'ing or Manchu dynasty it became the pasture land of both Mongols and Tibetans. It was then that the city of Hsi-ning 西寧, denoting Western Peace, was established, whose affairs were managed by a minister of state who governed the territory and the tribes.

In the 4th year of the Republic of China 1915, all this was changed and there was established the Kan-su border Ning-hai 寧海 protectorate; which included Ning-hsia 寧夏, and was guarded over by a Commissioner. It was then that the Mohammedans gained control of the territory and a Moslem general ruled ruthlessly over the Tibetans.

In the 17th year of the Republic, 1928, the province of Ch'ing-hai was created and parts of western Kan-su were incorporated into Ch'ing-hai.

Present Day Borders

In the north and east it borders on the province of Kan-su 甘肅, in the southeast it adjoins Ssu-ch'uan 四川, in the south Hsi-k'ang province 西康省 (Sikang), and in the south and west it is contiguous with Hsi-tsang 西藏 or Tibet. In the north it has a common border with Hsin-chiang 新疆 (Sinkiang) or the New Frontier, the latter adjoins it also on the west and to the north of Tibet.

Its capital and trading centre is Hsi-ning situated in the northeast of the province. There are 29 Mongol banners and forty clans or tribes of Hsi-fan 西番 or western barbarians as the Chinese love to term non-Chinese tribes, who are ruled by local chiefs or T'u-ssu 土司. The trade mart is also the seat of the government of the province. The diameter from east to west is about 600 miles, and its length from north to south about 450 miles, its area approximately 228.350 square miles. Its inhabitants number roughly 15,112,000 souls.

Present Political Divisions

The province of Ch'ing-hai is divided into 19 districts or magistracies. Nearly all the old names current during the Ch'ing or Manchu dynasty have been changed when the borders of the province were rearranged. I shall give here both the old and new names of the towns or villages which have been raised to magistracies, as well as of those which had been magistracies before the changes took place.

Hsi-ning or «Eastern Peace», the Tibetan Zi-ling, and Sining of western maps has been made the capital, and its old name has been retained. It is situated on the south bank of the Huang-shui formerly called the Hsi ho 西河 or West River, also Hsi-ning Ho 西寧河. The ancient name of the town was Huang-chung 湟中.

Huang-yüan 湟源 is the present name of the former Tan-ka-erh 丹噶爾 derived from the Tibetan name of the place; it is situated on the Huang-shui 湟水 and is 30 miles west of Hsi-ning and formerly belonged to Kan-su.

West of Huang-yüan is the magistracy of Hai-yen 海晏 or the Quiet of the Sea. To the north of Hai-ning is the Hsien district of Hu-chu 互助 or Mutual Assistance, formerly a mere village called Wei-yüan-p'u 威遠堡. Northwest of the latter is Ta-t'ung-hsien 大通縣 whose former name was Mao-pai-sheng 毛伯勝, it was also known as Pai-t'a-ch'eng 白塔城 or the White Pagoda City. To the north of it is the district of Wei-yüan 亹源 or the Source of the Wei or Hao-wei 浩亹 or the Vast Wei River, actually the upper Ta-t'ung ho. The old name of Wei-yüan was Pei-ta-t'ung 北大通 a mere empty shell of a place.

The easternmost district is on the south bank of the Huang-shui or Hsi-ning River, and is called Min-ho 民和, it is opposite the village of Hsiang-t'ang 享堂 where the Ta-t'ung Ho 大通河 joins the Huang-shui or Hsi-ning River, actually one third of a mile before reaching Hsiang-t'ang. The Yamen (official residence) of the magistracy of Min-ho was first at Hsia-ch'uan-k'ou 下川口 («Below the Mouth of the Stream») but was later removed to the ancient Ku-shan-p'u 古鄯堡 or Ku-shan-yi 古鄯驛 or Ku-shan post station, which is southwest of Hsiang-t'ang. Shang-ch'uan-k'ou or Above the Mouth of the River, is directly opposite Hsiang-t'ang.

Southwest of Hsi-ning is the new district of Huang-chung 湟中 or the Centre of Huang, where the great yellow Lama temple sKu-hbum, pronounced Kumbum, meaning Hundred-thousand Images, the Chinese T'a-erh ssu 塔爾寺, is situated, and where Tsong-kha-pa, the founder of the Yellow Sect was born. Adjoining the lamasery is the small trading town or village of Lu-sha or Lu-sha-erh 魯沙爾, a transcription of the Tibetan Klu-gsar, pronounced Lu-sar. Between Hsi-ning and Min-ho is the district town of Le(Lo-)tu 樂都 formerly known as Nien-pai 碾伯. South-southwest of it is the district of Hua-lung 化隆 the former Pa-yen 巴燕 or Pa-yen-jung 巴燕戎, more than halfway between the Hsi-ning River and Hsün-hua 循化, the seat of the Sa-lar Mohammedans. Southwest of the latter is the lamasery of Rong-wo, the Chinese Lung-wu ssu 隆務寺 created a hsien and called T'ung-jen 同仁. South of the Lake (Koko nor) is the district of Kung-ho 共和, and southwest of it the old hsien and city of Kuei-te, called Gus-mdo (Gü-mdo) in Tibetan, situated on the south bank of the Yellow River. Subject to it is the smaller hsien or district called T'ung-te 同德, actually only a lamasery called Ra-gya dgon-pa or La-chia ssu 拉加寺 on the Yellow River in the South. West of the Yellow River and almost opposite the brGyud-par (Gyü-par) Range is the long valley of Ta-ho-pa 大河壩, the Tibetan Hang Chhu or Hang River which flows into the Yellow River. The tiny group of huts located in the valley are also known as Ta-ho-pa or the Great River Bank or Plain. The Chinese with a look into the future, created of this wilderness a hsien or district called Hsing-hai 興海 or the Prosperous Sea; it is southwest of Kung-ho 共和. West of the Koko Nor is the little principate of Tu-lan ssu 都蘭寺 where there is a lamasery, it is called Dulan Hiid in Mongolian or the Warm Hermitage and marked on foreign (western) maps Dulan-kiit. It is also called Hsi-li-kou 希里溝, and is situated at the entrance to a narrow defile formed by the Dulan gol or Dulan River. Groves of *Juniperus tibetica* Kom., occur here. Near here is also a lake, the Dulan Nor or Tu-lan-hai on whose shore the local native prince had his camp. To the northwest is the small lake called Sirho nor, the Chinese Szu-erh hu 思爾湖 whence a trail leads to the salt swamps of Tsai-dam, from the Tibetan Tshwai-hdam, salt swamp, the Chinese Ch'ai-ta-mu 柴達木.

In the extreme central south is the district of Yü-shu 玉樹 or Jade Tree, whose ancient name was Chieh-ku 結古, called in Tibetan Khyer-dgun-mdo (Khyer-gün-do), also written sKye-rgu-dgon (Kye-gu gön) but the last is the name of the local monastery which controls the nomads of the region. The latter spelling of the name is according to Sir Basil Gould.[16] The Tibetan name has been transcribed by the Chinese Kai-ku-to 蓋古多. It was also the last Mohammedan outpost where the troops of Ma Chi-fu[17] had a fort and baracks. The altitude of the settlement is 12,928 feet according to George Pereira[18] and comprises 200 Tibetan families and 40 Moslems. The town consists of mud-built houses on the hillside north, and above the Pa Chhu or Pa River 巴河 or Chieh-ku shui 結古水, a tributary of the upper Yangtze. Northwest of Yü-shu is the district of Ch'eng-to 稱多 situated on the upper Yangtze, the Chin-sha Chiang 金沙江 of the Chinese or River of Golden Sand, and the Tibetan hBri Chhu (Dri Chhu) or Cow-yak River.

[16] Basil Gould, 1883-1956, diplomat, British trade agent and political officer in Tibet, Tibetologist.

[17] Ma Ch'i 馬麒 (1866-1931?) was governor of Ch'ing-hai from 1929 to 1931. He is mentioned in Howard L. Boorman: *Biographical dictionary of Republican China*. Vol. 2. New York: Columbia Univ. Pr. 1970, 474b (in the biography of his son Ma Pu-fang 馬步芳 who became governor of Ch'ing-hai in 1938)

[18] George Edward Pereira, 1865-1923, British explorer. See: *Peking to Lhasa*. The narrative of journeys in the Chinese Empire made by the late Brigadier-General George Pereira. London: Constable 1925. 293 pp.; H. Gordon Thompson: From Yunnan-Fu to Peking along the Tibetan and Mongolian borders, including the last journey of Brigadier-General George E. Pereira. *The Geographical Journal* 67.1927, 2-27; *Who's who in the Far East* 1906/7, 264.

The southernmost district is called Nang-ch'ien 囊謙 formerly known as Se-lu-ma 色魯馬, situated on the O-mu-ch'u Ho 鄂穆楚河, the Tibetan Ngam Chhu, which is the northeastern branch of the Mekong or rDza Chhu in Tibetan.

The northwestern and western parts of Ch'ing-hai are wild and uninhabited, and are mainly composed of salt swamps (Tshai-dam), barren mountains, waste lands, and snow clad ranges to the southwest.

The eastern part of the province is designated by the Tibetans as A-mdo and the region to the south of it Khams.

The Mountains of Ch'ing-hai

To give a detailed description of all the mountains and mountain ranges of the province is impossible as our knowledge of the area is insufficient and does not come within the scope of this account. Many of the ranges have been described by various travelers but the nomenclature is most confused. A detailed description is given in the accounts of the mountains explored by the author botanically, most of which were up to that time unexplored, as the eastern part of the Am-nye Ma-chhen, the Gyü-par Range, and the entire Min Shan 岷山 which belongs however to the neighboring province of Kan-su and not to Ch'ing-hai. The various ranges which comprise the Nan Shan in the extreme northwest have been visited by other travelers, notably Sir Aurel Stein[19], but few or no botanical collections were made.

Other areas explored by the author, and previously unexplored, were the gorges of the Yellow River, the mountains east and west of Ra-gya, south to Tsan-gar (monastery), and east of the same to Dor-gen-nang.

Each of these areas is described in detail, also the country traversed between them and La-brang.

As to the rivers of the province, the Yangtze, which the Chinese designate as the Ch'ang Chiang 長江 or the Long River, has its source in the K'un-lun 崑崙 mountain range, that is its northwestern branch called the Ch'u-ma-erh ho 楚瑪爾河; the longest branch, with its source in the extreme southwest of the province is the Wu-lan-mu-lun Ho 烏蘭木倫河 which joins the Mu-lu-wu-su Ho 穆魯烏蘇河, the Muru Ussu of the Mongols; united they form the T'ung-t'ien Ho 通天河 which further south becomes the Chin-sha Chiang 金沙江 or River of the Golden Sand.

The Yellow River is described elsewhere in detail. Next of importance are the Ta-t'ung Ho 大通河 and Ya-lung Chiang 鴉龍江; the former rises on the northern slopes

---

[19] Aurel Stein, 1862 – 1943, explorer of Central Asia; his important collections went to the Government of India and the British Museum. See Kazuo Enoki: Aurel Stein kei shôden. *Tôyô gakuhô* 1950:1, S.102-122; L. Rásonyi: *Stein Aurél és hagyatéka*. Budapest 1960 37 p. (Publicationes Bibliothecae Scientiarum Hungaricae 18); Publications of Sir Marc Aurel Stein. *Journal of the RAS* 1946, 86-89; C. E. A. W. Oldham: Sir Aurel Stein. *JRAS* 144, 81-86. *Aurel Stein on the Silk Road* / Susan Whitfield. London: British Museum Press, 2004. 143 pp. Among Stein's major publications are *Innermost Asia*: Detailed Report of Explorations in Central Asia, Kan-Su and Eastern Iran / Carried out and described ... by Sir Aurel Stein. With descriptive lists of antiques by Sir Aurel Stein. Oxford: Clarendon Press 1928. 4 vols. 4° – *Serindia*: detailed report of explorations in Central Asia and Westernmost China / Aurel Stein. Oxford 1921. 4 vols. – *Ancient Khotan*. Detailed report of archaeolog. explorations in Chinese Turkestan. Oxford: Clarendon Press 1907. 2 vols.

of the Ta-t'ung Shan, one of the parallel ranges comprising the Nan Shan, and the latter in the district of Yü-shu 玉樹縣.

The sources of the Mekong are in the south central part of the province, that is the two branches which unite at Chhab-mdo (Cham-do), the Chinese Ch'ang-tu 昌都, to form the Mekong or Tsa-ch'u Ho 雜楚河 in Hsi-k'ang, and known in Yünnan as Lan-ts'ang Chiang 瀾滄江. The upper part of it is called by the Tibetans rDza Chhu (Dza Chhu), and in Yün-nan Zla Chhu, pronounced Da Chhu.

Products

The main products are sheep wool, musk, felts, hides, furs and yak tails which were formerly exported to Japan to augment the hairdress of Japanese women; salt is exported from the salt lakes in the west, fish (mostly a species of carp) from the streams and lakes, borax, medicinal herbs, and the following minerals: silver, iron, copper, lead, zinc, petroleum, and gold which is found in the region to the northeast in the district of Wei-yüan.

Communication

There are many trails and caravan routes, but none that deserves the name of road in the modern sense of the word. The distances are reckoned in Chinese *kung-li* 公里, three kung-li being equal to one mile. In Kan-su and Ch'ing-hai the Li is a little longer than the Ssu-ch'uan Li.

From Hsi-ning to Lan-chou the capital of Kan-su is 275 li; south to Kuei-te 104 li; west to Tu-lan-ssu 357 li, reckoned from Huang-yüan; to Ta-t'ung northwest 63 li, and thence 52 li to Wei-yüan in the north; to Hu-chu northeast 52 li, southwest to Yü-shu 795 li, to T'ung-jen from Hsün-hua 83 li, to Hsün-hua from Hsi-ning 155 li, from Hsi-ning to T'ao-chou Old City via La-brang 344 li. One trail leads from Min-ho to Yao-kai 窰街 thence Lien-ch'eng 連城 to Yung-teng 永登 and north to Wu-wei 武威, the former Liang-chou 涼州, to Chiu-ch'üan 酒泉, formerly Su-chou via Shan-tan 山丹 and Chang-yeh the former Kan-chou 甘州 a distance of 947 li; the stretch from Wu-wei to Chiu-ch'üan or Su-chou could be accomplished by mule cart. As the streams have no outlet and disappear underground, the countryside becomes in places a quagmire which is frozen in the winter, and in spring when the ice thaws, carts often sink into the quivering earth.

At the time of my visit only caravan trails existed, but now a sort of motor road has been constructed to the larger places, and also to the Koko Nor (Lake).

Population

The main inhabitants of Ch'ing-hai may be considered Tibetans and Mongols followed by Chinese, Moslems and T'u-jen or aborigines. The Chinese and Moslems dwell mainly in the northeastern part as in and around Hsi-ning and Huang-yüan. The Mongols are mostly confined to the northeastern Koko Nor plateau, while the Tibetans occupy the rest of the country. In this book appear many Tibetan tribal or clan names, this does not indicate that they are not Tibetans for all speak the Tibetan language, yet they are distinct clans and each has its own chief who rules over each respective clan. They are further divided into sedentary Tibetans who live in villages, and nomads or hBrog-ba (Drog-wa) who occupy the grasslands. Each clan has its own winter and summer grazing lands, and they are nomads only to the extant that they shift their camps from one to the other, but never go farther afield. No clan would dare to move into the grazing lands of another. The encampments of lower altitude are reserved for winter occupancy, and those situated at higher levels for the summer camps.

No cultivation can be carried on on account of the height of these grasslands above sealevel, and the Drog-wa are thus dependent on traders who furnish them their tea and barley. These items are rarely if ever bought with silver, but given in exchange for sheep wool, the main product of the grasslands. Musk is a minor item, but hides furnish, next to wool, the most important article of export. The latter are usually brought into Hsi-ning or Huang-yüan and sold for cash.

The different Tibetan clans can be recognised by their dress, that of the men by their headdress, and that of the women by their ornaments worn on the back, on a long strip of cloth, suspended from their hair over the neck and back. These strips of cloth are of various colors, depending on the tribe, and so are the ornaments, either silver, amber, beads, coin, etc., and their arrangements on these strips.

The northwestern part of Ch'ing-hai is either waterless desert or salt swamp and not habitable, and so is the area adjoining it in the west, i.e., Tibet proper.

The Drog-wa live mainly on yak meat and mutton, buttered tea and Tsamba or roasted barley flour which is mixed with the buttered tea, kneaded with the hand into a dumpling and eaten. The Drog-wa will eat neither fowl nor fish, and no eggs. Potatoes or other vegetables are unknown and scorned. The only roots they eat are those of *Potentilla anserina* L., which they rob from the marmots who dig them, and store them for the winter in their burrows. One other item of vegetable is the fairy-ring mushrooms which spring up in the grasslands in the summer on darker green circular areas, but other vegetables they know not.

Sedentary Tibetans will often raise pigs and thus they have a more varied diet than the Drog-wa. Much butter, milk, and a sort of hard, dry cottage cheese are consumed; the latter is pressed dry, shaped into balls and left to dry near their stoves or on shelves around the tent, till they are brown from the smoke.

For fuel they use yak dung (argols) and sheep manure; the former is often gathered and stored for the winter by being arranged around their camps in the shape of a wall.

Distinct from the Tibetan nomad clans are the fierce mGo-log (Go-log), robber tribes, who live in the neighborhood of the Am-nye Ma-chhen, and south of the Yellow River. They are also divided into several clans who have each their own chiefs and tribal lands. That they are feared by all is needless to say. Large bands of them will raid the caravans on the highway between Hsi-ning and Lha-sa or rob neighboring encampments. They are jealous and superstitious and woe to the traveler who crosses their path, or enters their tribal lands.

Each clan will select a prominent peak or hill within its territory as the abode of its protective *sa-bdag* or lord of the earth, in other words mountain god, which each clan worships and makes offerings to each time a family partakes of tea; this is done by throwing out a ladle full of tea accompanied by quickly muttered prayers. Books could be written on Tibetan nomad clans, but this sketch will have to suffice.

One word must be said about the yak without which the nomad could not exist. The yak is everything, it furnishes milk and butter, meat which is cut up raw into long strips and hung up to dry in the wind and sun; its hide is used for shoes and pouches for bullets, tinder, etc., its horns for gun powder containers, its hair is twisted into ropes or woven into a coarse cloth out of which their tents are made; its dung furnishes fuel (argols), its tails are exported for fly wisps, and the living yak itself is used as a pack animal. Even the bones are found use for especially the shoulderblades, which when dry are inscribed with lamaistic prayers, tied one above the other to a yak hair rope and suspended in some conspicuous place where they will have to be moved, and in so doing the prayers are said for the party who hung them up. Or they will be collected and hung up after having been covered with prayers, under a specially constructed shelter outside some lamasery, hundreds of them, which are shaken, pile after pile, by pilgrims or worshippers after each completion of a pious tramp around the monastery, thus saying the prayers written on them.

As regards Moslems and their sects see population under Kan-su. They live mainly in suburbs of Chinese towns from Hsi-ning to as far south as Yü-shu. Mongols are also nomads, but unlike the Tibetans live in yurts constructed of sheep wool felt and are circular in outline. That they are much more comfortable than a Tibetan yak hair tent needs not to be emphasized. While Mongols will drink mare's milk no Tibetan will do so. Both Mongols and Tibetans are tall, well built and hardy, only the fittest survive in these high, windswept, inhospitable uplands.

The Region of the Koko Nor (Lake)

The name by which the lake is best known is the Mongolian one, namely koko – blue, nor – lake. In Tibetan it is called mTsho sngon-po (Tsho ngön-po) and in Chinese Ch'ing hai 清海, all meaning blue lake or blue sea.

As soon as one crosses the (pass) La-la ta-pan 拉拉達板, elevation 13,675 feet, west of Huang-yüan 湟源, the former Tan-ka-erh 丹噶爾, situated at 8,997 feet above sea level, one leaves behind paved streets, shops, temples, houses, and an agricultural population. A narrow ravine leads to the pass beyond which lies an altogether different world. To the west of the La-la ta-pan lie steppes, salt lakes without outlets, and salt swamps. Here are no villages and everywhere the eye looks there is grass even covering the rounded mountain ranges which encircle the basin in which this largest of Tibetan lakes is situated. The elevation of the lake is 10,700 feet above the sea.

Looking over the vast expanse of lake and steppe one has the impression that here is utter lifelessness, even the broad flat valleys are without a sign of a tree and lend their character to the surrounding mountains. Here silence seems to rule supreme, for man and beast seem to be non-existent. But this is in fact only a deception. The fierce nocturnal winds known as hei feng 黑風 or black wind, which sweep over the vast landscape do not permit the growth of bush or tree, and only grass or flat, rosette-forming plants, can flourish.

Looking south from the encircling ranges one beholds a veritable sea of mountains. In the north we have the parallel and interlocking ranges of the Nan Shan 南山 which extend from northwest to southeast, and culminate at wide intervals in high, snow-capped peaks. South and east of the Koko Nor the mountains are rounded, their crests dome-shaped and grass-covered, they adjoin the arid plain interspersed with swampy areas, gravelly beds and sand which extend to the shores of the brakish lake; at their bases, especially in the east, huge sand dunes are met with, but what little shrubby vegetation there is, one finds only near the foot of these mountains. The lake itself, according to Filchner, has a surface of 5,500 sqkm, and according to Przewalski is between 200-230 miles in circumference.

I personally visited only the southern and eastern shores, so had no opportunity to ascertain its exact size. The water, not actually salty but brakish, is absolutely clear so that every pebble can be seen at the bottom. The deepest part of the lake is 60 feet near its southern end. About the centre of the lake is an island called by the Chinese Hai-hsin 海心 or the Heart of the (Sea) Lake, on which a small Tibetan shrine or temple has been erected, inhabited by ten monks (at the time of my visit). No boats or canoes are known, and the only time the monks can reach the outside world is in the winter when the lake is frozen, from November to March.

Many streams flow into the lake of which the principal one is the Pu-k'o Ho 布喀河; the name is merely a Chinese transcription of the Mongol one, Bukhain-gol. It has two sources one north and near the western peak of the Ta-t'ung Shan 大通山, and another southeast of the Kara Nor or Black Lake. As the object of our exploration was a dendrological one we did not linger long in that vast grassy landscape, but visited its southern and eastern shores only. However the range east of the lake, known as the Potanin Range, proved of considerable botanical interest. It is at a notable distance from the lake, and its main valley extends from north to south, and is known as Ra-k'o gorge.

The Region between Hsi-ning and Huang-yüan

Hsi-ning 西寧 meaning Western Peace, the capital of the province of Ch'ing-hai is situated on the right bank of the Huang-shui 湟水, formerly called Hsi Ho 西河, or West River, also known as Huang-yüan Ho 湟源河, after the town 30 miles west of Hsi-ning, the last outpost of Chinese civilisation east of the Koko Nor. Hsi-ning is at an elevation of 7,800 feet, the streets are planted with willow trees and poplars, the mountains surrounding the town are bare and the whole scenery dreary and dusty. The plants cultivated are mainly wheat, sorghum, millet, potatoes and maize, while in truck gardens the usual Chinese vegetables are grown. Hsi-ning is the center of the wool trade, wool being brought from all over the Koko nor by yak, and thence loaded on camels is sent to Pao-t'ou-ch'eng whence it used to be shipped by rail to T'ien-chin 天津 (Tientsin) and from there by boat to the United States.

Only 30 li or ten miles southwest of Hsi-ning is the famous Tibetan monastery of sKu-hbum   (Kumbum) or 100,000 Images, called in Chinese T'a-erh ssu 塔爾寺, built on the spot where the founder of the Yellow Sect, Tsong-kha-pa was born. The lama-sery has been fully described by Wilhelm Filchner in his book *Kumbum Dschamba Ling*,

Leipzig 1933.[20] Adjoining the monastery is the Moslem trading town with caravan inns, called Lu-sha-erh 魯沙爾.

The path to Huang-yüan leads through the west gate of Hsi-ning and the suburbs due west, up the valley of Huang-yüan River whose waters are usually crystal clear. The valley is quite broad and so is the road which mule carts, heavily loaded with wool, four mules to each cart, frequent. The wheels of the carts are huge and the rims are of heavy iron with sharp prongs at short intervals, giving them the appearance of a cogwheel. These wheels useful on muddy roads, are also their ruin, cutting deep trenches into them. Here and there the road is planted to both sides with poplars, *Populus Simonii* and *Populus balsamifera*, also willows which have spread over the broad valley, especially along the borders of fields. Next to the crops mentioned previously are oats, beans, the *Vicia Faba*, and *Cannabis sativa* the Indian hemp from which the farmers weave their thick, rough, hemp-cloth; it is woven in narrow strips and then sewn together. The hills surrounding the valley are absolutely bare with not a sign of a tree, except those planted on the plain. The fields are extensive and very fertile, they are irrigated by the waters of the Huang-yüan River and its affluents, and belong to various villages. About 35 Li from Hsi-ning the valley broadens considerably, the hills are high, but bare. On this stretch the stream receives two affluents one coming from the south, and where the valley is exceptionally broad, one from the north. To the south, in the distance is visible the high range called La-ya Shan 拉鴉山 which separates the Huang-yüan stream from the Yellow River; it is composed of mica-slate and black dolomite and is about 15,000 feet in height, covered with snow for the greater part of the year. Directly ahead, west, there is a very narrow gorge from which the Huang-yüan stream issues. A mile beyond there looms up an exceedingly rocky range about 17,000 feet in height which extends from south-southeast to north-northwest, also bare of trees and covered with snow for about 2,000 feet; it is the eastern end of the Nan Shan 南山.

At the village of Ma-lung 馬龍 the stream issues from a gorge with the hillsides bare and rocky, being composed of granite, mica-schist with loose boulders hanging from above ready to fall. Through this somber gorge a trail winds following the meandering clear stream, here full of rapids. In places villages are situated in the gorge, while on the steep slopes groves of poplars, as *Populus Simonii* and *P. balsamifera*, enliven the otherwise dreary landscape. After a little over eight miles, the length of this gorge, the valley broadens and the trail ascends a small ridge whence the city of Huang-yüan can be seen, situated on the north slopes of the valley.

Outside the city, which is surrounded by heavy walls, was a Mohammedan garrison, the soldiers all dressed in black wearing peculiar angular caps drawn together by a cord on the top, like a lady's handbag. All the roofs of the houses are flat and resemble those of the middle east.

---

[20] Wilhelm Filchner: *Kumbum Dschamba Ling. Das Kloster der hunderttausend Bilder Maitreyas*. Ein Ausschnitt aus Leben und Lehre des heutigen Lamaismus. Mit 208 Abbildungen aus Kunstdrucktafeln nach eigenen Aufnahmen, 412 Skizzen des Verfassers im Text, einer Lichtdruck- und einer Buntdrucktafel sowie einer Klosterkarte. Leipzig: F. A. Brockhaus in Komm. 1933. XVI, 555 S. With prefaces by Berthold Laufer and Ferdinand Lessing. A large part of the book was authored by W. A. Unkrig (1883-1956) who, as a paid collaborator, is mentioned in the preface but does not show up as co-author.

It is a typical frontier town, but dirty to a degree. The valley in which it is situated is completely surrounded by snow-covered mountains which in the west separate it from the Koko Nor steppes and lake. The town is really beautifully located at an elevation of 8,997 feet, the snow-covered ranges encircling it and the clear stream rushing down the valley into the gorge below adding greatly to the beauty of the scene.

From Huang-yüan Across the La-la ta-pan to the Koko Nor
Leaving Huang-yüan by the west gate the road leads over a hillock down into the valley planted with the species of poplars and willows already mentioned, which with *Ulmus pumila* furnish the only available wood for the construction of bridges and houses. One poplar log about seven to ten feet long and ten inches thick sold for seven taels silver or about five Dollars U.S.

An old bridge spans the Huang-yüan Ho, the trail leading across, continues west around wheat fields up an affluent called the La-la Ho 拉拉河 whose source is in a pass of that name which separates the Huang-yüan district from the Koko nor. The La-la valley is hemmed in by grass-covered bare hills, and is named after the La-la village situated on the right bank of the stream. The only tree growth is to be found in the lower part of La-la valley where poplars are cultivated. At a crossing of the stream to the right bank is a tiny hamlet, not constructed of houses, but dug out of the loess walls of the valley, the villagers being actual cave dwellers. It is named Shih-yai-t'u 石崖土 or the Rock Cliff Place, after a rocky precipice a little beyond the village. The rock is here coarse granite and mica. A few li beyond the valley branches, the left branch leading to Sha-ra-khu-thul, the Chinese Ha-la-k'u-t'u 哈拉庫圖, ruled by a local chief, and the right, directly west, leads up to a shallow pass and a small hamlet to the left, at the head of the valley. Instead of following this valley to its head the trail leads up another valley to the left to a small Buddhist temple called Ta-tsang ssu 大藏寺; the valley is a rather long one, and the altitude at the junction of the two is 10,350 feet.

Here the valley floor is one mass of *Iris tenuifolia* Pall., with the prevailing scrub vegetation consisting of *Berberis diaphana* Max., very conspicuous on account of its crimson autumn foliage, and bright red, oblong fruits, and *Potentilla fruticosa* L. var. *parvifolia* Wolf., a yellow flowered form, the latter the most common shrub in this valley. On the slopes around 11,000 feet elevation grows *Caragana jubata* (Pall.) Poir., with thick, fleshy stems red when young, but it also extends higher up the steep grassy valley slopes. The buckthorn *Hippophaë rhamnoides* L., here a small shrub only about 1 foot tall, covers the steep slopes, like a carpet, whereas along the streambed it forms tall bushes, and when growing isolated becomes a small tree. A species of Swertia with ultramarine blue flowers thrives with *Gentiana Futtereri* Diels & Gilg, but does not extend as high as the latter which ascends to 12,000 feet. *Leontopodium lineare* H.-M., and the umbelliferous *Pleurospermum cnidiifolium* Wolff., only 2 feet high, with green flowers, extend from 11,000 feet to below the pass. *Aconitum szechenyianum* Gáy, and *Delphinium tatsienense* Franch., the latter with large blue flowers and very long spurs, are met with in gravelly patches on the grass-covered slopes, and are associated with the yellow composite *Tanacetum tenuifolium* Jacq., 1 foot tall. Another species of gentian, *Gentiana siphonantha* Max., forms fairly large sized clumps on rocky, exposed slopes with flowers of a purplish blue, while *Gentiana Futtereri* with rich blue flowers, striped yellowish, reminds of *Gentiana sino-ornata* of Yün-nan, mostly found in wet meadows. All the gentians seem to prefer the same type of situation for here occurs also *Gentiana straminea* Maxim., but instead of being prostrate grows over one foot in height, in long grass. In wet meadows at the head of the valley thrives *Saussurea stella* Max., forming flat rosettes with red leaves and reddish purple flower-heads. At 12,000 feet elevation and extending to near the very top of the pass 13,000 feet, grows *Saxifraga diversifolia* Wall. var. *Soulienana* Engl. & Irmsch., disporting yellow flowers.

Lower down at 9,500 feet occurs *Berberis Boschanii* Schneid., a shrub 3-4 feet tall with racemes bearing small red, transparent, elliptical fruits. It grows in thickets along the stream. This valley, as probably all others which extend east from La-la ta-pan were once forested with conifers as one lone tall *Picea asperata* Mast., indicated. The tree survived because of its great size and the difficulty of getting it out, once cut. On the grassy slopes at 9,500 feet a lovely species of *Gentianella* no 13347, not as yet identified, found a foothold also *Sorbus tianschanica* Rupr., further indicating spruce forest, on the outskirts of which it loves to grow; its brilliant red fruits make it conspicuous, but it does not reach its full height here, being only 5-6 feet tall. Here and there, scattered, occurs the very handsome *Lonicera syringantha* Max., which ascends to 10,000 feet. *Valeriana tangutica* Bat., with pale pink flowers adheres to the grassy exposed slopes at the same elevation. It is usually found in, or on the margins of spruce forests which the Chinese have now exterminated. A dragonhead *Dracocephalum heterophyllum* Benth., with creamy white, fragrant flowers keeps to the grassy exposed grades. This with a few willow bushes forms the plant growth near the head of this valley up to nearly 13,000 feet.

The summit of the pass is exactly 13,110 feet (hypsometer) and from it one can obtain a glimpse of one corner of the blue lake. The wind blows fiercely here from northwest, and freezing temperature occurs early in the autumn on La-la ta-pan as the pass is called. Ta-pan is the Chinese transcription of the Mongol word *daban*, in classical Mongolian dabayan, it is equivalent to the Tibetan *la* = pass of a mountain; this seems to be its principle meaning for the verb to pass or to cross over a mountain is *dabakhu* in Mongolian. Looking north from a peak to the right, still higher ones, with a few rock outcroppings form the continuation of the range which constitutes the divide between the Yellow River and the Koko Nor plateau whose rivers and lakes are without outlet. The otherwise rounded crest of the range is composed of grass-covered loess.

From the pass the trail descends into a broad valley flanked by bare hills, no shrub or bush, not even one foot high, grows here, all is bleak and dreary, short grass alone can endure the icy blasts which sweep over this region from the northwest. The valley floor on the other side of the range over which the trail leads in a westerly direction is 12,350 feet; looking northeast are light brown sand dunes extending into the actual Koko Lake, while a range, partly snow-covered, extends from south-southeast to north-northwest, behind which lies Chhab-cha, the Chinese Chia-pu-chia 恰布恰; it is the South Koko nor Range and seems not to be known by any specific name, it comes close to the southern lake shore, while the ranges to the north, east and west are considerably farther removed.

The vegetation along the shore mainly on sandy flats, as well as on the sand dunes at an elevation of 11,000 feet, is composed of the fleshy-leaved *Clematis tangutica* Korsh. var. *obtusiuscula* Rehd. & Wils., an erect shrub 2-3 feet in height; its fruiting heads are large, all of three inches in diameter, and a pinkish purple; on the drier rocky slopes near the foothills practically the only plant proved to be the chenopodiaceous shrub *Eurotia ceratoides* C. A. Meyer. On the sandy, wet grassy banks of the lake occurs the small, compact, rosette-forming *Potentilla bifurcata* L., often deformed by a gall insect; it is associated with *Aster Bowerii* Hemsl., with deep lavender flowers. Everywhere around the lake, at 10,700 feet flourishes *Gentiana dahurica* Fisch., with pale purplish blue flowers, and the prostrate rosette-forming thistle *Carduus euosmos* W. W. Sm. & Forr., its yellow flowerheads crowded in the centre. The Edelweiss *Leontopodium linearifolium* Hand.-Mazt., is partial to swampy areas near the sand dunes while *Pedicularis alaschanica* Max., *typica* Prain., is restricted to the meadows in the immediate vicinity of the lake where it disports its pretty yellow flowers in company with the rosette-forming purple legume *Oxytropis falcata* Bunge. Another thick prostrate rosette with deeply lobed leaves found here is *Saussurea Thoroldii* Hemsl., its small flowerheads packed into a dense purplish mass; it prefers the swampier meadows near the shore.

Wherever swampy meadows occur there are to be found the above composites associated with *Cremanthodium lineare* Max., with single yellow flowerheads and the flat, dark red, lanceolate entire-leaved, rosette-forming *Saussurea stella* Max., its roots deeply embedded in the mud. *Stipa conferta* Poir., which grows in masses and forms tall clumps vies with *Iris tenuifolia* Pall., and *Phragmites communis* Trin., for space.

Low flying birds are plentiful in this region and many can be killed with a stick. We bagged about 20 specimens representing a number of species among them two vultures *Haliaëtus leucoryphus* (Pall.), and a new species of owl or rather owlet *Athene noctua imposta*, later again collected by us in the grasslands east of the Yellow River between Ra-gya and La-brang.

On this vast plateau one is in a different world, away from the densely populated region of China, here reigns peace and beauty, and I cannot do better than to quote from my diary while camping on the lake shore.

September
«The sunset was most beautiful this evening. The sky took on the magic coloring of sunsets one beholds in upper Egypt as at Luxor across the desert. The mountains were a royal purple, the sky a pale blue which merged into yellow and green, black fleecy clouds hung over a pale, lemon-colored sky which turned a golden yellow over the deep red mountains behind which the sun disappeared. The placid lake turned a creamy white, and cranes flew lazily to their nesting places. Not a soul was visible, only our tents stood out against the lake with wisps of smoke from our camp fires encircling them. Soon the mountains turned a cold bluish grey, the plain turned brown, and the lake a dark blue towards the east. Yet twilight lingered in the west, a pale orange sky merged into green and then abruptly into a dark starry firmament, while the west end of the lake still retained its pale blue color.

About 2 p.m. a terrific gale started to blow which whipped my tent hither and yon; fiercer and fiercer became the gale so notorious on these plains: I tightened the tent-ropes, and as my tent was nearest the lake I feared to be blown into its angry waters: snap went one pole, and calls for help against the fury of the gale, remained unanswered till at last they heard me and came to my assistance; had I released my hold upon the tent poles there would have been little left of the tent. Mohammedan and Tibetan travelers never use wall tents, the canvas of their tents slopes directly to the ground so as not to allow the wind a chance to play with it. The nomad tent, an ungainly affair, is fastened to thirteen poles all arranged outside the tent, with one large pole in the centre, giving the appearance of a huge black spider. The fury of the wind pitched the lake into a high running sea whose breakers dashed against the very walls of my tent like the surf over an island reef.»

The wind which the Koko nor nomads call Black Wind is much dreaded and for this reason they keep their encampments near the foothills and in the mouth of the shallow ravines. We broke camp at 5 a.m. and continued our way along the lake towards Tu-lan ssu 都蘭寺 where forest exists of *Juniperus zaidamensis* Kom.

The terrific winds encountered in the Koko nor and which usually start about 2 a.m. blow constantly throughout the year, hence the region is one of the bleakest imaginable.

The plain around the lake rises gradually to the foothills to an elevation of 10,825 feet, whence a number of small streams empty into the lake. The lake recedes here, due to a huge landmass extending towards it from the mountains and describes a big bay, with a narrow sandspit extending some distance into the lake. The only plants encountered here are Stipa and the ever present *Iris tenuifolia* Pall., with the thistles and the bushy Clematis.

Towards northeast extends a huge dry brown plain with the snow covered mountains in the distance, the whole aspect is limitless and monotonous, everything is bare and bleak.

Our camp at 11,210 feet was a sheltered one and we hoped for a peaceful night, yet we were awakened after midnight when our soldiers opened fire at robbers who had sneaked up to our tents to rob us. Their method is to crawl noiselessly to the tents, cut the ropes so that one is caught like in a bag, but the would-be robbers were driven off due to the alertness of our Moslem soldiers.

Along the southern and eastern lakeshore at some distance are small fresh water ponds separated by sandbars.

The ground is here covered with coarse gravel and sand, the glare hard on the eyes and the heat unbearable at midday; a real desert country with loose sand dunes and shrub vegetation of *Clematis tangutica* var. *obtusiuscula* Rehd. & Wils., and a new variety of *Juniperus chinensis* var. *arenaria* Wils., with spreading branches which forms shrubs 1 foot high on the dunes at an elevation of 11,000 feet.

Here I shot a huge eagle and two species of sea gulls *Larus brunicephalus* Jerd., and *Larus Ichthyaëtus* Pall., as well as the gray goose *Anser anser* L., and a few small birds. The same vegetation as described previously extends to the foothills of the Potanin Range with the addition of *Oxytropis aciphylla* Led., also known from Mongolia and the Altai mountains, *Artemisia salsoloides* Willd., with rich dark green leaves and *Potentilla fruticosa* var. *parvifolia* Wolf., a shrub 2 feet high. On the grassy plain which seems endless occur huge raven, the *Corvus corax tibetanus* Hodgson, with long whiskers.

The Region East of the Koko Nor
Large sand dunes extend the whole length of the Potanin range which hems in the plain on the east. At the foot of the dunes are small, clear fresh water ponds, but the vegetation is extremely poor; here persist only thistles, Edelweiss, and *Stellera chamaejasme* L., with flowers purplish pink and white, this differing from the Yün-nan form which is yellow. The region here is most peculiar, exhibiting absolute desert conditions with high sand dunes adjoining fresh water ponds, then again dry grassy steppes joined by swamps with huge grassy hummocks, and beyond all that the brackish lake. Huge birds either eagles or vultures soar constantly at a great height over the plain, forever circling and on the lookout for dead yak or sheep.

The sand dunes were formerly the happy hunting ground of Tibetan bandits who waylaid travelers who, impeded by the sand, can only make slow progress, but Ma Chi-fu of Hsi-ning, with a ruthless hand exterminated them, chopping the heads off those who were not shot on the spot. In order to reach the Potanin Range, one must cross the dunes, a slow and laborious procedure as one sinks deeply into the sand, up and down at an elevation of 11,400 feet. *Artemisia salsoloides* Willd., a fleshy bush with divided leaves, and globose cushion-forming spiny *Oxytropis aciphylla* Led. are the only plants which find foothold on these dunes. The dunes give way to a grassy steppe, boggy in spots with large, grassy hummocks surrounded by standing waters. Huge lämmergeiers *Gypaëtus barbatus grandis* Starr., feeding on dead yak, and gazelles or antelopes, with a spread wings measuring ten feet from tip to tip, are common (see Plate **40**).

A gap through the range leads into a broad grassy plain which from the pass looks lifeless, but is inhabited by Tibetan nomads. To the south of the range at the southeast end, a spur extends directly south which is known as the Balekun Range, while the one on the east flanking the Koko Nor is named the Potanin Range after the intrepid Russian traveler.[21] By crossing the former range one comes to Balekun Gomi on the Yellow River west of Kuei-te 貴德, on the highway to Lha-sa. Directly east of the Balekun Mountains is the Chinese Jih-yüeh Shan 日月山 or The Sun and Moon Mountain, it is probably only an eastern extension of the Balekun mountain which derived its name from a little settlement to the south of it.

The lake is a deep and glorious blue and seems to merge with the horizon in the North. Here one meets with yaks loaded with wool, while sheep graze about the thousands. Every path leads to the lake over this vast expanse.

---

[21] Grigorij Nikolaevič Potanin, 1835-1920, Russian explorer of Central Asia. See Grigorij Nikolaevič Potanin: *Vospominanija*. Novosibirsk: Zapadno-Sibirsk. Kn. Izd. 1983. 332 p.; Conclusion: 1986. 339 p.; Potanin: *Piśma*. 1-5. Irkutsk: Univ. 1987-1992; E. Bretschneider: *History*, 1007-1033. Potanin undertook four major research trips through China and Central Asia; his main works are *Očerki sěvero-zapadnoj Mongolii:* rezul'taty putešestvija, ispolnennago v 1879 godu po poručeniju Imperatorskago Russkago Geografičeskago Obščestva. S. Peterburg, 1881-83. 4 vols. and *Tangutsko-Tibetskaga okraina Kitaja i central'naja Mongolija*: Putešestvie G. N. Potanina 1884-1886; Izd. Imp. Russk. Geogr. Obščestva. S. Peterburg: Tip. A. S. Suvorina, 1893. 2 vols.

Close to the lake shore, back of low mounds which here surround the lake, are small fresh water ponds. Waterfowl of many different kinds are here at home as the teal *Anas crecca* L., *Casarca ferruginea* (Pall.), the cormorant *Phalacrocorax carbo sinensis* Shaw, the redshank *Totanus Totanus eurhinus* Oberh., *Actitis hypoleucos*, the stint *Pisobia temminckii* (Leisl.), and many other birds. Of grasses around the lake *Stipa splendens* Trin., and *Stipa conferta*? Poir. or perhaps a new species (13383) are the most important, while others are the new *Brachypodium durum* Keng, *Poa attenuata* Trin., with the var. *vivipara* Rendle, and *Poa alpina* L., of wide distribution.

The lakeshore is black with ducks and various other waterbirds specially the large black-necked crane *Megalornis nigricollis* (Przew.) which breeds here. For a botanist or dendrologist the Koko nor holds out little but grass, the ever present Stipa and the Iris *(Iris tenuifolia* Pall.) which takes control of land sporned by the grasses, it extends everywhere around the lake. The region to the east of the lake, but still west of the Potanin Range, is called Ketteniha, the Chinese K'ai-teng 凱登, also marked K'u-t'u-erh 庫圖爾 on some maps, but the former is correct.

The Vegetation of Râ-hu nang or Ra-k'o Gorge

The Tibetan encampment at K'ai-teng belongs to the Sha-khur clan. Their encampments are on a beautiful undulating plain which emerges from a grassy valley spreading fan-like with low hills to both sides, while ahead in the east, a mountain range extends from north to south called the La-rtse Ri in Tibetan, and La-chi Shan 拉脊山 in Chinese. The wide expanse gradually merges into a broad valley with a stream called the Ch'ung-k'o Ho 沖克河 which flows south to Huang-yüan. *Potentilla fruticosa* var. *veitchii* Bean, with white flowers, grows here in the streambed with willows. Beyond the stream the plain extends to the foot of the above mentioned mountain chain, part of the Potanin Range. A trail leads parallel to the mountain range and thence turns northeast following a dry ravine to a pass at 11,500 feet elevation. Here the rocks are mica, schist and quartz.

From the pass the trail descends into a broad plain with fine pastures, a more or less broad flat valley bordered by a bare mountain range some 14,000 feet in height. Directly opposite the pass which leads into the plain, is a small lake called Ulan Nor, which I could not find on any map. Large herds of Huang-yang 黃羊, by which the Chinese designate the fawn colored gazelles frequent these plains, being attracted by the fine pastures. Beyond the plain, another high pass, 11,800 feet must be crossed into a narrow rocky ravine which debouches into the Ra-k'o gorge, the Tibetan Ra-hu nang transcribed by the Chinese as La-k'o Ho 拉科河 and on the Ch. G. S. L. S. marked as La-ch'u-kuo-lo 拉楚果勒 of which the first two characters represent a transcription of the Tibetan, viz. La the name of the valley, chhu = stream, and the last two, kou-lo, a transcription of the Mongol, word for river = gol. Thus it is not really the name of either the valley or stream. Chinese maps made in Nanking or elsewhere in China and not at the source, are unreliable as to names, as well as location of streams, mountains etc.; vide the location of Hsi-ch'ing Shan, a mountain range, the name of which occurs in the more than 2000 year old Annals of Confucius, but which until this day is still misplaced on Chinese maps.

Ra-kʻo Valley, at the time of our visit September – October was the encampment of both Mongols and Tibetans at an elevation of 11,000 feet, being sheltered from the cold winds sweeping the Koko Nor; Mongols however predominated. The first bushes encountered were those of *Potentilla fruticosa* L. var. *parvifolia* Wolf, with yellow flowers, *Berberis Boschanii* Schneid., and *B. dasystachya* Max. The forests were in the lower part of the ravine at an elevation of 10,000 feet. The vegetation is quite rich if not in species; at any rate it appears marvellous after a sojourn in the bleak wastes of the Koko Nor. *Picea asperata* with deep green glaucous needles is the prominent feature of the landscape; with it grows the new variety of birch *Betula japonica* var. *Rockii* Rehd., while on the outskirts of this forest thrived *Rhododendron Przewalskii* Max., and *Rhod. anthopogonoides* Max.; the former a shrub 4-5 feet, with thick, coriaceous, pale green leaves, bright yellow petioles with a fawn-colored indumentum on the underside, and whitish pink flowers, and the latter with very small bronze-colored aromatic leaves, and small white tubular flowers. These two species extend to the upper part of the Ra-kʻo valley to an elevation of 11,000 feet and also into the lateral ravines which open into the Ra-kʻo gorge proper.

The hillsides at our visit in the early autumn were a brilliant red from the *Berberis* bushes of which three species grow here, *B. Boschanii* Schn., *B. dasystachya* Max., and *B. diaphana* Max., all three being of about the same height, and golden from the foliage of the birches. *Cotoneaster tenuipes* Rehd. & Wils., a shrub 5-6 feet branching from the base, with black fruits prefers open rocky situation with full sunlight, and so does *Rosa bella* Rehd. & W., and *Rosa Swinzegowii* Koehne, the latter with large bladder-like fruits, and large oval, serrate leaflets, and the former with smooth, pyriform, scarlet fruits on long pedicels. *Sorbus tianschanica* Rupr., a small tree 5-20 feet, with brilliant red fruits in large cymes prefers the margins of the spruce forest, but occurs also singly on the hillside. The ever present buckthorn *Hippophaë rhamnoides* L., becomes here a tree on more exposed rocky situations, 15-20 feet in height with trunk of 1 foot in diameter. *Caragana brevifolia* Kom., forms spiny cushions on exposed rocky places near the upper margin of the otherwise bare slopes of the gorge, while below, it is a shrub 2-5 feet tall with suberect branches and minute leaflets, but always confined to dry situations. On the grassy hillsides flourish the fleshy *Caragana jubata* Poir., with rose-colored flowers, and along the streambed the lovely, fragrant *Lonicera syringantha* Max., reaching 5 feet in height, with scarlet subsessile fruits crowned by the calycine lobes. Herbaceous plants were at this season very few and were neither in flower nor in fruit except a species of *Gentianella* no 13662 as yet not identified; it had blue flowers and was less than an inch in height, occurring only at the higher elevation on the grassy slopes at an altitude of 11,000 feet. *Sorbus Koehneana* Schn., a much branched shrub 4-5 feet tall, with small leaflets and small pure white fruits in large clusters, accompanied its larger confrere.

Further down the gorge the forest becomes denser, and willows appear along the streambed. Birds are plentiful especially near the streambed in which the Ra-kʻo River roars, over large boulders, a regular mountain torrent. The gorge narrows and the vegetation becomes still denser till it abruptly debouches into a broad bare valley, on the left slope of which is situated the lamasery Ra-hu dgon-pa or La-kʻo ssu, while to both sides of the stream wheat is cultivated. Ravines debouch into the now broad Ra-k'o valley, and these are covered with *Berberis* and birches already enumerated.

The Forests of Kuo-mang ssu 過忙寺

The Huang-yüan valley is enclosed in the north by the Hua-shih Shan 花石山 over which a pass leads to the wooded valley of La-sa kou 拉薩溝 to where it widens at 8,650 feet at the village of La-sa.

The valleys and spurs which are interjacents between Huang-yüan and the lamasery of Kuo-mang ssu are all bare except the valley of La-sa kou which harbours a similar vegetation as Ra-k'o gorge. In the east are low hills which extend from north to south seeming to block the La-sa valley which empties its waters into the Ra-k'o gorge or here better called valley. The valley floor at the junction is 8,300 feet above sea-level. From a pass looking directly north a bare rocky mountain which I judged to be 11,000 feet in height, is visible and to the southwest a long snow range about 15,000 feet or higher separating the Hsi-ning 西寧 from the Yellow River valley; this is the Jih-yüeh Shan 日月山 or Sun and Moon Mountain of which a prolongation westwards is known as the Balekun Range.

From another pass 9,650 feet over the grass-covered loess which forms the eastern wall of the Ch'uan-lin valley a wonderful bird's-eye-view unfolds over the Pei-ch'uan ta-ho 北川大河 or the Great River of Pei-ch'uan (the North Stream) which flows from northwest to southeast past Mao-pai-sheng 毛伯勝 and enters the Hsi-ning Ho or Hsi-ning River just below Hsi-ning. Spur after spur follows intervening ravines, all are bare and grass-covered, the country being here cut up by a maze of valleys until the south gate or nan-men 南門 of Hsin-ch'eng 新城 or New City, at an elevation of 8,250 feet, is reached. A wall extends here over the hills and across the road, which is part of a southern branch of the Great Wall of China or Wan-li-ch'eng 萬里城 or the 10,000 Li Wall. The road leads to Mao-pai-sheng up the Pei-ch'uan ta-ho, the river flowing southeast. Here again limestone mountains have been pushed through the bed of schist and shale; the most interesting is Lao-yeh Shan 老爺山 crowned by a temple (Kuan-ti Miao 關帝廟), with its slopes wooded. Wherever temples are found on mountains, there the original vegetation has survived as the people are prohibited from cutting trees, etc.

This mountain is covered with a dense scrub vegetation composed of *Berberis vernae* Schn., and a new variety of *Rhamnus leptophylla* Schn. var. *scabrella* Rehd. var. nov., *Clematis brevicaudata* DC., which climbs over the *Berberis* bushes, *Cotoneaster*, willows, poplars, *Betula japonica* var. *szechuanica* Schn., and here and there a *Picea asperata* Mast. The temple is 1,500 feet above the valley floor, making the summit 10,900 feet in height. The ascent is fairly steep being a good hour's climb, the temple taking all available space on the top, leaving only a narrow path around it on the edge of the vertical walls which extend into the Pei-ch'uan ta-ho. A wonderful view is to be had from the summit; snow ranges surround the region on all sides, the great Nan Shan 南山 or South Mountains dwindle towards the east where they seem to merge with the Hsi-ning – Huang-ho divide; to the west is the snow range which encircles the Koko

Nor, while opposite Lao-yeh Shan, considered male, is his wife, another limestone mountain but whose name I could not learn; Chinese love to personify their mountains.

The Pei-ch'uan ta-ho receives here an affluent called the Hsien-min Ho 先民河 whose broad valley is planted with poplars (*Populus Simonii* Car.), while *Iris tenuifolia* Pall., is plentiful along the road and on the lower hillsides. A white Chorten (mchhod-rten), or reliquary shrine off the road indicates a lamasery in the vicinity. Kuo-mang ssu lies nestled in this valley inhabited by an aboriginal population known as T'u-jen 土人, against the hillside. The entire valley floor is here covered with *Iris tenuifolia* Pall., and *Iris ensata* Thunb., both low growing species. *Picea asperata* Mast., is a very variable species both in growth, color of foliage, and size and shape of cones, and undoubtedly could be split into two or more species or varieties. That the entire region was once densely forested there can be no doubt; the disappearance of the forests can only be ascribed to the Chinese, for the Tibetans in these regions do not burn wood but use either yak argols or sheep dung as fuel.

Kuo-mang ssu 過忙寺 is called in Tibetan by several names as bKra-shis sgo-mang dgon-pa; gSer-khog dgon-pa, and gSer-khog sgo-mang sgar-dang chos-bzang dgon-pa, pronounced Ser-khog go-mang gar-dang chö-zang gom-pa; it is famous on account of a lama who lived there known as the sMin-grol no-ma-han or the No-ma-han of the Min-drö-ling (Monastery), he was called Gong-mai lung-gis btsan-po or plain lama Tsan-po Gong-ma lung-gi i.e. of the superior instruction. He composed in 1820 the famous Tibetan geography of the world known as hDzam-gling-rgyas-bshad pronounced Dzam-ling gye-she.

Kuo-mang ssu is situated in a valley northeast of Huang-yüan, not far from the latter, and adjacent to Ra-k'o gorge but east of the Potanin Range. It is to the credit of the lamas of this monastery that the forests which cover the mountains opposite, facing north, have been left undisturbed. From almost the valley floor to the very summit of the range, a single species of spruce, *Picea asperata* Mast., forms pure stands. It seems to be the only conifer, except Juniperus of which several species occur, able to withstand the cold winters of this region. It extends north to the Nan Shan, west to the gorges of the Yellow River, and wherever it grows, its trunks rise from a deep carpet of moss. Once this moss is disturbed the trees succumb. It occupies northern mountain slopes where the snow lies for a long period, and the moss acts as sponge and holds moisture. The slopes facing south are either bare of trees and grass-covered or foster juniper forests. These have been depelted for the wood is used as incense by the Tibetans, trees are mutilated and deprived of their branches which are burned green, producing a white smoke, a pleasing offering to the Tibetan mountain gods.

The flora of this mountain is not rich and as the time of the year was advanced there were no herbaceous plants in flower, not even a gentian. The undershrub consists mainly of two species of Berberis, namely *Berberis kansuensis* Schn., a shrub 15 feet tall with long whip-like branches, large red fruits often only pale red, and oblong leaves, and *Berberis dasystachya* Max., which grows near and on the top of the mountain in clearings. This has orbicular leaves, but otherwise resembles its associate. *B. kansuensis* extends lower than its confrere and is more common at 9,800 feet elevation but it does extend up 10,300 feet. It is associated with *Lonicera nervosa* Max., a shrub 3-4 ft in height with black fruits, and the low growing 1 ft tall, large leaved *Rubus Przewalskii* Ilj., deeply embedded in moss. Its fruit are red and sweet.

The most conspicuous deciduous shrubs or small trees are *Sorbus tianschanica* Rupr., reaching a height of 15 feet, with large cymes of brilliant red fruits, and *Sorbus Koehneana* Schneid., 12-15 ft. in height and white fruits, both grow in moss forest up to the summit of the mountain 10,300 feet elevation. Another *Sorbus* as yet not named grew on the summit ridge only, it was 20 ft. tall, had ascending branches and pink fruits (no 13284). Lower in the Picea forest prevails *Rosa bella* Rehd. & Wils., a shrub 4 ft tall with red glabrous fruits on hirsute pedicels; *Cotoneaster acutifolius* Turcz., 6-10 ft tall with black fruits, and *Cotoneaster multiflorus* Bunge 5-6 ft tall with red fruits grow all in the mossy forest, while in clearings and less dense situations occurs *Euonymus Przewalskii* Max., with pinkish-red, pendant fruits; it is associated with willows and the yellow flowered *Potentilla fruticosa* L., *Leontopodium haplophylloides* H.-M. a plant 1-2 ft tall, and partial to the grassy slopes on the outskirts of the forest at 9,500 ft elevation. No Delphiniums nor Gentians could be found, the altitude being too low and there was also too much shade for such species to flourish. The forest faces north, the opposite side is practically bare save for a row of *Picea asperata* on the spur and to the rear of the monastery. Back of the monastery proper as well as along the stream in the valley *Populus Simonii* Carr., and *P. balsamifera* were cultivated, otherwise the loess hills were bare and grass-covered. The Picea trees reached a height of 80 feet, some perhaps 100 ft., one stump one and a half feet in diameter had 119 rings. One single Aconite and a *Paeonia* were observed but they were neither in flower nor in fruit.

Looking east from he summit, and back of it, is a deep valley densely forested and so is the range further east. In the distance could be seen a rocky range with sloping alpine meadows, then snow-covered.

As a whole the region is botanically poor and in fact becomes poorer until the Nan Shan is reached, the system of mountain ranges which face the plains of Kan-chou 甘州 and the Gobi desert. The Nan Shan is however only forested on the southern slopes up to an elevation of 12,000 feet which is the actual timberline. Forests can here exist in protected valleys and on mountain slopes not exposed to the dry winds of the Gobi.

The Region Between Kuo-mang ssu and O-po 俄博

The region between Kuo-mang ssu and O-po is composed of valleys flanked on the west by the Ta-pan Shan 大板山 which extends from northwest to southwest hemming in the Ta-t'ung Ho 大通河, and on the east by the Liang-chou Nan Shan, both extending parallel; the latter range is what Farrer terms the Ta-t'ung alps, a misnomer, for the Ta-t'ung Shan is far to the northwest and separated by the broad valley in which the Hao-wei River, i.e. the upper Ta-t'ung ho flows. The valley floors instead of being grass-covered are one solid mass of *Iris tenuifolia* Pall., mixed with *Iris ensata* Thunb., but cultivation is carried on, wheat being the principle crop next to *Cannabis sativa* L., from which the natives (Chinese) weave their hemp clothing; beans, peas, potatoes, oats, and mustard are also grown. *Populus Simonii* Carr., and *P. balsamifera* are cultivated in large enclosed plots in the valley, evidently belonging to the lamasery. The mountains are bare only the lower slopes of the valleys are covered with brush of *Potentilla fruticosa* L., Berberis, *Lonicera syringantha* Max., and willows, the latter being confined to near the streambed.

The valleys are literally alive with stone pheasants which no one seems to catch or shoot, shotguns are unknown and rifle bullets too expensive to waste on pheasants. Moslems usually hunt them with falcons as they can never eat a bird that has been shot dead, they must cut its throat before life has become extinct. Various hamlets are located in the valleys, the population being half Chinese half Tibetan. Certain villages are inhabited by T'u-jen or aborigines, similar to those found in the valley of Kuo-mang ssu. The women wear large trousers of blue cotton stuck into large, high boots which reach to the knees; for a jacket they wear red Tibetan *p'u-lu* or woolen cloth with sleeves of the same material. The abdomen is protected by a sort of shield which reaches from the waist to below the breasts, made of stiff red cloth, probably of several thicknesses, which is often embroidered. Necklaces of coral and other beads are worn around the neck and from the earlobes are suspended huge earrings three inches in diameter. Over all this they wear one long garment of blue cloth with a red border. On their head they wear a peculiar hat or cap in shape like that of an American sailor, it is made of yak hair or sheep wool felt. Two braids, one on each side, hang over the temples and ears to below the shoulders. The Chinese women of this region wear their hair in an oblong knob at right angles to the back of their head giving them a most peculiar appearance. That they are all but clean needs hardly to be stressed.

Beyond the villages Mongol and Tibetan nomads are encountered, the former using felt yurts circular in outline, while the latter use black yak hair tents, with their numerous poles as previously described, usually thirteen in number, arranged outside, around the tent to which the latter is tied. The Yak hair cloth being coarse, the tents leak the first ten years, but as the smoke and soot eventually closes the coarse meshes, they become waterproof thereafter.

The vegetation in the valley which leads to a pass over the Ta-pan Shan 大板山 is mainly composed of the brilliant red *Berberis diaphana* Max., *Berberis Boschanii* Schn., and *Berberis vernae* Schn., but no trees are visible. This pass used to be the happy hunting ground of Tibetan robbers but the stern rule of General Ma Chi-fu had cleared them out.

There are three approaches to the Ta-pan Shan, the best and easiest is the one from Kuo-mang ssu, and is known as the Hsia-ta-pan 下大板 or Lower Pass. The second is by Mao-pai-sheng 毛伯勝 which is very steep and rocky, and difficult to negotiate with loaded animals; the third is by a valley below Kuo-mang ssu. The rock outcroppings on the pass are blue slate and brown sandstone.

The vegetation on the Hsia-ta-pan is composed of the stiff *Caragana jubata* Poir., willows, *Potentilla fruticosa* var. *parvifolia* Wolf., the yellow flowered variety, while on the wet northeastern slopes grew three species of Rhododendron, viz. *Rhododendron Przewalskii* Max., *Rhod. thymifolium* Max., and *Rhod. anthopogonoides* Max., they occur only on the northeastern slope where the snow does not melt until summer, the prevailing winds being here from the northwest and west. The first Rhododendron was observed at 11,000 feet, but extended up to 12,400 feet elevation. This was the northernmost region where Rhododendrons were encountered. They were all shrubs 3-4 feet in height and formed dense clumps.

On the slopes of the pass 12,450 feet grow the large flowered *Delphinium Souliei* Franch., probably the northern limit of that species, and a Meconopsis such as I had not seen elsewhere. I surmised it to be a blue-flowered one. It was not higher than 10 inches, had 10-15 stalks rising from one rootstock, each stalk bearing one capsule at the apex, capsule and stalk being spiny, the latter down to the base. Seeds were collected under no 13306 but I never learnt where the seed was planted or what became of the plants which resulted from it. Other plants found on the summit pass are a species of *Saussurea* (no 13700), *Aconitum szechenianum* Gáy, and *Gentiana Futtereri* Diels, closely related to *Gentiana sino-ornata* Balf. f. As the season was too advanced, herbaceous plants had mostly been killed by the severe frosts, except the plants mentioned. The northeastern slopes of the pass were covered with snow, and when returning on November 14th over a month later, the entire mountain was deep in snow. Below flows the Ta-t'ung Ho 大通河 from north-northwest to south-southwest; a high bleak range forms the northeastern valley wall with large areas indicating vast snow fields; in the north the range seems to dwindle and emerges into the plain.

The valley floor of the Ta-t'ung Ho, is covered with *Iris ensata* Thunb., and *Iris tenuifolia* Pall., almost to the exclusion of everything else. Further up the valley the Iris gives way to cultivation, wheat being the only crop, the whole valley over one mile in width was one vast wheat field to beyond the walled town of Pei-ta-t'ung 北大通 whose name has since been changed to Wei-yüan Hsien 亹源縣, situated at an elevation of 9,600 feet. Several villages are located on the eastern bank of the river but more on the sloping plain towards the foot of the mountain. The immediate hills bordering the Ta-t'ung ho which meanders here at 9,100 feet in numerous branches in rocky beds, are composed of red sand stone with superimposed loess of considerable thickness.

In the distance, north, are visable the walls of Pei-ta-t'ung or Wei-yüan. Here the stream flows in many channels interspersed with islands which are covered with grass and *Myricaria dahurica* (Willd.) Ehrb., bushes and a few thistles as *Carduus euosmos* W. W. Sm. & Farr.; in the stream bed among gravel grows *Saussurea cana* var. *angustifolia* Ledeb., with rich pink flowers, and on dry gravelly slopes along the banks *Eurotia ceratoides* C. A. Mey., with *Hippophaë rhamnoides* L., also willows and *Berberis*.

There are six distinct branches to this river, only one of which was two feet deep; in the summer when the river is swollen, inflated skin rafts are used to set men and loads across, while animals have to swim. The banks of the stream are conglomerate.

Higher up the valley coming from the northwest a valley and stream debouches into the main valley, flanked by a high snow covered range. The stream, actually a roaring torrent, is the real Ta-t'ung Ho, but which the Chinese mark as the Hao-wei Ho on their maps, denoting the stream below Pei-ta-t'ung as the Ta-t'ung Ho. The range hemming in the river on the south is the Ta-t'ung Shan Shan-mo 大通山山脈, shan-mo meaning mountain system. By this lateral valley leads a trail to a place called Yung-an 永安 or Eternal Peace. Above the confluence of the two streams, the valley with its brook, for the word stream is no more applicable, the greater volume of water having been supplied by the main Ta-t'ung ho, is called Pai-shui Ho or the White-Water River, it flows in a bed of black sand and is not marked on any map. The elevation where the Ta-t'ung Ho enters the main valley is 9,875 feet, it issues from west-northwest.

Somewhat to the north Tibetan and Mongol nomad tents dotted the ever inclining plain. The Mongols denote the range which hems in the valley on the east as Donkhyr in Mongolian, and the high snow peak directly north Konkhyr, the latter Erich Teichmann[22] denied as existing. Weeks later we crossed a pass to the west of it. It is a formidable mountain which I estimated to be 20,000 feet in height. It is correctly located on Bretschneider's map of China[23], which in many other respects is unreliable and out of date. The mountain is also marked but not named, on Sir Aurel Stein's map Kan-chou, Serial no 46, Survey of India no 98, its height is given as 17,200 by a Clinometer reckoning. Teichmann states «such names are unknown to the people of this region». It must be realized that four languages are spoken here, Mongol, Tibetan, Chinese and Turki, the latter by the Moslems. My Moslem escort who knew the region well, immediately when asked the name of the mountain said Konkhyr. In Chinese it is transcribed Kang-ka-erh 剛噶爾.

The Pai-shui Ho flows at the foot of the western hills at an elevation of 10,350 feet debouching into the Ta-t'ung Ho. From here on the valley narrows, and grassy low hills extend westwards into the plain. Flocks of little partridges *Perdix Hodgsonii sifanica* Przew., called Karachi by the local people, enlivened the plain, while cranes (*Megalornis nigricollis* Przew.) fed in the remnant of a lake, which the Chinese call Kan-hai-tzu 乾海子 or Dry Lake, and huge eagles and vultures of enormous size circle at terrific heights over the landscape. The plain is here one vast sea of mole hills and holes, made by the hamsters *Cricetulus lama* and the short-tailed hamster *Cricetulus alticola* or related species, which make riding very difficult. The height of the spur is 10,820 feet and the valley floor one succession of low grassy hills, not a rock being visible anywhere. The plain rises to an elevation of 11,500 above sealevel, the mountains to west diminish in height and are without snow. From the pass a trail leads into a dry streambed which has its source in the mountains immediately to the east, and which extends southwest joining other streams coming from the northwest. Many Tibetan nomads are encamped here with their thousands of sheep and yak. Beyond, another pass 11,350 feet elevation, leads into a ravine with a stream called the Liu-huang Ho 硫黄河 or Sulphur River where Moslems wash gold in a very primitive manner; a village called Sha-chin-ch'eng 沙金城 or the Gold Sand City is situated on the east valley slopes of the Liu-huang Ho, apparently the stream is rich in gold.

The trail turns up a lateral valley northwest, the streambed is rocky and flows through a defile composed of schist and quartz, it is a lonely valley; the hills are bare but rock outcroppings of sandstone stand like sentinels all along the hillside guarding the bleak scenery. There are strata of yellow, red, white and grey loams adjoining each

---

[22] Eric Teichman (originally Erik Teichmann), 1884-1944, British diplomat and Orientalist. See his *Travels of a consular officer in Eastern Tibet*: Together with a history of the relations between China, Tibet and India / Eric Teichman. Cambridge: University Press, 1922. XXII, 248 pp. – *Journey to Turkistan* / Eric Teichman. London: Hodder & Stoughton (1937). 221 pp.

[23] Emil Bretschneider: *Map of China*. Second thoroughly revised and enlarged edition. St. Petersburg: A. Iliin 1900. 4 sheets.

other with here and there black earth. At an elevation of 11,850 feet the valley divides, and the trail follows a central ridge over grassy slopes; here we encountered the huge black vultures *Aegypius monachus* (Linn .), of which we later shot one for a specimen.

Over a gradual incline the trail leads to the top of the ridge and a pass called the Chin-yang ling ta-pan 金羊嶺達板, also written 景陽嶺 which would indicate that the Chinese is a transcription of either a Tibetan or Mongol name, but it may be that the pass has two Chinese names for the first means Golden Sheep peak, and the second Viewing the Sun Peak; ta-pan is superfluous for it is the Mongolian for peak, or pass. The spur is the divide between the Yellow River and the Central Asian basin, all streams flowing north of here have no outlet and loose themselves in the sands of the Gobi; the height of the pass is 12,500 feet. The descent on the western side is even more gentle than the approach from the east. Everywhere the eye beholds a maze of hills covered with snow. In the north-northwest is a high rocky range some 18,000 feet or more, but with little snow, this is the Nan Shan of which more later. While west in the distance is a fairly high isolated mountain I judged to be 17,000 feet in height. This is the Niu-hsin Shan 牛心山 or Bullock-Heart Mountain whose forests we later explored.

The Region of O-po
On the plain before us grazed large herds of gazelles, the Chinese Huang-yang 黃羊 or the yellow sheep, actually deer, but they are very pale fawn-colored and hardly yellow; they are very wary and difficult to approach. On the plain which is 1,500 feet elevation, lies the lonely, walled town called O-po-ch'eng 俄博城, also written 峨博城, situated at an elevation of 11,600 feet in a sea of grass and at the foot of bare grass-covered loess hills. Opposite, looking south, is, what I designate as the North Koko Nor barrier range but which is part of the Nan Shan 南山 whose ramifications are explained in the next chapter.

O-po, Obo in Mongolian, is surrounded by a high mud wall crowned with rock battlements, the gates are massive, built of fired brick, and of considerable thickness. On entering the place one is greeted by the same deserted appearance and emptiness as at Wei-yüan or Pei-ta-t'ung. The houses are low and entirely constructed of mud bricks, the streets dirty and dusty.

According to Chinese records it was impossible for any one to live here on account of constant raids by Tibetan bandits, thus forcing people to move elsewhere. The historical account of this place will appear in the Historical Geography of this region. The town boasts of two temples adjoining the north wall, but like the houses are in a terrible state of repair. O-po is the western border of the Wei-yüan district.

The Nan Shan shan-mo 南山山脈 or the Mountain System of the Nan Shan (South Mountain)

Nan Shan or the South mountain range owes its its name to its position for it lies south of the vast Gobi desert; otherwise the term would be a misnomer, for it is the last chain which skirts the Koko Nor province in the north. It forms the northern border of Ch'ing-

hai and the province of Kan-su which stretches like an arm between it and Ning-hsia 寧
夏, touching Hsin-chiang 新疆 or the New Frontier, on the west.

The Nan Shan reaches approximately 20,000 feet in Mt. Ch'i-lien 祁連山 in the
extreme western end of the range, and the same height in Mt. Konkhyr, the Chinese
Kang-ka-erh 剛噶爾, in its eastern end. The two peaks seem to be the only ones which
bear glaciers. To westerners the range is known as the Richthofen Range.

The Nan Shan actually comprises several ranges which extend parallel to each other
and are in some places connected by lateral spurs, or merge into each other towards the
west, with broad, almost plain-like, undulating valleys between them. They are so
intricately connected in their western end that it is difficult to determine the exact limits
of each. The multiplicity of names, not only those given them by the Chinese, but also
those applied to them by the Tibetans and Mongols who live among them, (the latter
more confined to the eastern end), make the problem still more difficult. The range
immediately south and parallel to the main one, extending from northwest to southeast
is called by the Chinese the T'o-lai Shan 托賴山 which must be a transcription of the
Mongol name, viz., to-le = hare[24], evidently because of the many hares which have their
retreats in these mountains, for the Chinese t'o-lai has no meaning. This range lies
between the Nan Shan and the Ta-t'ung Shan 大通山, and similarly extends from
northwest to southeast. A branch stretches south from where the Ta-t'ung Ho enters the
valley of Wei-yüan or Pei-ta-t'ung called, the Ta-pan Shan (see previous chapter). The
T'o-lai Shan my North Koko nor barrier range, is opposite or south of O-po; on Chinese
maps of Kan-su this range is also called the O-po Ta-nan Shan 峨博大南山 or the
Great Nan Shan of O-po. In this range further west, in the district of Ba-bo, the Chinese
Pa-pao 八寶 rises the beautiful, more or less isolated mountain called Niu-hsin Shan
牛心山 or Bullock-heart Mountain. All these ranges are snow covered but very few
extend to the perpetual snow line.

The Nan Shan, facing north, is divided by the Chinese into three sections; each
section is named after the town which lies to the north of it in Kan-su province, thus the
westernmost which faces Su-chou 肅州 is designated as the Su-chou Nan Shan, the
middle one facing Kan-chou 甘州 is designated as the Su-chou Kan-chou nan-shan,
and the easternmost facing Liang-chou 涼州 the Liang-chou Nan Shan. The names of
these towns have however been changed to Chiu-ch'üan 酒泉, Chang-yeh 張掖, and
Wu-wei 武威 respectively. The northwestern end of the Nan Shan has also been
named Ch'i-lien Shan 祁聯山 after the highest peak in the extreme northwest of it. K.
S. Hao[25] calls the eastern end the Mo-mo Shan, but I have never heard of such a name
or seen it on any map, foreign or Chinese. On Chinese maps the character mo 脈
appears very often as it is attached to the name of the range plus the character for shan
山 = mountain, both meaning mountain system or mountain chain; I believe Hao has
mistaken them and finding two mo characters close together took them for name, and
called it Mo-mo Shan. He could not have been among these ranges, for on his map,
which is completely out he marks the Kan-chou Ho as flowing between the Ch'i-lien
Shan and what he calls Mo-mo Shan in a plain covered with brushwood, while the Kan-

---

[24] John Hangin: *A concise English-Mongolian dictionary*. Bloomington: Indiana University, The Hague:
      Mouton 1967, p. 103: hare – туулай.

[25] Kinshen Hao (Ho Ching-sheng) 郝景盛.

chou Ho 甘州河 cuts through the Kan-chou Nan Shan and flows in a terrific and impassable, red, sandstone gorge about 4,000 feet deep. It is even difficult or next to impossible to approach the mouth of that gorge into which the river tumbles in terrific cascades. I shall come back to his map later when describing the Am-nye Ma-chhen. (K. S. Hao: Pflanzengeographische Studien über den Kokonor-See etc. in *Botanische Jahrb.* Bd LXVIII, Heft 5. 1938)[26].

What has been said about the names of Nan Shan, holds also good for the streams which have their sources in these mountains. The spur known as Chin-yang-ling Ta-pan which is the watershed between the Yellow Rover and central Asia, sends one stream south called the Pai-shui Ho 白水河 which joins the Ta-t'ung Ho further south near the village of Hei-shih-t'ou 黑石頭, both entering the Hsi-ning River or Huang-yüan Ho which empties into the Yellow River. Flowing northwest and west is the Hei-kou Ho 黑溝河 which passes by O-po and becomes the O-po Ho (see Aurel Stein's map Kan-chou). The former is the name given to it by the local inhabitants. V. K. Ting on map no 20 of his Atlas of China gives it as Fu-niu Ho 伏牛河, and the C. G. S. L. S. map gives it as Hei Ho 黑河 or Black River; locally it is known as the Black Valley River. This river flows from the northern slopes of the T'o-lai Shan west, to where it is joined by a larger stream issuing from a valley further south and west, separated by a spur partly forested with *Picea asperata* Mast. This stream is locally known as the Hei-ho 黑河. It skirts a rocky bluff and joins the Ba-bo or Pa-pao Ho 八寶河, identical with Hei-kou Ho; from their confluence the stream is called the Kan-chou Ho. On Chinese maps and Aurel Stein's map it is marked as Kan-chou Ho. The mountain spur is of red sandstone, as is the gorge in which the Kan-chou Ho flows through the immense Nan Shan. No human foot has entered the gorge and it will forever remain impassable.

The Forests of the Nan Shan or Richthofen Range

The District of Pa-pao 八寶 *or Ch'i-lien* 祁連
The region known to the Tibetans as Ba-bo is called Ch'i-lien 祁連 by the Chinese who have transcribed the Tibetan name to Pa-pao 八寶. It adjoins O-po in the west and much gold is mined in the district. The ancient name of the region was Pa-pao ssu or Pa-pao monastery.

The valley in which O-po is situated extends northwest. A high mountain juts out from the T'o-lai Range into the valley. It was not marked on any map, and Sir Aurel Stein does not mention it, nor does he give its location, although it is one of the most prominent landmarks of the region, it is known to all as Niu-hsin Shan 牛心山 or Bullock-Heart Mountain.

The entire valley floor is pitted or foveate due to thousands of rodents, mostly hamsters which have here their underground haunts. Nothing but grass is visible; over the entire valley floor we found abandoned mud stoves, indicating that Mongols or Tibetans had their summer camps here and had moved to their winter encampments at lower altitudes. The valley floor is here 11,000 feet above the sea but gradually declines

---

[26] pp. 515-668. This was Hao's doctoral dissertation and originated from his participation in Sven Hedin's Sino-Swedish Expedition.

towards the west. Niu-hsin Shan is actually a huge promontory extending north from the T'o-lai Range and only connected with the latter by a low ridge. I estimate the height to be 17,000 feet, but not much snow was visible on its southern slopes and it had no glaciers, but the mountains back of it were heavily snow-covered though appeared lower.

Fifteen miles from O-po a large affluent, called the T'ien-p'eng Ho 天棚河 or the Celestial Tent River is reached, probably so named after the large nomad encampments to be found here in the summer. This river called T'ien-t'ung Ho on Aurel Stein's map, has its source in the south, in a 15,660 feet high mountain of the T'o-lai Range. At the mouth of this river gold is washed in a most primitive way. It is said that the gold for the roofs which cover the main temples at T'a-erh ssu 塔爾寺 or sKu-hbum, near Hsi-ning, came from this region.

The broad valley is actually composed of ridges or eroded, and gentle, grassy slopes, partly forested with *Picea asperata* Mast., while the stream meanders through the intersected plateau which connects the T'o-lai Shan and the Nan Shan proper. The T'o-lai Shan itself, that is its northern slope, is bare, snow streaked and eroded, with the summit appearing truncate. At one place the range is pierced by a large V-shaped valley showing another but connecting higher range, in the background.

An affluent from the T'o-lai Range, called the Lo-t'ou Ho 駱駝河 or Camel River, issues from a deep forested gorge which extends for a few miles into the range. It was the first really forested valley we had seen since leaving Kuo-mang ssu. Alas the spruces further up the valley proved again to be *Picea asperata*, no other species of Picea nor Abies could be found. The rocky cliffs of this valley are covered with *Juniperus zaidamensis* Kom., it has flat, glaucous branchlets, and dark, bluish-black, globose fruits. It covers the rocky slopes to the exclusion of nearly everything else, but the forest was dying. The underbrush consists mostly of willows, *Caragana jubata* Poir., *Sibiraea laevigata* var. *angustata* Rehd. which Hao raised to a distinct species *S. angustata* (Rehd.) Hao, *Potentilla fruticosa* L., and the new grass *Brachypodium durum* Keng. Along the streambed with willows thrived *Hippophaë rhamnoides* L.

The Picea trees in this valley reached a height of from 80-100 feet with trunks of over two feet in diameter, the branches extended downwards at a sharp angle and hid the trunks; wherever the trees were healthy thick moss covered the ground, wherever the moss was absent the trees were dying. Here in this spruce forest, in the wilds of Central Asia, we collected a new bird *Prunella fulvescens Nadiae* B. & P.

The valley floor at 10,500 feet elevation was taken up completely with willows, and the ubiquitous Hippophaë, and *Sibiraea angustata* (Rehd.) Hao. Half way up the gorge Picea trees cease, and the slopes are partly covered with *Juniperus zaidamensis* Kom., with here and there a *Lonicera syringantha* Max.

Beyond the Lo-t'ou Ho we came to a tent lamasery, the first we had ever seen. It was siatuated on the western slopes of a ravine at the outskirts of spruce forest; the tents were the round Mongol yurts, of grey felt, and the lamasery was called Arig Ta-ssu (A-li-k'o Ta-ssu 阿利克大寺). Arig, here pronounced as if it sounded Arke, is a Mongol clan name, while ta-ssu 大寺 is Chinese and means large Buddhist monastery, or lamasery. Below the tent lamasery were the black yak hair tents of Tibetan nomads. The Tibetans cannot use wood in their tent ovens or stoves which are only fitted out for burning sheep manure, hence the forests have here survived, while if Chinese had occupied the region, there would not have been a single tree left.

The Forest of Niu-hsin Shan

The Picea forest which covered this beautiful mountain from almost the valley floor to its upper third was again composed of a single species namely *P. asperata* Mast. The streambed was filled with poplars (*Populus suaveolens* Fisch.), willows, Hippophaë, *Lonicera syringantha* Max., while on the outskirts of the spruce forest, densely carpeted with moss, grew *Rosa wilmottiae* a shrub 3-4 feet, with oblong to ovoid fruits and very small, to minute, calycine teeth; it also occurred with the poplars in the streambed.

The scenery was superb, lovely meadows were bordered by beautiful forests, and to the north, across the stream, a limestone range reared its snow covered summit into a blue sky; what the region lacked in plant species it made up in scenic beauty as the photos testify.

The Picea trees were of great size, the cones and foliage appeared different, but apparently *Picea asperata* is a very variable species. We had observed no Berberis since leaving the Ta-t'ung Ho, but here we found a few bushes of the small fruited species *Berberis vernae* Schneider, also met with on Lao-yeh Shan.

The streambed is here very broad and full of poplars, willows and the shrubs usually associated with them. Diagonally across rose a bold rocky range without a vestige of vegetation, and only long talus slopes or screes descended from the cliffs. Here the maps do not agree, the district is called Ba-bo in Mongolian and Pa-pao 八寶 in Chinese meaning the Eight Jewels, but I believe it to be a Chinese transcription of the Mongol name. The Chinese maps show a Pa-pao on the Ta-t'ung Ho or Hao-wei Ho 浩亹河 as the seat of a native official, it may be that the district extends across the T'o-lai Shan into the valley of the Ba-bo River.

The entire spruce forest of Niu-hsin Shan was devoid of undergrowth, thick moss only covered the ground. The forests extend to an elevation of 12,000 feet, apparently the timberline in this part of the Ch'ing-hai or Koko Nor province. Beyond 12,000 feet appeared scrub vegetation composed of willows, *Caragana jubata, C. pygmaea* (L.) DC., and *Potentilla fruticosa* L.; the Caragana had invaded however the spruce forest to half way up the mountain. Towards the 12,000 feet level spruce became stunted and a few extend above that elevation. There are still several thousand feet of mountain, but not a single tree or shrub could be observed.

The upper part of the mountain consists of crags and scree composed of gneiss, felsite and other metamorphic crystaline rocks. Undoubtedly in the summer time the upper slopes would repay the explorer with an abundance of herbaceous alpines, but in the limited time at our disposal and the long distance to be traversed to reach these regions, not forgetting the return journey of nearly a month to the base camp at Cho-ni on the T'ao river, made this impossible.

Birds were abundant in the spruce forests of Niu-hsin Shan, especially the Tibetan eared pheasant *Crossoptilon auritum* of a slate bluish grey color, with four long lyre-like, steel blue feathers in its bushy tail. It is smaller than its congener of Yün-nan and Hsi-k'ang in the south, but like it, roams the forests in groups of ten or more.

The natives steal its eggs and hatch them, selling the four lyre-like tail feathers which were then much in demand, but they never succeeded to breed the bird in captivity. Another pheasant which here roams the forests is the Blood Pheasant *Ithaginis sinensis-michaelis* Bianchi, it has a number of cousins in the forests of Yün-nan and in the mountains of Mu-li, but these belong to other races. We also shot here the three-toed woodpecker *Picoides tridactylus funebris* and many others. The blood pheasant, it may be remarked, differs also from those found in the forests of the Min Shan in Kan-su, in having a white crest of feathers on the heads of the male birds.

In the Ta-t'ung valley there exists a deciduous mixed forest with the same species of Rhododendrons as occur on the Ta-pan Shan, probably their northern limit, these forests of deciduous trees and their shrubby undergrowth are completely absent on the norther slopes of the T'o-lai Shan and on the southern slopes of the Nan Shan proper. Here we find only conifer forest composed of one species of *Picea* (*asperata*) and no undergrowth save two species of Caragana; the floor being covered with moss (*Mnium* sp.?). The vegetative zones are here very distinct. On the gravelly valley floor along the streams grow tall poplars *P. suaveolens*, *P. Simonii*, *Ulmus pumila*, and willows etc., while the slopes of the mountains up to 12,000 feet are covered with the coniferous forest mentioned, followed by a scrub vegetation mostly composed of willows. Alpine meadows are practically absent, the scrub vegetation being followed by rubble, boulders and scree covered with snow.

As we go further north across the Nan Shan, forests cease altogether, not a tree being visible anywhere except in narrow sheltered valleys. The lower slopes or lower mountain spurs are rounded superimposed with a thick deposit of loess and covered with grass, while *Picea asperata* trees grow on the crests. It is possible that they have once covered the slopes and have survived only on the summit.

Across the Nan Shan to Kan-chou 甘州 *or Chang-yeh* 張掖 *in Kan-su Province*
We decided to follow the Ba-bo Ho down stream to see if it was possible to follow it down the red sandstone gorge and through the Nan Shan to Kan-chou or Chang-yeh.

A short distance west, down stream the Ba-bo Ho receives an affluent called the T'ung-tzu Ho 潼子河; this river issues from the northern slopes of the T'o-lai Shan from a rather broad valley, with loess bluffs on both sides. In the flat rocky stream bed grew the usual poplars, willows, etc. Through the wide open valley steep snow covered peaks are visible and sharp ridges, enclosing near the summit, broad circular depressions full of snow, the beginning of a glacier. The height of these peaks ranges between 15,000 - 16,000 feet. The T'ung-tzu Ho flows into the Ba-bo Ho near Erh-ssu-t'an 二寺灘 or the Two Temple Rapid, its water being crystal clear; it is nameless on Aurel Stein's map. On the opposite side of the Nan Shan is loess covered in its lower slopes with vertical walls where streams have cut through them. From the main snow-covered backbone of the range rise two peaks of solid rock, apparently of the same composition as the Ta-t'ung Shan, as gneiss, felsite, and chlorite. The mountain is called Ma-lo-ho Shan 嗎羅河山 and is over 15,000 feet in height (see Plate **41**).

Beyond Erh-ssu-t'an the Ba-bo Ho (Hei-kou Ho) makes a bend and flows northwest through a bare rocky red sandstone gorge. One and a half miles beyond it is joined by the Hei Ho 黑河 or Black River issuing from the west behind a rather low red sandstone spur around which it flows, the spur jutting out into the Ba-bo Ho valley.

The mountains of the To-lai Range increase in height westwards, reaching 18,150 feet, less than twenty miles west of the confluence; the source of the Hei Ho being south of the peak Ch'i-lien Shan nearly 20,000 feet in height. At the confluence of the Ba-bo Ho and the much larger Hei Ho, which is at an elevation of 9,000 feet, is the hamlet of Huang-fan ta-ssu 黄番大寺 or the Yellow Barbarian Great Temple. The ridge which juts out into the Ba-bo Ho valley, separating the two streams, has its highest point about ten miles northwest as the crow flies, namely 17,080 feet, while directly opposite rises a higher peak in the T'o-lai Shan, 18,150 feet. The range which separates the Hei Ho and the Kan-chou Ho where it cuts through the Nan Shan is known as the Ch'i-lien Shan 祁連山. The gorge which dissects the Nan Shan and through which the Kan-chou Ho flows is still unsurveyed, as are the immediate mountains to the east and west of it. The gorge is impassable, an enormous volume of water rushes madly into it and continues between vertical walls of red sandstone. No human foot has ever ventured into that gorge, for there is no foothold anywhere, and as to the open valley and shrubbery through which Hao makes the river flow, and as he indicates on his map, this is nothing but invention. The stream emerges into a broader valley on the Kan-chou plain which is nearly always dry, as the river disappears underground, or what water does issue from the gorge is directed in ditches and channels for irrigation purposes.

The name Kan-chou Ho is applied to the stream formed by the confluence of the Hei Ho and the Ba-Bo Ho. The so-called ta-ssu or great temple is beyond the village, and is an abandoned empty lamasery. Here the valley is quite broad and so is the streambed; what is still extensively cultivated for a distance beyond the village. Immediately back of the village is a volcanic cone, of perfect proportions, deeply furrowed, and open on the southeast side, where it is cleft by a deep ravine to the very rim of the cone, it is a formation entirely foreign to the rest of the landscape. The cone is forested with *Picea asperata* Mast., on its northern slope. Large patches in the broad valley floor are covered with poplars, willows, *Hippophaë rhamnoides* L., etc. Five or six miles from Huang-fan ta-ssu the valley is wild and romantic, both sides being flanked by rugged red sandstone cliffs. On the western side is a peculiar cone of red sandstone the top of which had fallen in, leaving a black chimney-like hole. Niu-hsin Shan fills the valley looking south and appears from here as an isolated mountain mass.

On the sharper ridges to both sides of the valley are rows of spruces and on the steeper rocky bluffs grow *Juniperus zaidamensis* Kom. The river flows very swiftly along the foot of the vertical red sandstone wall making several sharp turns like the letter «S»; in the second turn there stands an island, part of the original valley floor which the river could not dislodge. In the north the valley is blocked by a snow streaked mountain about 14,000 feet in height. The stream enters a narrow defile and flows directly east into a terrific gorge, it is here where it pierces the great Nan Shan at an elevation of 8,500 feet, with the mountains rising thousands of feet above it, it drops 4,000 feet in a distance of about 55 miles as the crow flies to Kan-chou or Chang-yeh, which is at an elevation of 4,580 feet. The distance to where the stream reaches the plain is actually only about 40 miles, so the river drops an average of 100 feet per mile, but the drop within the gorge must be considerably more, and perhaps actual waterfalls exist, but to penetrate the gorge is absolutely unfeasable.

Counting the height of the mountain to both sides of the gorge the latter must be over ten thousand feet deep and thus vies with the gorge the Yangtze has cut for itself far to the south, through the Li-chiang snow range in Yün-nan. The Kan-chou Ho gorge is however only a fraction of its width of that of the Yangtze gorge in Yün-nan.

There is an ideal camping place under huge poplars on a splendid meadow which the river crosses in sharp curves ere it enters the gorge. The face of the huge mountain mass which blocks the river and forces it through the sandstone gorge is covered with junipers while forests of *Picea asperata* Mast., cover the slopes of the mountain higher up, but the sandstone walls are bare of any vegetation.

By following the Hei-ho upstream and crossing a pass over the To-lai Shan into the valley of the Ta-t'ung Ho (Hao-wei Ho) and thence over the Ta-t'ung Shan, one can reach Tu-lan ssu 都蘭寺 directly south, and west of the Koko nor (Lake). There are also several trails and passes over the Nan Shan north to the Kan-chou plain. We shot a number of interesting birds and great Ma-chi or *Crossoptilon auritum*, also a fat flying squirrel the size of a dog, quite different from those found on the Mekong in Yün-nan. They are said to be plentiful on the Nan Shan and in the forests of the T'o-lai range where we shot it.

Crossing the Nan Shan is not an easy undertaking, especially when one wishes to traverse it by unfrequented paths and through ravines where forests could be found. Engaging a Tibetan guide he led us up an unfrequented valley with a narrow trail to a pass 12,350 feet; this pass was actually a frozen bog; here we found still flowering *Incarvillea compacta* Max., 1-2 feet tall, the valley being known by the Tibetan name Nag-sha-thar which our guide pronounced Nashthar. It extends from east to west surrounded by grassy hills, on the slopes of which grew *Potentilla fruticosa*, the only shrub.

From the pass a trail descends into an equally barren valley called Ch'ing-tung kou 青硐溝 or the Green Cave Valley, turning north, the valley being surrounded by rocky snow capped peaks. In the middle [---][27] of the valley willows and *Caragana jubata* formed the only shrub vegetation. Crossing lateral ravines and several passes, one finally reaches the [---] pass of the range. It is called Chhar-lo Nye-ra or Rain (and) Lightning Pass, it is 13,350 feet above sea level. From here a splendid view could be had over the snow range, especially noteworthy [---] the peak Konkhyr in the east.

The valleys, gorges, ravines across the Nan Shan through which the Kan-chou plain is reached, harbor the same vegetation as the valleys south [---]. The north face of the Nan Shan, facing the desert is absolutely barren of tree growth. For herbaceous plants the season was too far advanced.

Loocking north from an elevation of 9,800 feet from the lower spurs of the Nan Shan, we could behold in the hazy distance the plain of Kan-chou bordered in the north by the rather low range called Ho-li Shan a bare reddish range which on the east is adjoined by the Lung-shou Shan 龍首山 or Dragon Head Mountain. The highest point of the former is a peak of 12,000 feet directly north of Tung-lo Hsien 東樂縣 now called

---

[27]   Passages (usually 1 word) not legible in copy.

Min-lo Hsien 民樂縣. The height of the range diminishes towards the west, the last peak being 7710 feet. Beyond the range is the Gobi desert and the province of Ning-hsia.

To the east is a huge plain which adjoins the horizon, and there lies the cart road through dust to Liang-chou 涼州 now called Wu-wei 武威.

From Kan-chou to Su-chou now called Chiu-ch'üan 酒泉 or Fountain of Alcohol is a six to seven days journey along the highway to Chinese Turkestan or Hsin-chiang, the New Frontier. *Picea asperata* forest is encountered en route at the end of the second stage at Ma-chia fan-ti 馬家番地, at the third stage at Lo-erh-chia 羅兒家, and then three miles from the latter place at Pa-ko-chia 八個家.

On the plain of Kan-chou *Elaeagnus angustifolia* L., occurs both wild and cultivated, while in the desert areas around the city the composite *Echinops Turzaninowi* Ledeb., is quite common in company with *Tribulus terrestris* L. Of *Ulmus pumila* L., large trees 60-80 feet in height with large spreading and ascending crown are plentiful at the foot of the Nan Shan and on the Kan-chou plain and so is *Populus nigra* L.

The eastern end of the Nan Shan is disected by a famous gorge the Pien-tu-k'ou 扁都口 which leads directly to O-po. Other valleys lead into the Nan Shan but not across it, as the Li-yüan kou 梨園溝 or Pear Garden Valley whose condition however belies its name, for there are neither pears nor any other vegetation except *Nitraria Schoberi* L., a widely distributed shrub which extends from Central Asia to north China, and its constant companion *Zygophyllum xanthoxylum* Maxim., which is at home from Mongolia to Chinese Turkestan.

The valley represents arid conditions; the spurs are composed of red eroded sandstone, sculptured to an unbelievable degree. On the valley floor along a small stream grow large willow bushes with very fine branches and very narrow leaves, as yet undetermined. Under them a type of stone pheasant *Phasianus colchicus sohokhotensis* Baturl., played hide and seek, while the new *Ianthocincla elliottii perbona* B. & P. hopped among its branches.

From Li-yüan-k'ou a trail leads to the top of a plateau through a corkscrew-like ravine where three species of Berberis, *B. caroli* Schneid., *B. vernae* Schn. and *B. Boschanii* Schn., flourished. Large fat hares were abundant and scurried about by the hundreds, stalked by the Tibetan lynx which the Chinese call She-li. Other plants were *Sorbus tianschanica* Rupr., and *Potentilla fruticosa* L.; feeding on the fruits of the former we found the very rare rose finch *Erythrina rubicilloides rubicilloides* (Przew.) B. & P.

From the top of the much eroded loess plateau at an elevation of 8,600 feet a wonderful view can be had over the entire Nan Shan or Ch'i-lien Shan, the range which extends beyond Su-chou or Chiu-ch'üan. The plateau extends gradually up to the snow range. The latter is very precipitous in its southeastern end, but towards the center much broken up, with two prominent rocky peaks in the northwestern limit, after which the range dwindles. At the southeastern end it terminates abruptly leaving a broad gap, on the other side of which another range commences with a gradual slope. Behind the gap is visible a high snow range parallel to the northern one, but it is not here where the Kan-chou River cuts through the range but much further east.

At the yellow monastery of K'ang-lung ssu 康龍寺, hidden among the valleys of the Nan Shan are extensive forests of *Picea asperata* Mast., which cover the mountain slopes, while *Iris ensata* is the sole occupant of the valley floors. The forests extend for over a hundred miles to other monasteries as Ch'ang-kou ssu 長溝寺 or the Long Valley Lamasery, Shui-kuan ssu 水關寺, and many others. There are in all seventeen lamaseries scattered among the valleys along the Nan Shan or Ch'i-lien Shan, all of them surrounded by Picea forest.

Of all the conifers introduced by me to America this species of spruce has proved the hardiest. It has been planted on the coast of Massachusetts where it has endured intense cold and the furious gales which blow from the sea, where other trees succumb. Its hardiness accounts for its wide distribution in the most inhospitable areas of Ch'ing-hai province (Koko Nor) where no other conifers save junipers can thrive, and finding no competition became the sole occupant of those cold and bleak regions. That it can thrive at sea level in the New England States, when it rarely goes below 9,000 feet in its native home, is proof of its adaptability.

As to the Pien-tu-k'ou gorge which leads across the eastern end of the Nan Shan to O-po, suffice it to say that it is rocky in the extreme, the only tree we encountered were a few stunted *Picea asperata* trees and *Juniperus zaidamensis* Kom., the former only four to five feet tall, associated with them and scattered throughout the gorge were *Caragana jubata, Potentilla fruticosa*, and *Potentilla salesoviana* then in fruit, which until then had not been recorded from the Koko Nor or the Kan-su province. It is a handsome shrub with large white flowers and large leaves. It was previously known only from the Himalayas whence it extends to the Altai mountains.

From half way through the gorge south, all ligneous vegetation ceases, and the hills and slopes become grass-covered. A pass is reached called O-po-ling-tzu 峨博嶺子, elevation 11,400 feet, the divide between Kan-chou and O-po. A short distance beyond the pass, the gorge merges into the grassy waste of O-po.

The Region Between Hsi-ning and Hsün-hua

The land south of Hsi-ning is an undulating plateau mostly bare or grass-covered, and intersected by arid loess ravines and valleys of an average altitude of 8,500 feet.

The inhabitants are mostly Moslems who live in mud villages with usually a mud fort to which they can flee in case of danger from marauding Tibetans.

Some of the passes across the spurs which separate the valleys reach a height of 11,000 feet and more, especially those further south.

The trail leads at right angle to the valleys, forcing one continuously to descend and ascend. Southwest are peculiarly created mountains which form the Hsi-ning – Yellow River watershed. One of the spurs or passes encountered is the Hung-t'u ta-pan 紅土達板 or Red Earth Pass elevation 11,650 feet. The land is much broken up by valleys with a long range and several snow peaks in the background called the La-ya Shan 拉鴉山 probably identical with Hao's Lagi Schan, the latter name is neither Chinese nor Tibetan. A pass 12,335 feet in height leads over it. In a northwesterly direction extends another range called the Chiao-p'en Shan 校盤山 whose highest peak is 12,630 feet. After the village of Tsa-pa 雜巴 the valleys extend from north to south towards the Yellow River and the whole landscape is dusty, the mountains are eroded and clouds of dust envelope the traveler. Hua-lung 化窿, the former Pa-yen-jung 巴燕戎 is situated on a spur at an elevation of 10,250 feet, with a mountain range to the east extending north called Pai-lu-ling 白鹿嶺 or the White Deer Range. The surrounding hills are loess and absolutely bare except on a small ridge on which a lamasery is situated called P'i-chia ssu 皮家寺 and where groves of junipers and spruces could be seen. A stream leads here to the Yellow River; the former has its source in a 10,900 feet pass. The surrounding hills are cultivated while the rest of the country is absolutely bare of vegetation. Hua-lung is inhabited by Chinese and Moslems, the latter being the overlords, the Chinese living in constant fear of the Mohammedans who retained there a fair-sized cavalry-garrison.

Beyond Hua-lung Tibetan villages are encountered with the surrounding hills terraced and cultivated. In the south a long snow range extends from east to west with a maze of intervening valleys between Hua-lung and the former. The geological formation is red sandstone or red conglomerate superimposed with grey loess. From a pass of 10,550 feet one descends to a flat spur whence an extraordinary view can be had. There are thousands of grass-covered knolls or knobs the country having the appearance of a rough sea; the knolls are abrupt and eroded while the whole is surrounded by walls of red sandstone, weirdly sculptured. The land is cut up into loess ravines and innumerable trenches, the surface washed away and only skeletons of the surrounding hills remain, every bit of surface soil, loess or whatever the covering might have been washed away into the Yellow River. The red gravel mountains have more the appearance of conglomerations of Buddhist stupas, towers or temples. Everywhere is dust and gravel, the bad lands of northwest China . The region is called La-mu Shan 拉木山, elevation 8,000 feet, as is a village at the foot of the eroded plateau at an elevation of 7,600 feet. The Yellow River is visible from the plateau, its valley extending from west to east. From here on only Salar Mohammedan villages are encountered till the village of Kan-tu 甘都 mainly inhabited by Salar and Chinese, is reached. The latter is situated near the mouth of the valley which debouches from La-mu-shan into the Yellow River a distance of seven miles. To both sides of the valley are weird mountains recalling skeletons of dinosaurs, extending into the Yellow River; the eastern one is called Tung Shan 東山 and the western one Wa-chih Shan 哇只山 with a deep valley west of the latter called Chieh-t'ang kou 乣塃溝.

Below the western cliffs is a lamasery with numerous caves or holes dug into the gravel for Buddha niches, the monastery is called Kan-tu ssu 甘都寺. The village itself is situated on the north bank of the Yellow River at an elevation of 7,300 feet, but quite a distance from the latter and near the foot of the eroded mountains, the gravelly strata of which is of a pale brick red and superimposed with a pale yellowish, drab-colored loess, The region itself is absolute desert inspite of the streams which are diverted however into irrigation ditches and give life to the land.

Here the Salar are ignorant of the Salar language and live in peace with Tibetans and Chinese. Chinese weaping willows are planted everywhere within the village.

Beyond Kan-tu the desert, a dry loess plain, is intersected by deep channels, these originate with a single hole in the ground where the loess is apparently less densely packed, the water boring down to the base of the plateau leaving standing columns of loess 50-60 feet high, and often 40 feet in diameter, a regular labyrinth of underground channels, the detached loess blocks leaning and ready drop, form a regular forest of loess columns carved out of the plateau.

Three miles from Kan-tu is the ferry across the Yellow River, its waters a yellowish green in the winter. On the south bank is situated the Hsien city of Hsün-hua 循化 the seat of the Salar Moslems, surrounded by desert. The wall of the town is in good condition and the interior clean, with broad streets and gutters to both sides something unheard of in Chinese towns. The eastern end of the town is empty, the yamen is in the middle of the otherwise forlorn village. The Yellow River valley is here broad, the enclosing walls are red sandstone superimposed with loess. Habitation exists only where lateral valleys empty their streams into the Yellow River, the latter flowing too deep for irrigation purposes, pumping machines being unknown.

The Region Between Hsün-hua and La-brang
Three miles beyond Hsün-hua is the village of Shih-hang 石巷 situated on a bluff overlooking the Yellow River which makes here a sharp turn, the walls are steep, whole gravelly islands are in the center of the streambed. The Salar men all dress in black, also their women folk, except young girls who wear either pink or green clothes and usually a green turban like Chinese women in Kan-su, they had bound feet.

Nightly high winds blowing down the Yellow River valley enshroud the country in dust so that breathing becomes very difficult and grit between one's teeth is constant.

From Shih-hang a trail leads south up a broad valley flanked by phantastic mountains and cliffs robbed of every bit of soil making them appear like huge gothic cathedrals one joining the other, a succession of nature's master pieces of erosion. The stream which here issues is known as the Ch'ing-shui Ho 清水河 or the Clear Water River. The villages, including the larger one called Yai-man 崖漫 are all inhabited by Salar Moslems, at the latter a Salar garrison being stationed.

Between Yai-man and La-brang is wilderness and no villages are encountered. R. C. Ching denies that there is a trail, but a trail exists, although rather difficult and the happy hunting ground of robbers. The vegetation is first a xerophytic one, Berberis bushes forming the main vegetation with Caragana, *Potentilla fruticosa* L., and here and there a conifer. At 11,000 feet Incarvillea becomes very common. Beyond the valley a pass 12,250 feet elevation leads to the grasslands where nomad tents make their appearance of the black, yak-hair variety, and down a ravine wooded with Berberis etc. The ravine merges into a grassy plain surrounded by bare hills and mountains. In a grayish cliff, peculiarly sculptured like the naves of a gothic church, prehistoric remains of man were said to occur, but the presence of robbers did not permit lingering in the region. At the foot of the cliffs on a grassy slope is a small lamasery, the Pai-shih-yai ssu 白石崖寺, or White Rock Monastery, it could not have been erected in a lonelier place. Cold winds sweep the plain from the north. There is not a shrub or tree, the only fuel being dung.

The little brook, a sheet of ice meandering like a white snake through the waste of grass, such is the aspect of this inhospitable country. From Pai-shih-yai ssu it is 14 miles to La-brang, the latter being separated from the grassy plain by a spur over which a pass leads at 11,500 feet.

La-brang itself is the filthiest monastery I have ever encountered. Frozen dead dogs lie about in the middle of the roads, and dead birds, sheep wool, stagnant blood, coagulated lumps, and large pools of blood from slaughtered animals, entrails and legs of yak and sheep, dirt and rubbish in which chickens and dogs rummaged, littered the landscape. In the village the Moslem were slaughtering animals and buckets of blood were dumped into the middle of the street. The fences or enclosures are of the bones of slaughtered animals; added to all this fierce gusts of wind which envelop the place in clouds of dust and one has a picture of La-brang, whose elevation is 8,585 feet above sea.

The Yellow River

The two lakes which form the source of the Yellow River had been explored by d'Anville[28], rediscovered by Przewalski, and visited by many travelers including Sven Hedin, Filchner and others. The actual source of the river is considerably farther west than the two lakes which are fed by springs of the Odontala or Starry Steppe thus called by the Mongols, sKar-ma-thang or the Starry Plain by the Tibetans, and Hsing-su hai 星宿海, Constellation Sea or Starry Sea by the Chinese, on account of the innumerable lakelets which cover that vast area. This marshy tract has a circumference of about 100 miles, its waters unite to form a stream called Altyn-gol by the Mongols; this feeds the first of the two lakes called Cha-ring nor, the smaller of the two; the eastern one, the O-ring nor, their names appear thus on western maps. The lakes are situated at an elevation of 14,000 feet, and are connected by a channel. Into this channel empty two streams, the longer called the Djaghing gol, is 110 miles long and extends from southwest to northeast into the channel; the other is a shorter one coming from the south, and empties near the eastern end of the channel into it. The Tibetan names of the two lakes are sKya-rings mtsho, pronounced Kya-ring Tsho, and sNgo-ring mtsho, pronounced Ngo-ring Tsho. The Chinese transcribe the names phonetically Cha-ling Hai 扎陵海 and O-ling Hai 鄂陵海, respectively. The river which issues from the northeast end of the O-ring Nor flows east as the rMa Chhu of the Tibetans, to become the Huang Ho 黃河 near Kuei-te 貴德 where it enters China proper.

We explored the unknown gorges of the Yellow River from the mouth of the rDo-rgan nang (valley) which the Tibetans call the sGo-chhen or the Great Gate, southeast of Ra-rgya monastery to beyond the mouth of the brGyud-par River, pronounced Gyü-par (Chhu) River, southwest of Kuei-te. All the land which lies between Ra-rgya Gon-pa on the right bank of the Yellow River to, and including the northern slopes of the Gyü-par Range. Crossed all of its tributaries which debouch into it from the east, and explored botanically and ornithologically all the large valleys that are wooded in their lower reaches.

---

[28] Jean Baptiste Bourguignon d'Anville, 1697-1782, French geographer and outstanding cartographer. Here reference seems to be made to his *Nouvel atlas de la Chine*. La Haye: Henri Scheurleer 1737. 11 pp., 42 maps. The atlas assembles maps from Jesuit sources, previously included in Du Halde's *Description ... de la Chine*. 1736. Anville never visited China.

The grasslands which extend to the gorges of the Yellow River (north of Ra-gya) represent undulating plains of an average height of 12,000 feet, disected by a few long valleys extending from east to west, while others extend from south to north and vice versa, with passes between them reaching altitudes of over 14,000 feet. The larger valleys are reversed, i.e., their heads are wide and flat, having their source in the undulating grasslands; they becoming deep narrow, impassable gorges as they near the Yellow River. Grass covers deep loess deposits, while the underlying strata is either schist, slate, sandstone, and conglomerate which is visible in the lower sections of the valleys and in the gorges of the Yellow River proper.

At Gochhen the elevation of the Yellow River is 10,308 feet, and in the north at the Gyü-par valley it is over 9,000 feet, while its source is at 14,000 feet elevation, a drop of nearly 5,000 feet in a distance of about 400 miles, measured directly west as the crow flies.

Its largest tributary on its right bank is the hBaa, pronounced Bâ River, which flows the whole length of the Gyü-par Range some three miles to the south of it, in a more or less shallow valley being about 650 feet, in its central part, below the loess plain through which it has cut its course; it debouches into the Yellow River a little below the Chhu-ngön River coming from the west.

The rival of the Bâ River is the Tshe Chhu, the second largest tributary of the Yellow River west of the T'ao River. The Tshe Chhu has its source in the grasslands of the sBa-bo-mar (Ba-wo-mar) and the Rong-wo tribes; the largest of the branches, known from its source as the Tshe Chhu, issues from the northeastern (Tshe-sde-ra elevation) one which flows parallel before uniting to form the main Tshe Chhu, is called Chha-shing Chhu. The two streams are shallow and flow east in marshy grassland which they enclose; here are two grass-covered mountains, the westernmost the Ma-mo-ren-chhung-ba (Ma-mo ren-chhung-wa) and the eastern one the Na-mo-ri-on-rdza-sde (Na-mo-ri Ön-dza-de) [Plate **42**], while the plain itself is called the Na-mo-rgan-thang, pronounced Na-mo-gen-thang; it is the territory of the Hor-pa or Hor tribes. Here is neither tree nor shrub, nothing breaks the monotony of the landscape except the black tents of the Hor tribes. We saw no antelopes, but wolves going singly in search of food, while huge eagles and Lämmergeier soared at great heights, or hovered near a camp waiting for the departure of a caravan.

The two main branches of the Tshe Chhu unite north of a mountain called the Sa-ri-mkhar-sgo, the stream flows in a curve southeast, where it strikes a hill called Seng-ge khang-chhags, there it makes a sharp curve around it, and is forced south by another mountain to the west of it, the Am-nye sgar-dang. It flows directly south when three mountains necessitate it to flow west, leaving two within the knee called collectively the Yur-rgan-ri; here are also the remnants of a lamasery called mGur-sgar (Gur-gar) while the third mountain to the south of it, forces the river to flow between them, the third mountain rising from the flat plain is called the Ma-mo shar-snying (Ma-mo shar-nying). Here there is a ford, the stream turns east-southeast, cutting deeper into the plain as it approaches the Yellow River, and finally debouches into it through a narrow gorge, as do most of the streams which have their sources in the grasslands.

The Tshe Chhu receives a number of affluents which rise within the quadrangle the river forms, the longest, having its source on the southern slopes of Sa-ri mkhar-sgo, is the Chhu-nag nang or the Black River which flows due south and parallel to the Tshe Chhu. The affluents flow in shallow valleys separated by rounded spurs over which passes lead westwards. The Tshe Chhu enters the Yellow River south of a mountain called the A-rig dzo-rgon-ma.

The next in size is the Gyü-par stream which flows from southeast to northwest, having its source below the Tho-thug pass, elevation 13,900 feet. It debouches into the Yellow River opposite A-tshogs dgon-pa, pronounced A-tshog Gom-pa, a lamasery on the left bank of the Yellow River. It flows to the north of the Gyü-par range and receives several small affluents from the range on the left, and from the plain to the northeast, the largest being the hJu-chhung as seen by us, but according to the name, which means the Small hJu stream, there must be a still larger one or hJu-chhen, the Large hJu, but which we could not locate.

Beyond the hJu-chhung stream, in a northeasterly direction, is the waterless plateau called Ma-la-dge-thang (Ma-la-ge-thang).

One stream which has its source west of the Tho-thug pass, called the Mu-gyang, debouches directly west, cutting through the loess into the Yellow River valley.

South of the Bâ valley are two tributaries, the northern one the hJang-chhen, and the southern one the hJang-chhung, i.e., the Large and Small hJang (River), respectively. Both have their sources west of Dzo-mo La or the Half-Breed Yak Cow Pass. There are still other smaller ones of which we could not learn the names.

South of the Jang-chhung, is the sTag-so-nang, Tag-so Valley, carrying a stream from east to west. To the southeast of the Tag-so stream is the Long Gold Valley or gSer-chhen nang (Ser-chhen Nang) which has its source in the Tshe-sde-ra pass, pronounced Tshe-de-ra, elevation 13,550 feet, it debouches into the Yellow River west of gTsang-sgar (Tsang-gar)(monastery) where the Yellow River makes a bend flowing directly west towards its source. All the other smaller tributaries will be described in the travel narrative and botanical account of the region.

On the left or west side of the Yellow River the largest tributary is the Chhu-ngön or the Blue River, it carries the greatest amount of water to the Yellow River or rMa-chhu (Ma-chhu). It comes from the western side of the Am-nye Ma-chhen, but flows first in a wide curve northwards following the common direction of the rocks, and debouches into the Yellow River in an east-southeasterly direction a little above the mouth of the Bâ River. It flows in a deep gorge with precipitous walls cut into the slate and silicious schist; nowhere is it fordable owing to the rapid fall of the water. Like the Yellow River it has immense cataracts but no actual waterfalls. In parts it is forested with *Picea asperata* Mast., the spruce so common in this region.

North of the Chhu-ngön is the Hang Chhu, the Chinese Ta-ho 大河, in which a small Mohammedan garrison is, or was stationed, in its upper course where the Hsi-ning – Lhasa trail crosses it. The new name which the Chinese have given the place (the old one was Ta-ho-pa or the Great River Plain) is Hsing-hai 興海. It flows in a rather broad valley and debouches into the Yellow River 2° south of west, coming from west-northwest. There are terraces on both sides of the stream, with some forest and willow shrubs along its bank.

South of the Chhu-ngön is a smaller tributary called the hBrong-sde-nang pronounced Drong-de nang (valley) or the Wild Yak Valley.

Coming from the eastern slopes of the Am-nye Ma-chhen is the large Tshab Chhu which is formed by Ye Khog and Yün Khog (q.v.) respectively; before entering the Yellow River it receives an affluent from the south called the mGu-chhu or mGur-gzhung. Other, smaller tributaries will be given in the description of the region west of the Yellow River.

The Great Loop of the Yellow River

Within the great knee of the Yellow River, here called rMa Chhu, there stretches a long mountain range uninterrupted from nothwest Long. 100° 15' to southeast 101° 30', which has been termed the Sarü-Dangerö Gebirge by Karl Futterer[29]. He saw the range from south of the Yellow River which he says flows in a 10 km broad valley. He also states that he thinks the Sarü-Dangerö range certainly forms a part of the «Amnematschin» Range.

Now as to his name Sarü-Dangerö for the range. I have already remarked that it is not the name of that long mountain chain; his «reü» and «rö» are identical with the Tibetan ri mountain peak. The range has two high peaks, one called Sa-ri at 101°15' Long., and a northwestern one Dang-ri at 101° Long., these astronomical positions are approximate only.

Many tributaries of the Yellow River, flowing north and south within the Yellow River loop, have their source in that range for which we were given the name Sha-ri-yang-ra, it is the Chinese Ch'ang-shih-t'ou Shan 長石頭山 or the Long-Rocky Range.

Futterer states that the snow-capped peaks rise to over 5,000 meter (16,400 feet) and 1,650 m (5,413.3 feet), above the river, and that the broad river is bordered by rounded, graceful spurs above which rise the steep and rugged cliffs of the high range and its peaks. The upper third of the mountains was covered with snow but no glaciers were visible. The range is too low for glaciers, only mountains over 18,000 feet possess glaciers in this region. In the west the range becomes higher and towards the east it dimishes in height. Nowhere was there visible a breach which would indicate a large valley, or an intersection in the crest of the range. At about Long. 100° 15' it makes a turn north and is joined by another spur which extends from 100° to about 100°35' Long. It is hemmed in on its southern slope by the main branch of the rGu-gzhung (Gu-zhung) which has its source in the spur connecting the two ranges. On the other side of the transverse range flows the hDu Khog which must have its source in the western end of the main range. The largest tributary of the Yellow River flowing north, with its source on the northern slopes of the range is the Shi Chhu, in the large valley of which, the stream flows east and then northeast into the Yellow River. The Ngu-ra tribal lands commence, and extend north of the range to approximately 101°45' Long., to an isolated

[29] Karl Futterer, 1866-1906: *Geographische Skizze von Nordost-Tibet. Begleitworte zur Kartenaufnahme des Reiseweges vom Küke-nur über den oberen Hoang-ho durchs Thao-Tal nach Min-tschôu.* Mit zwei Karten. Gotha: Justus Perthes 1903. 66 pp. (Petermanns Mitteilungen. Ergänzungsheft 143.) – *Durch Asien: Erfahrungen, Forschungen und Sammlungen während der von Amtmann Dr. Holderer unternommenen Reise*; mit Unterstützung des Grossherzoglich Badischen Minist. der Justiz, des Kultus und Unterrichts und des Naturwiss. Vereins in Karlsruhe hrsg. von K. Futterer. Berlin: D. Reimer 1901-1911. 3 vols.

mountain called Am-nye Ngu-ra, the mountain god of the Ngu-ra tribe. The head of the Shi Chhu consists of many branches, the southern one, the mKhar-sgo Nang cutting deeply into the range causing it to be considerably constricted.

The largest tributary with its source on the extreme southern slopes of the range is the Kho Chhu; it receives many affluents near its head where dwells the Tshang-hkhor-zhug-ma tribe, and whose encampments extend into the headwaters of the Gu-zhung River. In the central part of the Kho Chhu dwell the sGog-rnam Tibetans (Gog-nam), and in the lower part, to its confluence with the Yellow River and beyond, south of the river, into a southern tributary called the hJig Khog, the Khang-gsar mGo-log (Khang-sar Go-log) have their encampments. To the east of the Go-log territory is that of the lNga-ba (Nga-wa) tribe called lNga-sde (Nga-de) or lNga Khog or the Nga Valley, ruled by a chief. The daughter of the chief of the Nga-wa is the wife of the chief of the Bu-tshang Go-log both of whom we met at Ra-gya Gom-pa.

A trail extends at the foot of the southern slopes of the range crossing the head waters of all the southern tributaries over passes ranging from 13,654 feet in the west to 11,802 feet in the southeast. There is one stream, a northern tributary of the Yellow River, the Dwang-chhen Chhu or the Great Dwang River which has its source south of the range and disects it about the center. This is however not visible from south of the Yellow River; it is the only tributary which pierces the range from southwest to northeast. A trail crosses its head waters south of the range. South of the central part of the range dwell the Mu-ra Tibetans whose encampments are bordered on the east by that of the Shag-chhung Clan.

Another tributary flowing partly parallel to the southern bend of the great loop and debouching in the Yellow River in a southeast and south direction is the gSer-zhwa Khog which flows at an altitude of from 12,451 feet to 11,800 feet; this stream receives an affluent called the Gong Khog, west of which, and north of the gSer-zhwa khog, dwell the sGar-tham Tibetans. The Gong Khog flows southwest of the last prominent peak in the easternmost part of the main range, the Lha-ri or the peak in the easternmost part of the main range, the Lha-ri or the Peak of the Gods. East of the latter are grasslands which extend to near the north bend of the Yellow River and isolated mountains of which the Yo-dar-tho-yi-ma is the most prominent. A trail passes to the east of it coming from two lamaseries to the south, the Shar-shu-ri-khrod (Shar-shu ri-thrö), and the mTshan-grags dgon-pa (Tshen-drag Gom-pa), and leading to the northern bend of the great loop of the Yellow River. Another trail branches off to the left (west), to Ngu-ra where there is a ferry across it.

Between Long. 100°30' and 100°45' the southern loop of the Yellow River turns north at an elevation of 11,639 feet (boiling point) where it is the recipient of two small tributaries, the mGo-mang, on the west and the A-ser-nang on the east.

In the extreme east end of the loop, but within it, is a large tributary called the gLang Chhu which flows from southwest to southeast through a great marsh known as the gLang Chhu mtsho-rgan (Lang Chhu Tsho-gen), and directly east of the two lamaseries previously mentioned. Above the marsh on the west bank of the river, are extensive sand dunes.

About 30 miles south of the Yellow River, between longitude 100°45' and 101° (approximately) and 33°30' and 33°45' n. latitude there rises a most magnificent mountain mass called by the Tibetans gNyan-po-gyu-rtses-rdza-ra (Nyen-po-gyu-tse-dza-ra) reaching a height of 20,000 feet. It is in the territory of the Khang-sar and Khang-gen Go-log tribes whose sa-bdag or mountain god it represents.

Unlike the Am-nye Ma-chhen, Nan Shan, Gyü-par, and Sha-ri-yang-ra ranges which are composed of schists, sandstone, conglomerate, soft phyllite, marble and quartz, this previously unrecorded range is composed entirely of limestone. The mountain has seven glaciers which radiate like the spokes of a wheel from below the main summit. The streams issuing from four of these glaciers form beautiful lakes around the northern base of the mountain which enhance its beauty. This mountain was discovered by William E. Simpson an intrepid young American missionary intensely interested in geography. He was later murdered by Moslem bandits.

The Yellow River receives two more tributaries of note, both having their sources southeast in Ssu-ch'uan 四川, one northwest, and the other west of the Kung-ka-la 貢嘎拉 of the Ssu-ch'uan Min Shan 岷山; the first which is the larger one is called sMe Chhu (Me Chhu) by the Tibetans, and Hei Ho 黑河 or Black River by the Chinese. It enters the Yellow River at the northern end of the knee where its flows west towards its source; the second, the shorter one, is called the dGah Chhu (Gâ Chhu) by the Tibetans, and Pai Ho 白河 or White River by the Chinese. This tributary enters the Yellow River at the southern end of the bend, below the lamasery Sog-tshang dgon-pa (Sog-tshang Gom-pa). In the valley of the northern tributary dwell the mDzo-dge hBum-tshang Tibetans, that is in the plain flanking the river on its right or northern bank, while on its southern bank the Gur-sde Tibetans have their encampments. The confluence of the Me Chhu and Yellow River near which is a ferry, is called Mar-me hdren-mtshams (Mar-me dren-tsham).

A range to the north of the Me Chhu extending from east to west, is the Yellow and T'ao River watershed; a trail leads from sTag-tshang Lha-mo in Ssu-ch'uan across the foothills of the above mentioned grass-covered divide, where, at the southwestern foot of it, and west of an affluent called the hBrong Khog (Drong Khog) are many fresh water springs. Here a trail leads north from the Yellow River; a lower pass over the divide is 11,118 feet, and a second pass over the main watershed, the hBrong Khi (Drong Khi), is 12,824 feet above sea level. This trail leads north to the Shi-tshang dgar-gsar (monastery) on the T'ao River.

Chinese maps mark the northern end of the wedge which Ssu-ch'uan pushes between Kan-su on one hand, and Ch'ing-hai on the other, far too much south, it extends to north of the bend of the Yellow River and not south of it.

The Am-nye rMa-chhen Range

The Name of the Range
On all the maps published in the west, the name of this range is still given as Amne Machin which is incorrect, various alternatives also occur. The first explorers to

approach the range were Russians, viz. Przewalski, Roborowski, Kozlow[30] to name the most important. To the Chinese the range was known from the immemorial; it occurs on their maps as Ch'ing-shih Shan 碃石山 but in their books more correctly as the Chi-shih Shan 積石山.

In Tibetan classics, its Tibetan name, always without the prefix Am-nye, is rMa-chhen spom-ra, or rMa-rgyal-chhen-po spom-ra or the Great Ma-spom-ra and the Great One King Ma-spom-ra, respectively. The word rMa, pronounced ma, is also the Tibetan name of the Yellow River viz. rMa Chhu; chhen or stand for great or the great one, respectively, and spom-ra is the second part of the name pronounced pom-ra.

Am-nye also written A-mye, meaning ancestor, grandfather, is an honorific and only used in the spoken language, never in texts. As the mountain is, or represents, the mountain god of the Tanguts and Gologs who dwell in its environs, when speaking of the mountain will always prefix the name with the honorific Am-nye, similar as westerners would prefix the name of one canonized with «saint». Thus has the term Am-nye, which has nothing to do with the actual name of the mountain or mountain god, come into geographical nomenclature.

In Chinese works the Tibetan name, plus the honorific, has been transcribed commonly as A-mu-ni Ma-ch'an mu-sun 阿木你麻續母孫 to which is added Ta-hsüeh-shan 大雪山 or Great Snow-Mountain. The Chinese name is merely a phonetic transcription of the Tibetan name and has no meaning. During my stay in the Lamasery of Cho-ni I obtained a copy of the classic pertaining to this mountain god, which classic is chanted when the god is worshipped and offerings made to him by the lamas. In this Tibetan classic which has been translated by Dr. J. Schubert[31], a full description of he mountain god is given, also the Tibetan names mentioned previously. This translation will be published with the Historical Geography I am writing of the entire region.[32] It is needless therefore to further discuss why the name Machin or Amnyi Machin, as occurring on western maps, cannot be retained. The name of the mountain must be given as Ma-chhen pom-ra.

In the geography entitled hDzam-gling rgyas-bshad[33] written by the famous Gong-

---

[30] Petr Kuźmič Kozlov, 1863-1935, Russian explorer of Central Asia, best known for his research on Karakhoto. See *Russkie voennye vostokovedy. Biobibliografičeskij slovaŕ*. Moskva 2005, 114-116; he started his career by accompanying Prževal'skij on two trips.

[31] Johannes Schubert, 1896-1976, librarian, then professor of Tibetan and Mongolian at Leipzig University. See *Asienwissenschaftliche Beiträge. Johannes Schubert in memoriam*. Herausgegeben von Eberhardt Richter und Manfred Taube. Berlin: Akademie-Verlag 1978. 201 pp., 28 ill. (Veröffentlichungen des Museums für Völkerkunde zu Leipzig 32.); Eberhardt Richter: Johannes Schubert (1896-1976). *Namhafte Hochschullehrer der Karl-Marx-Universität Leipzig* 5.1984, 66-74, portrait; H. Walravens: Briefwechsel Johannes Schuberts mit Ernst Schäfer und Bruno Beger. *Nachrichten der Gesellschaft für Natur- und Völkerkunde Ostasiens* 175-176.2004, 165-224. – The mentioned translation of the text was not published as the editor, Giuseppe Tucci, considered it not up to standard.

[32] *The Amnye Ma-chhen Range and adjacent regions*. A monographic study. By J. F. Rock, Honorary Research Associate, Far Eastern and Russian Institute, University of Washington. With 82 plates and 5 maps in color. Roma: Is.M.E.O. 1956. IX, 194 pp. (Serie orientale Roma 12.) – The structure and contents is very similar to the one of the present book but more focused on the historical geography.

[33] Turrell Wylie, 1927-1984: *The geography of Tibet according to the 'Dzam-gling-rgyas-bshad: text and English translation* / Btsan-po. Roma: Istituto Italiano per il Medio ed Estremo Oriente 1962.

mai-lung-gis bla-ma btsan-po of sMin-grol-gling, the name of the mountain also appears as rMa-chhen-spom-ra. The work is famous and was written in 1820.[34]

In a work composed at the order of Ch'ien-lung 乾隆 (1736-1796) called Ch'in-ting Hsi-yü t'ung-wen-chih[35] 欽定西域同文志 chapt. 15, fol. 18a, Gazetteer of the Western Region giving the names of mountains and places, etc. in six languages, the name of Am-nye Ma-chhen is given in Chinese as A-mi-yeh-ma-le-chin mu-sun o-la 阿彌耶瑪勒津木孫鄂拉 which is a transcription of the Mongol name, the first six syllables represent the name, and the last four a transcription of Mu-sun-u-la the Mongolian for Ice Mountain. And so is the Tibetan name occuring in the same work a transcription of the Mongol name, viz., A-mye-mal-chin-mu-sun-u-la and not the actual Tibetan name. Under the Mongol name is the explanation that Ma-le-chin means «old man's head, and mu-sun ice, i.e., it resembles the bald head of an old man; the top of the mountain is shining and pure».

Location of the Range on Existing Maps

The Am-nye Ma-chhen Range is actually an isolated mountain mass, not within the knee of the Yellow River, but as shown on Roborowski's map accompanying his work[36] Vol. I, part III, which my findings also corroborate. The range is northwest of Ra-rgya Gom-pa and directly opposite a pass east of the Yellow River and north of Ra-rgya where I first saw it. The pass is called Mo-khur Nye-ra and is 12,800 feet in height; from a peak to the northwest of the pass, at an elevation of 13,220 feet, I was directly east of the Am-nye Ma-chhen and approximately at 35° latitude.

On all existing maps the Am-nye Ma-chhen range is shown to extend into the knee of the Yellow River, which it decidedly does not; it extends from northwest to southeast, and as seen from Mo-khur Nye-ra falls within 268.5° to 277.5° of the compass. The first pyramid is at 270°, the actual peak Am-nye Ma-chhen from which the range derives its name at 272° and the huge dome, the northernmost at 275° of the compass. This is the real Am-nye Ma-chhen or rMa-chhen spom-ra, to give it its proper name.

Other maps in Roborowski's work are not correlated with the map showing that range, and were made by someone else who apparently thought he knew better, and changed the name of the range which Roborowski definitely called the Amnye Machin, and made the range to the west of it the Amnye Ma-chhen. Sven Hedin copied the error and published it in *Petermanns Mitteilungen*.

XXXVII, 286 pp. (Serie Orientale Roma 25.) – Wylie: *A Tibetan religious geography of Nepal / Btsan-po.* Roma: Istituto Italiano per il Medio ed Estremo Oriente 1970. XVIII, 66 pp. (Serie Orientale Roma 42.)

[34] An earlier translation was by Vasil'ev: *Geografija Tibeta.* Perevod iz Tibetskago sočinenija Mińčžul-chutukty. V. Vasil'eva. Doloženo v zasědanii Istoriko-filologičeskago otdělenija 2 maja 1889 g. S.Peterburg: Akad. nauk; I. Glasunov, Eggers & Co.; K. L. Rikker; Riga: N. Kimmel in Komm. 1895. [4], 95 pp. 8°

[35] A Japanese facsimile edition, with index, was published Tôkyô: Tôyô Bunko 1961-1964. 4 vols.

[36] Trudy Ėkspedicii Imperatorskago Russkago Geografičeskago Obščestva po Central'noj Azii soveršennoj v 1893-1895 / pod načal'stvom V. I. Roborovskago. Izdanie Imp. Russkago Geogr. Obščestva. Č. 3: Naučnye rezul'taty ėkspedicii V. I. Roborovskago: s kartoju i 19 planami. S.-Peterburg : M. M. Stasjulevič 1899.

From high peaks both east and west of the Yellow River no other higher mountain range is visible beyond the Am-nye Ma-chhen, had any range higher than the latter been immediately west of it, it would certainly have been seen.

The Am-nye Ma-chhen range must however be of considerable width, and spurs of notable length must extend west; it is probably these spurs rather than the actual mountain which certain travelers approached, like Clark and Hao. The latter says in his botanical paper that he spent three days at the Am-nye Ma-chhen and marks an area covered on his map as being his travel route; this does not only encircle the mountain, and also crosses it in the center where there is no pass, but extends far within the knee of the Yellow River; such a journey would take at least twenty days. To circumambulate the range starting from its eastern foot takes a week without stopping to do botanical work. He gives no rivers, except the name of one, which he obtained from a French map of 1902.

All early and even recent maps, both Chinese and foreign of this region are notoriously wrong. It seems further strange that Hao should have to copy the name of the range from a French map of 1902, which name does certainly not apply to the Am-nye Ma-chhen. The whole western region and a great part of the eastern side of the range are still unexplored, and for a Chinese who claims to have climbed it only to an elevation of 5,200 m, or about 16,200 feet, to have to fall back on a French atlas fifty years old, seems certainly strange. His photograph shows a valley purporting to be in the Am-nye Ma-chhen with a snow-covered spur, which may be anywhere. He does not give us a photo of the range nor of the actual mountain mass Am-nye Ma-chhen. Had he actually covered the area or gone around it, as marked on his map, he would have been able to add considerably to our geographic knowledge, but this cannot be done in three days, especially when one collects plants and climbs around on the higher slopes. I leave it to any one who is familiar with travel in the Tibetan highlands, to judge for himself, and try to reconcile his statement of having spent three days at the Am-nye Ma-chhen with the area he marks on his map as having been covered by him; the map is drawn at the scale of 1: 5,000,000. Nothing more need be said.

We observed the Am-nye Ma-chhen also from the highest peak of the brGyud-par, pronounced Gyü-par or Jü-par range, from an elevation of 14,546 feet, but saw no range to the west of it. From the highest peak of the Jü-par Range, brGyud-par mtshar-rgan = Jü-par Tshar-gen, the Am-nye Ma-chhen Range came within 219° and 231° of the compass. The compass indicated 228° for the centre of the dome, the northernmost and highest peak of the range. It forms a semi arc, and is undoubtedly connected with the range which extends southeastwards into the knee of the Yellow River, but that range, although snow covered, and below the perpetual snow line, is known as Sha-ri yang-ra according to my lama guide, but the name does not occur on any map. In this range are two peaks which stand out above the others of the range, the Sa-ri to the east, at about 101°15' longitude, situated between two valleys, the western one called the Chhu-har Khog whose stream flows north and enters the Yellow River opposite Thar-lung dgon-pa, and the one to the west, the Dwang-ri, pronounced Dang-ri, with the large valley carrying the Dwang-chhen Chhu (River) to the west, and the smaller Dwang-chhung Chhu to the east. Futterer in his work *Durch Asien*, vol. I, opposite page 364 gives a long folded plate depicting this mountain range within the knee of the Yellow River.

This range he calls Sarü-Dangerö Gebirge; apparently when asking a Tibetan the name of the range, the Tibetan knowing only the two peaks, which are sacred and represent mountain gods, replied Sa-ri (and) Dang(Dwang)-ri, this Futterer took for the name of the range. Both peaks are in Ngu-ra Tibetan territory.

This range was known as the Ch'ang-shih-t'ou Shan 長石頭山 or Long Rocky Range to the Chinese and so appeared on Chinese maps. For some reason or other, the name Chi-shih Shan 積石山 has been substituted, which is the actual name of the Am-nye Ma-chhen.

The two ranges are distinct, although connected by lower mountains. General Pereira in his diary which he kept on his journey from Peking to Lhasa wrote under May 26th: «crossed Churi or Chida pass through the Chulung Range 14,500 feet. The going was stony and boggy, descended through gloomy valleys to the sandy Luan-ch'uan plain, here saw Amne Machin peaks possibly 70 miles away to the southeast from an altitude of 13,000 feet.»

May 28th: «Crossed the Tungri pass 13,867 feet (this pass is over the range which separates the To-so Nor from the Chu-ngön River) from the top of the pass obtained view of To-so Nor. (This range Roborowski calls the Ghirun-tun, *italics* author's).»

May 29th: «Passed through gap in the Ch'ang-shih-t'ou range which appears to connect up with the range in which the Am-ne Machin is situated.»

June 2nd: «Crossed west Malayi pass 14,500 feet and later forded the Yellow River, 30 yeards wide, 2-2 ½ feet deep.»

Pereira believed the range to be higher than Everest, but after careful calculation I have come to the conclusion that it does not reach 21,000 feet. The perpetual snow line commences at 17,000 feet; the range being much glaciated thanks to its height.

Description of the Range
The Am-nye Ma-chhen is a huge mountain mass extending over nine degrees of the compass, and the highest peak is the huge dome to the north, it is little sculptured, but forms a huge massive dome and is called dGra-hdul-rlung-shog pronounced Dra-dül-lung-shog or Enemy Subduing-Air (Wind) Wing, but the last syllable may be written gshog, meaning to pierce. The next highest peak is to the south and forms a rounded pyramid, it is named after the god of mercy of which the Dalai Lama is an incarnation, namely Chen-re-zig, written sPyan-ras-gzigs. The actual peak from which the entire mountain mass derives its name is a lower one, between the southernmost and northernmost mentioned above, and is Am-nye rMa-chhen or rMa-chhen spom-ra, pronounced Machhen Pom-ra, the second «h» in the second syllable indicates that the «ch» is pronounced aspirated.

Ma-chhen pom-ra is a sa-bdag or earth owner and therefore a mountain spirit, or god of the land or ground. He is supposed to be an angry, jealous being of terrifying appearance, to whom sacrifices are offered in the shape of a fire of green juniper boughs, the god delighting in the fragrant smoke of the juniper.

In the west the mountain drops steeply to the plain, but in the east, its preliminary ranges which gradually decline towards the Yellow River gorges, are cut up by many valleys, one coming from the northern slopes is called the Ye Khog or Right Valley, written gYas-khog, and the southern one, the Left Valley or Yön Khog, written gYonkhog. These two valleys unite in front, about centre of the eastern slopes of the Ma-chhen pom-ra and united they flow as the Tshab Chhu east into the Yellow River. It is one of the largest tributaries of the Yellow River in this region, except the Chhu-ngön, written Chhu-sngon or Blue River, Tafel's Tschürnen. The latter has its source to the west of the Am-nye Ma-chhen Range, makes a large bend to the north, and flows around the northern end and finally debouches into the Yellow River almost opposite the Bâ River. The Chhu-sngon also pronounced Chhu-ngön is called Churmyn by the Russian travelers and is so marked on their maps. The correct name is Chhu-ngön. The Chinese write the name phonetically Chu-erh-men-ch'u Ho which, as can be seen from the transcription, was taken from foreign maps and not from the Tibetan name. Another affluent of the Tshab Chhu is the hDom Khog, or the Valley of Assembly. It is in this valley that the Tibetan nomads hold their gatherings and discuss their differences. Another notable tributary is the Gu Chhu or Gur-zhung, see journey west of Yellow River.

These are the main valleys east of the Am-nye Ma-chhen Range. Between the Tshab Chhu and the Chhu-ngön River is the Drong-de-nang or Drong-de Valley which sends a small stream into the Yellow River, the name is written hBrong-sde-nang, hbrong meaning wild yak, sde a part of, as the nang or valley, apparently part of a valley where wild yak roam.

North of the Chhu-ngön is the Ta-ho-pa 大河垻 which the Tibetans call the Hang Chhu, and to the south of it is the Nga-thang written rNga-thang or Camel Plain.

The Pilgrims Trail Around the Am-nye Ma-chhen

The greatest number of pilgrims circumambulating the Am-nye Ma-chhen is in a horse year, when more than 10,000 Tibetans visit the mountain or walk around it. It has been known that Tibetan women have measured the distance around the mountain with their bodies in penance or religious zeal, taking two months to do it. Otherwise the circuit can be accomplished in seven days. No one rides, not even the high incarnations, but all walk. There are many places arranged for the burning of incense around the mountain and the pilgrims, while circumambulating either chant or pray. The trail encircling the mountain commences at the mouth of the Yön Khog, a valley which extends to a pass on the southern point of the range, called rTa-mchhog-gong-ma or the excellent superior horse, but it is also said to be written bDe-mchhog-gong-ma meaning the superior bDe-mchhog, the latter is a Yi-dam or protective tutelary deity of the lama church. Beyond the pass one descends to the southern foot of the range to a large plain called Ngang-gi-shog-hdebs or the spread-out wings of the wild goose. On this plain roam many wild yak and wild asses, the Tibetan *rkyang*, pronounced Chiang. From this plain a place is reached called Bye-ma-hbri-sde, pronounced Je-ma-dri-de, where there are many yellow sand dunes which stand like offerings to the god; they are conical in shape and their bases are grass-covered, but their apices are bare. The place derives its name from these sand dunes, ja-ma (bya-ma), meaning sand. We are now on the west side of the mountain and reach a place called Gos-sku-chhen-mo (Gö-ku chhen-mo) or the great painting. The name is derived from the varicolored rock-cliff of which the steep slopes of the Am-nye Ma-chhen are there composed; the rocks are of all colors giving the mountainside the aspect of a great painting. Near the cliff are two conical hills, one is called Mo-pa, in Amdo pronounced Mo-wa = the diviner, and the other the gtor-ba, or tor-wa, he is the one who throws tor-ma or tsamba offerings to the gods into the fire, or merely throws them out; tor-ba means to strew, to scatter. The place or hill is so called because around the hill are many rocks representing the offerings which the tor-wa has thrown out.

Beyond the trail leads up a pass called the rGal-thung la (Ge-thung La) or the Short Crossing Pass. Near the pass is a rock with the imprint of a lama's hand; there is also a projecting rock, about which legend relates Seng-chhen rgyal-po (Seng-chhen Ge-po) or the Great Lion King tied his horse to, while resting on the pass. He is none other than the mighty Ke-sar, of whom Tibetan bards recite innumerable epics of his many heroic deeds.

From the Ge-thung La one descends into the rGal-thung nang or Ge-thung valley which merges into another valley, at the mouth of which is a huge boulder with an Obo or cairn on the top. This white boulder is called Nu-bo dGra-hdul rlung shog (Nu-wo Dra-dü lung shog) and represents the younger brother of Am-nye Ma-chhen pom-ra. Nu-bo meaning younger brother, dgra = enemy, hdul = conquer, subdue, rlung = wind, shog = wing; it may however be written gshog, also pronounced shog, meaning to pierce, to split, and this may be the more correct rendering having reference to his conquering his enemy like a piercing wind. This place is to the northwest of the Am-nye Ma-chhen. From here the trail arrives at the bank of the Chhu-ngön or Blue River which debouches into the Yellow River opposite the hBâ River q.v. without crossing the Chhu-ngön one ascends a valley whose stream empties into the former river, the valley is called the Brag-stod (Drag-tö) or the upper rock. At the head of this valley is a level place called Ri-gur-stong-shong literally meaning mountain tents one thousand elevated plain, having reference to the flat head of the valley where a thousand tents can be pitched. A shallow pass leads over the ridge separating Drag-tö valley from the long valley called the Ye Khog or Right Valley down which the trail leads (in contradistinction to the Yön Khog or Left Valley), to a place full of junipers (*Juniperus tibetica*). The Ye Khog valley narrows to a veritable gorge and is here joined by two other valleys with constricted exits; these three gorges are called collectively mDzo-mo rgod-tshang (Dzo-mo gö-tshang) or the lair of the wild halfbreed yak cow. The three valleys being designated, one as the gong-ma or upper, the middle one as ü-ma written dbus-ma or middle one, and the gzhug-ma read zhu-ma or the lower.

The Ye Khog and Yön Khog united beyond Dzo-ma go-tshang, and from their confluence the river is known as the Tshab Chhu to where it debouches into the Yellow River. That part of the valley is densely forested with junipers (*Juniperus Przewalskii* Kom.), and spruces *(Picea asperata* Mast.). Near the mouth of the Yön Khog are three thrones built of river stones and rocks, they are simply known as bZhugs-khri (Zhu-thri) or the thrones. One is for the La-brang (bLa-brang) incarnation, i.e., for the highest incarnation of that largest of all lamaseries in China, just within the borders of Kan-su, hJam-dbyangs-bzhad-pa (Jam-byang zha-pa) or the Laughing Jam-byang; one of the highest incarnations of Ra-rgya dgon-pa (Ra-gya Gom-pa), situated on the Yellow River, q.v., Shing-bzah Pandita (Shing-za Pandita), and one for the highest incarnation of the Rong-wo (Rong-bo) lamasery. When any one of the three incarnations circumambulates the mountain he will mount his particular throne, to meditate and pray, and to accept homage from the pilgrims.

Journey to the Am-nye Ma-chhen from Cho-ni 桌尼 via bLa-brang (Hsia-ho hsien 夏河縣) and Ra-rgya dgon-pa (T'ung-te hsien 同德縣)

The region between Cho-ni and La-brang or Hsia-ho Hsien
The region between Cho-ni and Hei-ts'o 黑錯 is a high grassy plateau of an average elevation of 10,500 feet, intersected by shallow, grassy valleys with grass-covered passes. Out of the loess in places project limestone crags or individual mountains of the same rock formation as Pai-shih-yai Shan, to the north of T'ao-chou Old City. Still farther north rises the higher Ta-mei Shan which separates the Rong-war plateau from the Hsia-ho valley. In this valley we find rock outcroppings but these are shale and schist, indicating that the mountain has been forced up through beds of schist and shale similar to the Min Shan. Forest is only found in the vicinity of Hei-ts'o and in the valleys to both sides of Ta-mei Shan, as at La-brang proper where *Picea asperata* occurs in company with *Picea purpurea*, the latter's northern limit.

The main stream which waters this region is the Ta-hsia Ho 大夏河 or the Great Summer River and its affluents. It has its source to the west of Hsia-ho hsien or La-brang, and southwest in the grassy high plateau of which the pass called Tshâ-a-mi-kha, elevation 12,020 feet, is the divide; it is part of the system which forms the Yellow River – T'ao River watershed and separates the Hsia-ho territory from that of the A-mchhog tribal lands. The passes vary from 11,400 feet in the east, to 10,140 and 12,020 feet in the west. The A-mchhog country is watered by the rDog Chhu which flows from west to east for a considerable distance when, after having passed the lamasery called hBo-ra dgon-pa, it flows south-southeast into the T'ao Ho 洮河. It receives several affluents from the north which have their sources in the Tshang-dkar and rTse-ü passes, east of the hJog-khyi pass. These affluents surge through valleys which cut the grassy range at certain places, that otherwise forms the rDog Chhu-Hsia Ho divide. The whole region is a maze of valleys with interjacent spurs over which passes lead from one valley into the other; all are of loess formation and grass-covered, some of the valley floors being swamps with protruding grassy hummocks. Further east towards Lin-t'an the valleys become drier. All the spurs are intricately connected and their highest crests indicate the ancient level of the plateau.

The vegetation in this area is very poor in species, and with the exception of the forests indicated, is treeless.

T'ao-chou Old City is the first town encountered after leaving Cho-ni. In the temple compounds of T'ao-chou *Viburnum fragrans* a shrub 15 feet tall, with pale pink, fragrant flowers is cultivated as an ornamental while on fields and waste places *Aconitum gymnandrum* Max., with dark bluish-purple flowers, flourishes like a weed. Everything else had not yet awakened from the winter's sleep. The surrounding hills are loess and grass-covered; the town itself is gray, dreary and spells desolation. The roads leading north of T'ao-chou are lined with poplars, and the hills of red clay are terraced and cultivated to near their summits.

Here in the grassland we shot a golden *Aquila chrysaëtos daphanea* Menzl., and *Circaëtus gallicus* (Gm.) a rather rare bird for West China.

The highest pass between T'ao-chou and Hei-ts'o is 11,400 feet, the grassy plateau embodies shallow depressions and short valleys, making travelling comparatively easy. The soil varies from grey to red loess and black loam. Although nomads are encamped everywhere, there is still an odd Tibetan village here and there, also a lone monastery called rJe-tshang dgon-pa or She-tsung ssu 舍宗寺 on the hillside, at an elevation of 10,500 feet.

From a rocky pass where a lone hermit had built a stone dwelling, depending on travelers for his food, a high, snow-covered range, which I judged to be between 15,000 and 16,000 feet in height is visible; it is in a northeasterly direction from Hei-ts'o and is called Am-nye gnyan-chhen. To the Chinese it is known as the T'ai-tzu Shan 太子山 or the Mountain of the Heir-Apparent. This range is the evil counterpart of Am-nye Ma-chhen. While the latter is a protector of the lamas and is considered a beneficient sa-bdag or earth-lord, Am-nye Nyen-chhen, as the name is pronounced, is a malevolent spirit and considered the protector of thieves and robbers. It is therefore spoken of as the Nag-phyogs-pa pronounced Nag-chho-pa or the Black Side, while Am-nye Ma-chhen is known as the dKar-phyogs-pa (Kar-chho-wa), or the White Side. Anyone possessed by a demon is considered to be of the Nag-chho-wa or Black Side. Nyen-chhen is an epithet denoting the great cruel one, or fierce one, while the former means the well disposed one.

Large groves of conifer forest extend to almost the summit of the range which stretches from west-northwest to south-southwest. It is a formidable range, and branches or affluents of the Ta-hsia Ho 大夏河 have their source on the northern slopes of it. United they discharge north into the Yellow River at Yung-ching 永靖, the former Lien-hua Ch'eng 蓮花城 or Lotus City. The range is apparently of limestone and seems to be a northern extension of the Hsi-ch'ing Shan 西傾山. Southwest of it and north-northeast of the monastery of K'a-chia ssu 卡家寺 is another high mountain called Liao-chia Shan 廖家山.

Hei-ts'o ssu 黑錯寺 or Hei-ts'o Monastery called in Tibetan Tse-ü Gom-pa, written rTse-dbus dgon-pa or the Central Peak lamasery, is situated on a grassy hillside in a shallow valley at an elevation of 9,205 feet, and belongs to the Yellow Sect. It contains about 200 compounds or houses, possesses a nine storey building called dGu-thog or «nine storeys» a description of which will appear in the Historical Geography of the region to be published. It had 350 residing monks, but before the Moslem-Tibetan war in 1925, it had 993 monks. It was partly destroyed and entirely looted by Mohammedan soldiers who also put more than twenty lamas to the sword.

Opposite the lamasery is a small forest of the ubiquitous *Picea asperata* Mast., associated with *Betula* sp? Berberis, Spiraea etc.

The lamasery was destroyed over 100 years ago by the Chinese and then again rebuilt. The monks and lama officials are the most haughty and unfriendly of the entire region. Teichman who came to Hei-ts'o several years before me was not received and had to put up elsewhere.

Not far from Hei-ts'o, in another valley is situated the yellow sect lamasery called in Tibetan Kha-rgya-sgar-rnying and K'a-chia ssu 卡加寺 in Chinese, while near a stream, in another valley is the little monastery bKra-shis dgon-pa, pronounced Tra-shi Gom-pa, or Monastery of Happiness, which the Chinese transcribe Cha-shih ssu 扎仕 寺. From this lamasery a trail leads west over a rocky spur and into a narrow valley up the stream, northwest. This valley is or was wooded, the trees were again *Picea asperata* and Betula; all the large trees had been cut and thousands of saplings were piled up along the streambed. This is not done by the Tibetans but by the Chinese Moslems who, like the Lolos of Yün-nan, and the Chinese in general, are the arch enemies of the forests.

On our return journey the ground had come to life and there was a profusion of rich green grass studded with *Pedicularis decorissima* Diels, a new species with remarkable flowers, the bright blue *Delphinium Henryi* Franch., *Pedicularis semitorta* Max., the new *Leontopodium haplophylloides* Hand.-Mazt., *Ligularia virgaurea* (Max.) Mattf., the globose bluish-white *Delphinium albo-coeruleum* Max., the campanulaceous *Adenophora Potanini* Korsh., with blue bells, and its congener *A. Smithii* Nannf., an herb with dark blue flowers.

Three miles or ten li beyond is another monastery called Rong-war written Rong-bar dgon-pa at an elevation of 9,500 feet with a village of the same name one li distant. It is larger than Hei-ts'o and controls the latter. On the loess banks at Rong-war flowered *Eurotia ceratoides* C. A. Mey., a shrub 3-5 feet in height with blooms of a grayish-pink. At the head of the valley is a snow-covered range, the lower slopes of which appeared black, indicating forests. This is the Ta-mei Shan 大煤山 or the Great Coal Mountain.

The northern slopes of a valley descending from the above mountain support lovely forests of *Picea asperata* Mast., here associated with the tall and stately *Picea purpurea* Mast.; the slopes facing south are covered with *Juniperus glaucescens* Florin; not a single juniper could be detected among the spruces nor a spruce among the junipers, they kept strictly apart, each to its favorite exposure, the spruces clinging to the northern slopes where the snows melt late, and the dampness favors the growth of Mnium moss, whereas the junipers love dry feet, and no undergrowth whatsoever, not even grass.

The streambed is lined with willows, both shrubs and trees, among which grows the Kan-su gooseberry *Ribes stenocarpum* Max., the latter is often planted by farmers as hedges among their fields. Berberis are also plentiful.

Higher up on the valley slopes the spruces cease and a shrubby juniper takes their place, while the latter gives way to *Potentilla fruticosa* L., this in turn relinquished the ground to grass.

The pass over the Ta-mei Shan is called in Tibetan Dam-mai-la-rgan which the Chinese phoneticised Ta-mei-la 大煤拉, to which they add the syllables shan-k'ou 山 口 or mountain pass. It is the steepest pass in this region, is 11,675 feet above sea level, and extends from south-southeast to north-northwest. When returning over this pass early in August we found various plants in bloom such as *Delphinium tatsienense* Franch., which, with its deep sky blue flowers occupied the grassy slopes at 10,500 feet elevation, and *Aconitum laeve* Royle, a plant 5-6 feet tall, displaying pale pinkish-lavender flowers. A little lower occurred *Dianthus superbus* L., and at 9,000 feet *Thymus serpyllum* L. var. *mongolicus* Ronn.

The valley which leads down from the pass is called the Dam-lung and extends from south to north, its floor is at 9,500 feet above sealevel and its stream debouches into the

Hsia Ho 夏河 or Summer River, on the north bank of which the great La-brang
monastery is situated. Flowing northeast by Ho-chou 河州, now called Lin-hsia 臨夏
it enters the Yellow River at Yung-ching 永靖 after crossing a deeply eroded, loess-
covered plateau called Pei-yüan 北塬, the name being applied particularly to that part
which extends west of the river. Following the Hsia Ho west up stream, the trail leads
over rocky cliffs and grassy banks till La-brang is reached at an elevation of 9,585 feet
above sea level.

Hsia-ho hsien or La-pu-leng ssu 拉卜楞寺

Hsia-ho hsien was established in the 17th year of the Chinese Republic (1928), and was
so named on account of its situation on the Hsia Ho or Summer River. Its local name is
La-pu-lang or La-brang (bLa-brang). It is next in importance to Kumbum or T'a-erh ssu
in Ch'ing-hai province, while Hsia-ho Hsien is in Kan-su. It controls in all 108
lamaseries in its neighborhood, and a monk population of more than 300,000 lamas.

La-brang monastery is known as bLa-brang bKra-shis-hkyil, transcribed in Chinese
La-pu-leng Cha-shih-ch'i 拉卜楞扎仕溪. Since our visit in 1928, the monastery, the
largest in China, harbouring usually between 4000-5000 monks, has been created a
hsien or district as stated above. It is a regular city with many large stone buildings. A
cavalry barack was across the stream from it where a Moslem general ruled with a cruel
hand. He was later replaced by the so-called Christian, and later Red, general Feng Yü-
hsiang, and the district incorporated into Kan-su province. A complete description of
this monastery will appear in the Historical Geography of the region.

The Region Between La-brang and Ra-rgya dgon-pa or Ra-gya Lamasery

The region between La-brang and the Yellow River is one vast swampy plateau, with
valleys and high passes, and long plains. Most of the region is an actual morass with
stagnant water and hummock-forming grasses; it is only near the great gorges of the
Yellow River and the mouths of its tributaries that basic red sandstone cliffs are
revealed which give way to shale, mica-schist and a purple conglomerate in the valley
of the Yellow River proper.

The valley of the Hsia Ho, called the Sang-khog (bSang-khog) in Tibetan, in which
La-brang is situated on the left or north bank of the stream, is rocky, and covered mostly
with *Caragana tibetica* bushes on its slopes facing south. The hills facing north are
covered with conifer forest of two species of spruces *Picea asperata* and *Picea
purpurea*. In the winter and spring fierce west winds blow carrying clouds of dust with
them, enshrouding the entire valley and making life disagreeable.

The monastery controls most of the Tibetan clans who have their encampments west
of La-brang and near to the Yellow River, but not the monastery of Ra-rgya on the
Yellow River.

The flora of the region is poor in species and only when we approach the valleys
which join the Yellow River from the east do we find a more interesting type of
vegetation. Of conifers *Picea asperata* is the only representative besides *Juniperus*

which genus is represented by several endemic species. We encounter a meadcow flora restricted to the narrower parts of the valleys, near where they debouch into the Yellow River, and in the gorges of the latter. The alpine flora is confined to the high passes which lead over the high intersecting spurs from one valley to another.

R. C. Ching in his account of the itinerary during his botanical exploration in Kan-su says about the La-brang Ho as he terms the Hsia Ho: «The clearness of the water suggests the existence of immense forests at its source, far up in the Tibetan country.»

That no forests of any kind exist at or near its source we ascertained, for wc followed the Hsia Ho its whole length to its source, northeast of the 12,020 foot pass Tshâ-a-mi-kha, nor are there any forests beyond, but vast swampy grassland till, as already remarked, we approach the gorges of the Yellow River. The Hsia Ho has many affluents on both sides, all the northern ones of which we crossed (the trail leading at right angles to the affluents), on our journey west to the Yellow River and Ra-rgya dgon-pa.

The Journey to Ra-rgya

West of La-brang the Hsia Ho stream emerges from a narrow valley known as the bSang Khog (Sang-khog). Beyond the defile the stream divides and flows in a broad valley which resembled more a plain than a valley. Here took place the last fight between the Moslem army and the Tibetan nomads of the entire region, and from even west of the Yellow River.

A long valley leads from northwest into the Sang Khog, it is called mDâ-nag Khog. Its source is northeast of a mountain called dBang-chhen-shar-snying (Wang-chhen shar-nying) and partly also in a mountain to the north of the latter called Am-nye gnyan-ri. It describes a long curve from southwest to east-northeast flowing practically parallel to the Sang Chhu, but separated from it, and its smaller affluents by a long grassy spur; in this spur is a prominence which rises to about 12,000 feet called the Chhong-rtse in which the second, right lateral valley, the Chhong-rtse nang has its source. The valley of the Dâ-nag Chhu (mDah-nag Chhu) is the camping ground of the Tibetan clan hJo-rong-og, while north of the divide is the encampment of the mDo-ba (Mdo-wa) clan of the Tibetans. The Sang Chhu valley is flat and broad and rises almost imperceptibly. In the triangle between the two streams (v. s.) grow many willow shrubs, also *Hippophaë rhamnoides* L., and *Potentilla fruticosa* L., the latter scattered throughout the valley. The hills are loess, bare and eroded, while those flanking the Dâ-nag Khog are of red sandstone. The spurs flanking the main valley are terraced, the remains of ancient cultivation. Owing to the wild A-mchhog Tibetans who dwell to the south, and whose main occupation is robbing, no settlers can exist here.

The elevation of the valley floor is here 9,550 feet. Opposite the Dâ-nag Khog a short valley opens out called the Ngor-chhen, here a trail leads across a spur separating the latter from the next valley the hJog nang in which it continues southeast over the hJog Khyi (pass) to A-mchhog. The scenery is wild in the extreme, herds of yak and horses graze peacefully, searching for grass beneath the snow.

Not far beyond is another afflurent also from the south, it is the dGu-dgu Chhu (Gu-gu Chhu); across from the pass whence the Gu-gu stream has its source, another valley of the same name extends southeast. The Tibetans have the practice of calling two valleys which have their sources in the same mountain or pass, but on opposite sides, by the same name, therefore when speaking or writing of a valley one must be careful to indicate which one is meant.

The trail follows at the foot of the spur flanking the Sang Chhu on the south (left), below the trail are a number of springs called Chhu-ngo bkra-mdog (Chhu-ngo tra-dog) the last two words meaning many-colored (variegated). The broad valley of the Sang Chhu is now abruptly narrowed by two projecting ranges. South is a valley called dBen-chhen drag-las (Wen-chhen drag-le); the spur or hill flanking it to the right is named dBen-dkar (Wen-kar) and that to the left hJab-gyâ-nag (Jab-yâ-nag). The valley is thus narrowed by these two prominent hills beyond which it again widens, the ground is soggy and black and is really a marsh.

Our camps were always arranged in a circle with loads and horses in the center; the latter had to be hobbled and padlocked on account of possible marauding Tibetans, it has even been known that robbers file the chains and thus make off with horses which are highly prized. Even when only resting for lunch horses have to be hobbled with yak-hair ropes, one front leg to a hind leg, so that they cannot stray too far, and to prevent mounted bandits to charge and drive them off. For the securing of yaks for the night a long rope is staked into the ground, and each yak is tied to it with another rope which is fastened to the nose ring, they are thus attached at regular intervals, often, when the caravan is a large one, in several parallel rows.

Our Arig nomads used argols for fuel which they fanned into red heat by means of a sheepskin bellow to which a crude iron tube is attached at one end, it is quite a knack to manipulate these bellows as they are open all round only the neck of the sheepskin being tied to the iron tube. Pheasants were calling and hares jumped about in the Potentilla bushes. Several Tibetan mastiffs were watching our camp and during the night Tibetans took turns to be on the alert for bandits. The nomads, as soon as their tea had boiled, take a ladle full, place a little butter in it and throw it out of their tent while yelling a prayer to the mountain gods, This they perform several times, but only from the chief's tent. Along the stream were many willows, also *Myricaria dahurica* (Willd.) Ehrbg.

The vegetation in the valleys is mainly grass and *Potentilla fruticosa* said to be the yellow-flowered variety. The ground is water-logged, the water clear, but brownish.

At 10,790 feet a valley opens with a rather steep incline called Rang-chhog gzhug-ma (Rang-chhog zhu-ma), up which a trail leads to the Ngu-ra encampment, six to seven days journey distant by Yak caravan, and four days by horse. The yak caravan advanced slowly up the rocky, boggy valley, and had to keep close to the rock cliff to avoid the marshy ground. Here the hills flanking the Sang Khog decrease in height and the trail emerges again into a broad valley to the right, in a west-northwesterly direction is the source of the Sang Chhu which at La-brang becomes the Ta-hsia Ho 大夏河 and empties into the Yellow River. There are two valleys on the south side of the Sang nang, the Rang-chhog dbus-ma (Rang-chhog ü-ma or the middle Rang-chhog), and the Rang-chhog gong-ma (Rang-chhog gong-ma or upper Rang-chhog), the valley is here one great bog and one of the affluents contains quicksand. Water stands in pools above the grass and the horses sink deeply into the black mud.

At the head of the valley is the gravelly pass Tshâ-a-mi-kha, elevation 12,020 feet above sea-level. the valley beyond discharges into a larger one called Chhu-nag nang or the Black Stream valley, which receives here another branch named dBang-ra-rgan (Wang-ra-gen). The latter has its source in an isolated grassy mountain dBang-chhen shar-snying (Wang-chhen-shar-snying) whose height is probably near 15,000 feet.

Beyond is a seven mile long plain called the hDar-tshogs-thang (Dar-tshog-thang). It is on this plain where the Sang-khog Tibetan clan, which inhabits the valley of the Sang Chhu in the summer, holds its annual horse races on scrawny, grass-fed horses. It is one of the most forsaken and dreary-looking places imaginable.

The long and wide, snow-covered plain, framed by bare low hills produced a picture of desolation and utter loneliness. We made our way across this weird plain like a row of ants up a pass over the low ridge flanking it. We thought we were alone, but somewhere Tibetans were hidden, they had followed us and had been watching for an opportunity to rob us.

This opportunity soon presented itself. One of my men, not used to horseback riding, was thrown from his horse which had been frightened by a hare, the horse galloped off with saddle flying, and before we had realized what had happened, a group of horsemen appeared, and drove the frightened horse which we never saw again. This taught us to be on our guard. As we did not have extra horses the poor chap had to walk till it was possible to procure another horse from a nomad camp. Although Mr. Simpson and a number of armed Tibetans of our entourage followed, and searched for the rest of the day, the horse and marauders were gone.

The pass is called the Dar-tshog nye-ra and leads down into another plain, the Wog-chha-thang, which extends toward the east. Crossing another pass south-southwest, the trail descends into a still broader valley known as the Gan-dmar gzhung (Gen-mar-zhung). The whole landscape has an indescribable stamp of desolation. The entire region as far as traversed is one vast morass or marsh. At a sandstone cliff our Tibetans gathered in groups around yak dung fires and were boiling their everlasting tea. To me it was appalling to be buried in these weird marshes, crossing ridge after ridge and swamp after swamp with nothing except yellow-brown grass, no sign of a shrub or tree. Wherever one looked one beheld bare hills and water-soaked plains with a temperature which certainly did not suggest the month of May. The valley is at an elevation of 11,350 feet. We crossed the trail to the mouth of a valley opposite, the rNam-nang (Nam-nang), but instead of ascending it we continue up a valley to the left, the two valleys being separated by a low spur; this valley is called the mKhas-chags (Khe-chhag) and is nothing more than a continuation of the morass, it leads to a pass of the same name. Before its termination into the pass Khe-chhag Nye-ra it receives three small valleys each one called hBor lung or Bor Valley, they are termed the (first nearest to the pass) gong-ma or upper, the second ü-ma or middle, and the last the zhu-ma or lower. The broad head of the upper valley is at an altitude of 12,132 feet boiling point. As the yak were tired from crossing the swampy plains we did not think it wise to cross the Khe-chhag Nye-ra (pass) more than a thousand feet higher than our camp. Here in this valley we found *Rhododendron capitatum* Max., *Rhododendron Przewalskii* Max., and *Rh. thymifolium* Max., all shrubs 2-3 feet tall, only the first and last were in flower. Near the top of the pass a dwarf form of *Primula fasciculata* Balf. f. et Ward, formerly known as *Primula reginella* Balf. f., was already flowering in the surrounding swampy meadows of this bleak, cold region, the little pink flowers with their yellow eyes brightening the drab, yellow grass, covered with patches of snow here and there. There were leaf-rosettes of *Meconopsis punicea* Max,. but no sign of inflorescences as yet, while of *Meconopsis integrifolia* (Max.) Franch. and other herbaceous plants only the dead flowering stalks of the year before remained. Undoubtedly in the height of summer this would prove an interesting region botanically. The growing season is however very short in this part of the world.

The rawness and inclemency of the weather was well matched by the hostility and churlishness of our nomad yak drivers, they were as unfriendly as the landscape was bleak, and I could not help but think that it was due to the environment which had fostered such an inimical character, and never one produced a smile or a friendly word. There was an old man, but could not have been more than fifty, people age quickly here being continuously exposed to the elements, who was antagonism personified, he was the very image of a churlish beast, he had participated in many a fracas for his face was cut in all directions, and so were his lips which had the aspect of multiple harelips. Even his nose[37]

were nearly blown off our horses by the 50 mile gale, at a temperature of 15°F., this on the 8th of May. The descent was much worse than the ascent, the trail was regular chute, the gravel frozen hard, while the snow whirled around us as we descended, slid down is the better word, into the teeth of a gale. Arrived at the foot of the pass we gasped for breath, and I felt my cheeks freezing, also the skin around the eyes where my spectacles touched it. The valley beyond is also called Khe-chhag-nang and as lonely and desolate as its other half on the other side of the pass.

Here we found abandoned encampments, the mud stoves of the nomads on which they boil their tea, burning sheep manure; also square blocks of yak dung piled up to three feet in height with a mud frame, these served as altars on which the nomads burn juniper twigs as offerings to the mountain gods. Of nomads there was no sign. The gale did not abate, but if anything increased in fury. We halted here for some of our yaks had given out and could hardly move, even without a load. We allowed them to graze and gather strength, although yaks like the nomads seem to be immune to hardships.

Rather than wait in the icy wind we decided to go ahead leaving a few Tibetans with the played-out yak. The trail leads from east-northeast to west-southwest, till we came to a plain known as Khe-chhag thang. The sky was clear and the west wind this time brought no clouds in its wake. While crossing another plain called the sTeg-sgam thang (Teg-gam thang) with a fair sized stream in its center, we shot some snipes and ducks, in the vicinity of a nomad encampment whose fierce dogs attacked us. Climbing out of this amphitheatrical depression we descended into a long valley, the Ma-mo gzhung (ma-mo zhung), the winter encampment of the Sog-wo A-rig tribe to which our Tibetan yak drivers belonged.

---

[37] Page 290 is missing in the typescript. It is not clear whether it is a recent loss – now also p. 289 is absent from the files of the Royal Botanical Gardens Edinburgh but, fortunately, preserved in the xerox copy.

The wind or gale continued in all its fury but glad to have arrived at this encampment we pitched our tents at an elevation of 11,449 feet, just far enough away from the tents of the nomads so as to be out of reach of their dogs (see Plate **43**).

We rested a day at the Sog-wo Arig camp and also arranged for a change of yak to replace those which were nearly all played out. Winter time or early spring is a bad time to travel with a yak caravan as the grass is poor and young grass not yet sprouted, the hardships are too great from cold, snow and gales. There are actually only two seasons in this part of the world, summer and winter, and the former is very short, as one can expect snow storms even in July and August. It may be remarked that there is not a single frost free night throughout the year in these bleak uplands.

Near the camp of the Sog-wo Arig we shot several ducks and the blacknecked crane *Megalornis nigricollis* (Przew.), a male, for it had a bright red cap on its head.

The boys also amused themselves catching the large marmots *Marmota himalayana* so common in this region. They tied them to stakes on a long rope, and whenever one approached they raced around in a circle, but more often raised themselves on their hindlegs and bared their long teeth. We made arrangement with the nomads to take us from here to Ra-gya, but this they refused, although they had promised to do so in La-brang; furthermore they announced that before they would move they would have to be paid in full in advance in lump silver. As there was nothing to be done, we had to acquiesce in their demands, but they agreed to take us only as far as the lamasery of gTsang-sgar (Tsang-gar), two days east of Ra-gya. There we would have to make other arrangements to be taken to the Ra-gya monastery, but this did not prove difficult, in fact we were glad to get rid of these rude, and churlish Mongolized Tibetans, with their lamb-skin caps and long red tassels. However the Sog-wo Arig chief's son, the chief lay dying of paralysis in his tent, became more civil on acquaintance, and the men did not mind to have their picture taken, but refused to have their womenfolk photographed, saying that the women would not like it. Although they spoke no Mongol, and were indistinguishable from Tibetans, they all lived in Mongol yurts, much more comfortable than the black, yak-hair Tibetan tents which leak for years when new, and only become waterproof when the soot of years has filled the coarse meshes of the cloth which resembles burlap than any other weave.

We had a peaceful windstill Sunday at this camp and entertained many of the nomads who wanted to see the wonders of an Urussu camp, Urussu meaning Russian, the nomads knowing no other nationality, and every foreigner (white man) is called Urussu. Some of the nomads especially the son of the chief had distinctly negroid features, as large projecting jaw, huge thick lips and a negroid cranium. They are an independent, absolutely fearless lot, and always go fully armed with sword and rifle even at their own encampments. They are of a swarthy brown complexion, age early, on account of their primitive existence, forever exposed to the unfriendly raw elements.

The only money in use here is lump silver which has to be chopped with chizel and hammer and then weighed, that everyone has his own scale which dips to his advantage need not to be wondered at. For minor purchase such as milk, cheese or argols, one barters, using needles and foreign thread which are much in demand, also cotton drilling of either blue or brilliant red color; the fixed rate is usually one square of this cloth for the hire of an animal or man as guide. These they patch together to make trousers, or give their women folk for jackets. We carried two yak loads of goods for barter, and presents for chiefs and incarnations and their stewards, for the chiefs we had pieces of black sateen, and violet purple brocade with gold thread designs for the stewards, the bright red or yellow brocade was reserved for the higher incanarnations, as were watches, while needles, thread, tea and salt served in lieu of small change.

The 10th of May found us on the way at 6 a.m. after a cold brilliant starlit night, without a breath of wind. It was a glorious morning, and what under the circumstances one would here call «balmy» atmosphere. From our camp the trail led us to the slopes of the ridge on which the nomads were encamped, northwest, then directly north to the valley in which the Tshe Chhu flows, one of the larger tributaries of the Yellow River. This river has one of its sources, the southern branch, in the Tshe-sde-ra (Tshe-de-ra) pass, and the northern in a grassy spur of the plain called Ma-mo-ren-chhung-ba (Ma-mo-ren-chhung-wa), which is also the name of an isolated grassy hill around which a southern and northern branch flow. It makes a right angle around a mountain called the Seng-ge khang-chhags and flows directly south to where it receives the Ma-mo-zhung at the Sog-wo Arig encampment. Here it makes another right angle and flows west-south-west between hills, two on its northern bank and one on the southern one. In the triangle formed by the confluence of the Ma-mo-zhung and the Tshe Chhu is a conical hill with a cairn or Obo on top, this is the southern hill, the Ma-mo-shar-snying. On the opposite bank of the stream is a single house, the embryo of a lamasery which the nomads had been trying to build with great difficulty owing to the entire absence of timber. The mountain against which the would be lamasery nestles is called Yur-rgan-ri, the lamasery-to-be was already known as mGur-sgar (Gur-gar). Owing to the swamps the trail which keeps to the left flank of the valley, leads to the ford of the Tshe Chhu at an elevation of 11,250 feet. Here at the river bank were many ducks and white terns, of the latter we collected specimens, it proved to be a variety of the common tern of both coasts of the Atlantic *Sterna hirundo tibetana* Saund.

West-south-west rises a mountain range, quite conspicuous in the landscape, it has five peaks of which two, the central ones are prominent and about 15,500 feet in height. This range is called A-rig dzo-rgon-ma (Arig dzo-gön-ma); almost at right angles to the range is a less high, about 14,500 feet, whose name I could not learn; the Tshe Chhu flows in a deep gorge between these two ranges and debouches into the Yellow River beyond them. No one has yet been at the confluence of these rivers. The trail continues west-north-west, crosses the affluent O-man-hde (Wo-men-de), and arrives at another encampment of the Sog-wo A-rig in the valley of the second affluent from the north, called the Chhu-nag Chhu or the Black Water River which has its source on the southern slopes of a mountain called Sa-ri mkhar-sgo (Sa-ri khar-go) in the north, and within the bend of the Tshe Chhu. At the banks of the Chhu-nag we decided to pitch our camp at an elevation of 11,450 feet. Here we found the encampment of an incarnation over 80 years old who was traveling, like we ourselves, to the lamasery of Tsang-gar to which he belonged. His name was Lags-kha-gtsang and later in the day we made his acquaintance.

We considered that day's temperature sultry and hot although the thermometer registered 52°F. at noon. At 6 p.m. the sky became overcast, a wind began to howl and rain rattled on our tents. We were in the middle of the vast grassy plain where the wind had full sway. During the afternoon we went to the nomad encampment and called on the old incarnation who received us well. He camped in a comfortable yurt or Mongol felt tent which we entered on the left, that is turned left after entering, as custom describes, while a Tibetan tent is entered going right. He sat on cushions surrounded by reliquary shrines and silver charm boxes, a prayer wheel and a few painted scrolls, the latter hanging from the roof of the yurt. I presented him with the usual silk scarf, pictures of the Dalai and Pan-chhen Lamas which he raised to his brow in blessing. After that we repaired to the tent of the local tribal chief where we were «entertained» to tea. A fire roared in the mudstove in the middle of the large tent which was crowded with nomads, men, women and children. They were all husky and well built, but dirt begrimed. An old woman tended the fire and tea kettle. Producing from a pile of sheep manure wooden tea bowls, she dipped them into the ground up sheep dung and with her dirty black hands began to wipe them, after which she proceded to wipe them again with a black filthy rag. They were now «clean» and ready to be filled with tea. It is true everyone should carry his own wooden tea bowl, among many other things, in his voluminous garment, but as I did not live or dress Tibetan style, they furnished me with a bowl of their own, «cleaned» it in the way described above. A small wooden box was then set before us with three equally large compartments, the first contained ground roasted barley flour, the middle one butter in which yak hair bristled and covered with a layer of dust from powdered sheep manure; like a mold it exhibited the imprints of five fingers with which the last nomad helped himself to the butter. The last compartment contained salt. So as not to offend these simple, friendly people I took a sip, but just a tiny sip. Thus ended the tea party in the chief's tent and we returned to our camp.

Next day the caravan went in three relays of twenty yak each over the same monotonous, grassy, rolling plain between low hills, the same kind of landscape we had seen since leaving La-brang. The grass flora is made up of the following genera and species: *Stipa mongolica* Turcz., *Poa attenuata* Trin., *Poa flexuosa* Wahl., *Poa* aff. *arctica* R. Br., *Deschampsia cespitosa* (L.) Beauv., *Elymus sibiricus* L., *Trisetum* sp ? *Festuca ovina* L., *Koeleria argentea* Griseb., the new species *Koeleria enodis* Keng, and the new variety *Elymus sibiricus* var. *brachystachys* Keng, to mention the most important.

In the summertime many herbaceous plants grow among these grasses which lend color to the otherwise drab landscape.

It was difficult to know whether one followed a plain or a large valley so vast is the landscape; as it was we had followed the Sha-bo (Sha-wo) valley west-northwest over a gradual incline at an elevation of 11,500 feet. Sha-wo nang has its source in a hill or mountain to the north called hDam rdzab-ri (Dam-dzab-ri) which rises high above the surrounding grassland. A trail leads across the intervening spur, only 250 feet above the level of the valley, into another valley known as Gan-dmar khog (Gen-mar Khog) on account of its red clay wall which flanks it on the northeast. Antelopes, the Chinese Huang-yang 黃羊 are common here and so are wolves of which we met many. Of the former we bagged one for the pot for they are fat and most excellent eating. The Gen-mar stream is of considerable length and flows in a southeasterly direction into the Tshe Chhu. The Gen-mar valley was dotted with black tents of the nomads and thousands of yak and sheep grazed over the broad landscape, the floor of the valley being only 90 feet below the pass. Gen-mar khog has three upper branches which unite to form the Gen-mar stream, the shorter ones from the northwest are the sPhyi-sgar (Chhi-gar), and the Lar-sgol (Lar-göl), while the third flowing from west to east, the longer one, is called the Rag-chhung nang. Here the grass was very poor not even an inch high having been overgrazed by the herds of yak and sheep of the nomads who belonged to the Tso-khar tribe. On and on we went, when all at once the pleasant balmy atmosphere changed in to a chilly wind, and we were bombarded with snow pellets for a considerable time; changes in temperature follow each other in great rapiditity. After traveling for about sixteen miles we decided to call a halt and finding the grass better on the northern bank of the stream pitched our tents near the encampment of another Tibetan clan called the Rong-po (Rong-wo) subject to the Rong-wo lamasery situated on the road to Kuei-te 貴德. The Rong-wo people are easily distingished by their felt head gear which resembles the cover of a chafing dish. Our camping place was designated by the name Ru-nag or the Black Horn. Here in the shallow Gen-mar stream at 12,100 feet we discovered fish, my men catching 48 by diverting the stream into a dry depression five feet deep. The nomads were astounded and asked what we were going to do with these worms. When told that we would eat them they looked disgusted, for no Tibetan of this region would ever touch fowl, fish or eggs.

A clear sky and a glorious morning found us next day ascending the first, lower part of the Nyin-zer La, elevation 12,520 feet; from this first pass there is a slight depression and a gradual incline to the main pass 12,650 feet. From the summit there is visible a black range, then partly snow-covered of approximately 15,000 feet in height, it extends from southwest to northeast, the Yellow River flowing back of it. In front of it, and separated by a valley called the gSer-chhen Nang or the Great Golden Valley, is a parallel range but much lower and grass-covered, whose name we could not learn. From here the descent is gradual over a somewhat rocky trail, the hill-side to the left was covered with bare bushes that could not be identified. A very short ravine leads from the pass into a long, narrow valley which stretches south-southwest to the Yellow River, it is known as the rDo-rgan Nang (Dorgen Nang) and is 12,220 feet above the sea.

Crossing its stream diagonally the trail leads between the grassy hills over a small pass known as sKyod-dmar (Kyö-mar) elevation 12,420 feet, and descends into a broad valley bordered by low hills; this is the sGo-shub Nang (Go-shub Nang) composed of much tilted slate and shale, in which Tsan-gar (lamasery) is situated.

On the rocky slopes we shot quite a number of beautifully marked partridges *Alectoris graeca magna* (Przew.) who, with rabbits or hares and marmots, were most common. Not far above Tsang-gar (monastery) there is a phantastic red sandstone cliff, pitted with holes caused by the action of the water. On this cliff grew beautiful, rich green, junipers (*Juniperus glaucescens* Florin), the elevation at the foot of the cliff being 11,2000 feet. A few more bends down the narrowing valley and we reached Tsang-gar (monastery). We were well received and shown an old red house and court where we could stay as long as we wanted. In the court yak manure was spread out a food thick, to dry, and into this our unruly nomads, on arrival dumped our loads most unceremoniously. There was not one who came to say good bye, they had been paid in advance and there was no need to stand on ceremonies. I was real glad to see the last of them. The first lap of our journey came thus to an end with only the loss of one horse.

Tsang-gar (monastery) buried in this far and out of the way glen, the Go-shub-nang, is built on the right slopes of the valley, the stream flowing deep below in a narrow trench. The lamasery is over 250 years old, but built a little later than La-brang Tra-shi-khyil, and houses about 500 lamas and a high incarnation which at our time of visit was the fifth. The first incarnation or rather the first generation, who later became reincarnated, came originally from the Pan-chhen Lama's domain in bKra-shis lhun-po (Tra-shi lhün-po), southern Tibet. The present incarnation Tsang wan-dita who is a brother of the steward of the lamasery, is from a tribe west of the Yellow River. The lamasery has three chanting halls or temples, and many small buildings housing the monks; it also has fifteen incarnations of various ranks but lower than Tsang Wan-dita, the first of whom was the founder of the monastery.

Tsang-gar is situated at an elevation of 11,000 feet, with the mountains which hem in this valley, rising high above it.

During our stay here of several days we visited the lamasery under the guidance of the steward who himself had built the latest chanting hall, a very substantial structure constructed entirely of rock. The main brass image of bTsong-kha-pa (Tsong-kha-pa) the founder of the Yellow Sect had been brought from Lhasa. The large chanting hall, built nearly sixty years ago, contained eighty pillars (supporting the roof) of tall spruce trees. In the vestibule were the usual frescos of the four guardians, Lokapalas, but in addition there were two others, one representing rMa-chhen spom-ra the mountain god of the Am-nye Ma-chhen Range, and the other that of their own mountain god of the region the local sa-bdag sGo-chhen after whom the mouth of the Do-gen Nang valley is named, Go-chhen meaning the «Great Gate». He is pictured riding a yak and brandishing a sword.

The tribe which inhabits the region is called the Yu-ngok?; their eastern boundary is the Nyin-zer pass (q.v.)

The gorges of the Yellow River south of Tsang-gar

While our caravan rested at Tsang-gar we took the opportunity to explore the Go-shub Valley and that of the Yellow River which in this region, prior to our arrival, had never been visited by white men. The lamas told us that there were forests in the lower parts of the Go-shub valley and in the Yellow River gorges proper, whence the timber came for the building of their lamasery. [Plate **44**]

On May 13th we left Tsang-gar following the Go-shub valley down stream which became so rocky and narrow that it was impossible for horses to pass; we therefore crossed the valley and climbed the spur which separates it from the Zher-shib valley, both merge a short distance before reaching the Yellow River. As these valleys were too narrow to permit reaching the Yellow River, we went, at the suggestion of our lama guide, southeast, and south to the mouth of the Do-gen Nang called sGo-chhen or the Great gate.

The rocks in the Go-shub and Zher-shib valleys are slate and shale, very much crumpled and tilted. On the upper slopes grew bushes of *Juniperus glaucescens* Florin, with semi-globose fruits. The tops of the spurs hemming in these valleys are composed of locss and are grass-covered. From the Zher-shib valley we climbed to a bluff

overlooking the Yellow River which flowed 600 feet below us in a rocky gorge of slate
and sandstone. Here we shot a new bird *Prunella fulvescens Nadiae* B. et P., and took
photos both up and down stream of the Yellow River (see Plates **45-47**) from an
elevation of 10,690 feet, looking northeast up a stream, and northwest down stream. The
gorges of the Yellow River are here wooded with *Picea asperata* Mast., *Betula japonica*
var. *szechuanica* Schneid., quite tall trees which grew to the water's edge. Willows
abound along the streams such as *Salix myrtillacea* and *S. Wilhelmsiana* M. B. = *Salix
taoensis* Görz n. hybr., *Salix rehderiana* Schn. var. *brevisericea* Schn., *Juniperus
glaucescens* Florin, *Cotoneaster adpressus* Bois, which tightly covered boulders, at the
bases of which flourished the fern *Cheilanthes argentea* (Gmel.) Kze. On dry loamy
banks and rocks grew the pale yellow, rather sad looking, *Primula flava* Maxim.

The Yellow River flows here very swiftly in its prison of slate, shale and sandstone.
From the bluff we descended over a very rocky path into the narrow mouth of Go-chhen
in the Do-gen valley, the mountain walls flanking Go-chhen facing north are covered
with forests of tall *Picea asperata* Mast., mixed with the Betula and willows above
mentioned. Here along the banks of the stream we shot a troglodyte wren which only
frequents the rocks along the very banks of streams, *Nannus troglodytes idius* (Richm.)
The Yellow River which flows here at an elevation of 10,200 feet, is about 150 yeards
broad, very swift and full of swirl pools. Directly opposite Go-chhen there is an
enormous rapid in which nothing could live or survive. Further up stream, in the Go-
chhen valley, we found several willows as *Salix wilhelmsiana, Salix myrtillacea*, 12 and
8 feet in height respectively, the new hybrid *Salix taoensis, Salix rehderiana* var.
brevisericea, 80 foot tall spruces *Picea asperata*, with short and descending branches
and deep green foliage. On the rocky slopes grew *Caragana brevifolia* Kom., *Potentilla
fruticosa, Juniperus glaucescens* Florin, and herbaceous plants not then in flower. The
only Primula that braved the cold proved to be *Primula flava* Max., it was the first to
flower, and occured in most of the valleys tributary to the Yellow River, as well as in
the Min Shan to the southeast.

From Tsang-gar to Ra-gya Gom-pa
On May 15th we said farewell to our lama friends, and with a caravan of sixty new yak
we left Tsang-gar, retracing our steps up the Go-shub valley to below the red sandstone
cliff and red clay, and turn up a small ravine of slate, shale and schist, also sandstone,
especially in the upper strata. A short distance higher up in the valley, the schist gives
way to sandstone and red clay altogether. The valley extends west-southwest and
gradually becomes shallower and grass-covered, with the slopes low and gentle, as do
nearly all valleys sending tributaries to the Yellow River. At their source they are wide
and merge into the grasslands, but gradually cut into the loess and through the
underlying red sandstone below which we find the schist, shale, and slate through which
the Yellow River also found it easy to cut its way.

Crossing a small pass, elevation 11,388 feet we descended into a gentle sloping
grass-covered valley. From now on it is continuous traversing of valleys and their
intervening spurs, the trail leading at right angles to them. Weird sandstone bluffs flank
them, the red cliffs terminating in castle-like towers and battlements, in the shallow

recesses of which small groves or forests of junipers with rich green foliage have taken a foothold. *Juniperus glaucescens* Florin, is the only species found in the valley of Go-shub. In the mouth of the valleys are spruce forests facing north, while the dry south-facings slopes are covered with Berberis, *Potentilla fruticosa* L., and specially *Berberis tibetensis* bushes. A red, clayey, muddy trail brings us to the long Great Gold Valley or the gSer-chhen Nang (Ser-chhen Nang). Where the trail fords the stream the valley is quite broad, but north of it, it narrows into a red sandstone gorge with vertical cliffs honeycombed with caves, and partly covered with *Juniperus glaucescens* Florin, which seems partial to red sandstone. On the west side, a short distance down stream, there is a huge long cleft in the cliff through which has oozed, as one might say, a pillar of ice, spread at its base like the train of a wedding gown (see Plate **48**). the upper half of the valley is again red sandstone while the lower half is composed of schist and slates. Deciding to explore this valley further we continued down stream, where we encountered one single nomad tent hidden in a juniper grove in this lonely gorge. We were not aware of any present till we were suddenly attacked and nearly mauled by fierce mastiffs whose habit is to make for a man's throat. The valley became narrower, and about three miles or ten li beyond, it was joined by another smaller one called the Small Gold Valley or the gSer-chhung Nang bringing red muddy water, while that of the larger valley was crystal clear. As it was impossible to continue in the valley due to its narrowness, we climbed the right flank of the ravine over loose shale covered partly with grass and with here and there a spruce tree. Before us lay the gorge of the Yellow River, and looking down from a bluff, elevation 10,300 feet, I photographed the River which here flowed southwest towards its source.

The river is here only about 80 yards wide and flows swiftly some 200 feet below the bluff. The gorge itself is not very deep, but its slopes fall steeply, and in places vertically go into the river, leaving no banks along the water, and this at low water mark. The drop in the Yellow River from the mouth of the Go-chhen Valley to that of the Ser-chhen is exactly 100 feeet. In the sheltered gorge *Picea asperata* established itself also *Betula japonica* var. *szechuanica* Schneid., and junipers, but the latter kept together facing south.

The upper part of the gorge is flanked by huge vertical cliffs 600-1,000 feet high, which still further up end in a grand amphitheater with a majestic sweep; here eagles had their nests and the tops of the pinnacles were covered with junipers. Icicles, like mighty stalactites, hung from the walls of caves and their ceiling; it was a weird and wild setting, yet peaceful. The streambed was filled with willows as *Salix pseudo-wallichiana* Görz a shrub 10-15 feet tall, x *Salix taoensis*, *Ribes vilmorini* Jancz. four feet high, while on gravelly flats grew large *Juniperus glaucescens* with trunks 3-4 feet and more in diameter indicating an age of 500 to as thousand years. Here among the junipers and willows we shot *Phylloscopus (Motacilla) affinis* (Tickell), but other birds were scarce, except on the grassy summits of the spurs where Tibetan eared pheasants, *Crossoptilon auritum*, roamed, always flocks of eight to ten. Having followed to the end of the gorge, we returned to our trail and climbed the spur at an incredible angle to an elevation of 10,820 feet whence we continued west-southwest over grass-covered loess hills, then northwest, the Yellow River flowing in a gorge to our left. Going now west, we descended into Ser-chhung or the smaller Gold Valley where we met the same type of red sandstone, the same junipers and willows, the latter in flower but almost leafless. Ser-chhen valley is the divide between the Yu-ngok? tribe under the rule of Tsang-gar and that of the sGar-rtse tribe (Gar-tse) to the west under Ra-gya monastery.

From the Ser-chhung valley the trail climbs a narrow ravine west-southwest to a pass 11,280 feet called A-ra-u-lag, a Mongol name. As the trail leads at right angles to the tributaries one is obliged to continuously descend deep valleys cross streams, and passes, such as the Lung-dmar-kha at an elevation of 11,090 feet, with the Lung-dmar or Red Valley below it which sends a red torrent to the Yellow River. Its sides are covered with willow bushes and treelets of *Hippophae rhamnoides*, *Sibiraea angustifolia* (Rehd.) Hao, Betula and *Juniperus glaucescens* Florin. Here we encountered and shot many of the Tibetan eared pheasants *Crossoptilon auritum*, always partial to wooded areas.

Instead of following Lung-mar valley, we climbed the right hand bluff as steeply as we had descended, up to an elevation of 10,300 feet. here, below us, flowed the Yellow River, angry and turbulent surging against its prison of red sandstone. Opposite to where the Red Valley empties into the Yellow River, a short valley called the dGâ-khog (Gâ-khog) joins the latter.

The Yellow River valley is here sprinkled with spruces, junipers, willows etc., loess covers the upper slopes with the red sandstone beneath; this formation can be found nearly everywhere as in the Hsi-ning Valley in the north, but it becomes only visible in deep valleys, where grass covers the loess above and the red sandstone is exposed under its thick layer.

Endlessly up and down the trail leads across several small valleys and their intersecting spurs till we emerge into the valley of the yellow River proper, descending steeply to its banks. Around a bluff, on a gentle-sloping meadow lies the Monastery Ra-gya dgon-pa, at the foot of a high cliff resembling a huge bird with outspread wings, it is the redeeming feature of the otherwise bleak landscape.

Ra-gya gom-pa and T'ung-te hsien 同德縣

The district of T'ung-te was established in the 24th year of Chinese Republic, 1935, and is actually ruled by Kuei-te hsien 貴德縣 situated on the south bank of the Yellow River several days journey to the northeast. It is a trading centre where cloth, salt, sugar, barley flour, tea and iron pans, etc., are exchanged for wool and musk. Only Moslem traders, and mainly those of the New Sect who have their headquarters in T'ao-chou Old City, come to trade with the nomads from west of the Yellow River. The New Sect Moslems were the only ones who could trade with impunity with the Go-log within the loop, and south of the Yellow River.

Ra-gya dgon-pa (Ra-gya gom-pa) and its neighborhood
Ra-gya monastery situated on the right bank of the Yellow River, consists of two sections separated by a shallow ravine, the larger section being west of the ravine. It nestles at the foot of a peculiar purplish-red conglomerate sandstone mountain (see Plate **49-53**), the south face of which drops vertically; the top is cleft, and its lower

lateral rock walls give it the appearance of a huge bird hovering, or ready to take flight. It is this aspect which apparently has suggested to the lamas or Tibetans the name Khyung-sngon (Khyung-ngön) or the Blue Garuda, the great mythical bird, the enemy of the Nâ-gas or serpent spirits. It is the mountain spirit of the region and especially of Ra-gya lamasery, whose festival falls on the 11th day of the fourth moon, which in 1926 fell on the 22nd of May.

Everywhere within a radius of a mile around Ra-gya monastery roam hundreds of blue sheep, the *Ovis burrhel*. They are sacred and hence unmolested and unafraid, one can approach them and photograph them. East of Ra-gya in Lung-ma valley they are wild and numerous (see Plate **54**).

The monastery comprises eight major temples of which two have more than two storeys. It contains also two large square buildings, one in each section; the one in the north section being much larger, houses the printing establishment of an edition of the bKah-hgyur or the translated word (Kanjur) the Tibetan classic or Buddhist canon of 108 volumes; it was not previously known that Ra-gya monastery had a printing establishment; it was not feasible then to obtain a set of the books, although this edition is unknown in the western world. In addition to these large buildings there are hundreds of small mud houses, the abodes of the 800-900 lamas or rather monks, who dwell here, plus the La-brang or Palace of its highest incarnation.

It boasts of one high incarnation who at our visit was a man of 20 summers; he was the incarnation, strange as it may sound, of the mother of Tsong-kha-pa the founder of the Yellow Sect who was born at Tsong-kha = the Onion Bank, where the famous monastery of sKu-hbum (Kumbum) now stands, some ten miles from Hsi-ning.

His incarnation is known as bLa-ma Shing bzah Paṇḍi-ta after the mother of Tsong-kha-pa Shing-bzah a-chhos. His name as transcribed in Chinese reads Hsiang-tsa la-ma 香咂喇嘛. Under him are two high incarnations and eighteen minor ones, incarnations of lamas who in past existences led saintly lives, but were no outstanding personalities as Tsong-kha-pa. The latter had declared on his death that he would not be reincarnated, and hence naturally the lamas never looked for his reappearance. Instead they found the incarnation of his father, who, during our stay at Kumbum, was absent, visiting Pei-p'ing (Peking) while his mother's incarnation was found in the boy of Ra-gya. The latter's dwelling is the highest located, back of the large printing establishment.

Although there is not a single Chinese living at Ra-gya and none would have dared across the grasslands for fear of the Go-logs, they established «on paper» a magistracy and called it T'ung-te 同德 or United Virtue. The only non-Tibetan people who come to Ra-gya are Moslem traders who live like Tibetans, dress like them, travel like them, and unlike the Chinese, need no rice for their subsistence.

The only reason I suppose to establish a magistracy (on paper) and the giving of a Chinese name to the place was for the sake of face. And inspite of of the red regime which is most anxious to establish its authority to the farthest ends of their realm, and what is not their realm, they will find it difficult to control the lawless tribes of the grasslands, for no Chinese be he red or white, can or will, care to endure the hardships of a nomad's existence. A Chinese can only live by agriculture, and nothing can be grown at the inhospitable heights on which the nomads live, so for Chinese to settle in the grasslands west or south of Ra-gya, is to be a discounted possibility.

After one day's rest at Ra-gya where Hsiang-tsa had assigned us commodious quarters southeast of the lamasery, the property of his uncle an incarnation who was then absent in the Go-log country, we called on Hsiang-tsa or Shing-bzah Rin-po-chhe as he was respectfully called by the Tibetans, to deliver to him a letter of introduction from the great incarnation of La-brang Monastery hJam-dbyangs-bzhad-pa (Jam-byang zha-pa or the laughing Jam-byang) whose real name was Ngag-dbang Ye-shes rgya-mtsho (Nga-wang Ye-she gya-tsho), then a boy of ten summers; both have however passed away since, like all mortals.

We presented Shing-za Rin-po-chhe with a white silk scarf, the Tibetan calling card, a watch, ten squares of gold brocade and a twenty dollar U.S. goldpiece, and suitable presents to his steward and other ranking lamas. We told him that we had come to explore the Am-nye Ma-chhen range and that we hoped he would help us attain our project, and to forward three letters we had been given by the La-brang incarnation to the three ruling Go-log Chiefs. They were to grant us protection especially the most powerful chief of all the Go-logs, the Ri-mang tribal chief, Ri-mang sprul-sku ldan-brag (Ri-mang trül-ku den-drag) who was also considered an incarnation or Trül-ku, and the chiefs of the two less powerful tribes the Khang-rgan (Khang-gan) and Khang-gsar (Khang-sar) respectively. In reply he, or rather his steward, made the following speech: Ever since 1921 when general Ma Chi-fu of Hsi-ning had attacked the Go-log, after having requested Ra-gya gom-pa to forward Ma's letters to all the Go-log chiefs demanding their submission, the Go-logs declared themselves enemies of Ra-gya for forwarding the letters, and were at dagger's point with the monastery. They had returned the letters and bluntly refused to submit. He stated that he personally was only on friendly terms with the chief of the Khang-gen tribe but that the Ri-mang chief could not be trusted even if he agreed to give protection. That there were three roads or trails around the Am-nye Ma-chhen, one fairly good one with few streams to cross, but that that road was full of robbers, and that only shortly before our arrival an incarnation from his lamasery had been robbed there of forty horses and four hundred sheep, in fact of all his belongings. The other roads were difficult, as one had to ford a very swift stream, a dangerous undertaking. That the best way would be to make a quick dash to the Am-nye Ma-chhen on horseback without any pack-animals, and that very soon, before the Go-logs became aware or heard of our presence at Ra-gya. To this I demurred as our object was not to make a quick trip, but to collect plants, take photos and explore the country more leisurely. He then agreed that the best way was to send the letters with suitable presents to the Go-log chiefs, and that he personally would write individual letters to them, and to await their replies. Furthermore, that in case the Go-log chiefs should fail to reply (which in fact they did, except one), we should go to the Ong-thag tribe (it was this tribe which attacked Roborowski) who lived northeast of the Am-nye Ma-chhen, and to whom he belonged, his elder brother being chief of the tribe. That their encampment was seven days journey by yak from Ra-gya, and that he would request his brother to have his men escort us around the Am-nye Ma-chhen, for, as he remarked, the Ong-thag tribe had never been defeated by the fierce Go-log, and were not afraid of them. He also told us that there was more forest in the gorges of the Yellow River, and in the estuaries of its tributaries than at the Am-nye Ma-chhen proper. We thereupon thanked him and returned to our quarters.

As there remained nothing else to do but to await the replies of the Go-logs, which could not be expected in less than twenty days, we made plans to explore the region around Ra-gya, the gorges of the Yellow River to the north of Ra-gya, and the brGyud-par Range (Gyü-par or Jü-par) in the north, all regions previously unexplored.

The Mountains Around Ra-gya
The mountains enclosing, or forming the Yellow River Valley at Ra-gya are all loess and mostly grass-covered, beneath the loess the strata is either red sandstone or sandstone gravel conglomerate with the lower third or less, shale and schist and some quartz. The mountains are bare, except to the south of Ra-gya, facing north, where spruce forests cover the steeper slopes, with junipers on the opposite side of the river near Ra-gya proper, facing south. The elevation of the mountains south of Ra-gya reach a height of over 13,500 feet and in May were still snow-covered.

The spruce *Picea asperata* Mast., the only conifer outside junipers, does here not reach the height as in the deep valleys further north, but is still a formidable tree of 50-60 feet with trunks of two feet in diameter indicating an age of 250 years or so.

The undergrowth is mainly thick moss (Mnium) in which little can grow except the white-flowered *Rubus idaeus* L. var. *strigosus* Max., which loves shade. On the outskirts of these forests we find the yellow-flowered *Thermopsis lanceolata* R. Br. a widely distributed plant from Siberia and Altai to Northeast Tibet. It takes here the place of Piptanthus which grows in similar situations in southwestern China as on the Li-chiang snowrange; willows are most common and while they are scattered all over the hillsides below the spruce forest, they are happier near the streambed and brooks of the valleys. *Salix myrtillacea* Anders., branching from the base grows to a height of 10 feet, and is associated with the new species *Salix Rockii* Görz of equal height; this latter species is also found on exposed rocky slopes among *Juniperus tibetica* Kom., and *Salix rehderiana* var. *brevisericea* Schneid., 6-8 feet in height. In the same locality occur two species of Ribes, *Ribes Meyeri* Max., and *R. Vilmorini* Jancz., the former also known from the Altai mountains and north China; a third very common species but frequenting open rocky situations is the Kan-su gooseberry *Ribes stenocarpum* Max., with fairly large but sour berries, first known from western Kan-su province, now reckoned as belonging to the Koko Nor or Ch'ing-hai. Hybrid Salix occur such as x *S. taoensis* Görz, *Salix juparica* Görz x *S. sibirica* Pall., the new variety *Salix oritrepha* var. *tibetica* Görz, and *Salix paraplesia* Schneid. Among rocks to both sides of the Yellow River grow the two inch tall *Ephedra monosperma* C. A. Mey, at an elevation of 11,000 feet, and the lovely *Androsace Mariae* Kan. var. *tibetica* (Max.) Hand-Mzt. with pinkish-white flowers, resembling *A. spinulifera* from Yün-nan. Another new hybrid *Salix myrtillacea x Rockii* Görz, frequents northern exposures with spruces, but usually along streambeds.

Of honeysuckles which prefer to grow alone on the upper margins of grassy spurs, the finest is *Lonicera syringantha* Max., with rose red, fragrant flowers, forming beautiful globose bushes, with *Lonicera microphylla* Willd., first known from eastern Siberia with yellow flowers and glaucous foliage; like its congener it forms lovely clean round bushes.

Among boulders in exposed places flourish *Caragana Maximowicziana* Kom., *Berberis Boschanii* Schn., also *Spiraea alpina* Pall., a shrub 3 feet tall with yellow flowers, and on the grassy slopes and bluffs the blue to purple flowered *Iris tenuifolia* Pall. holds sway in company with the prostrate, white-flowered scrophuraliaceous *Lagotis brachystachys* Max.

Associated with the spruces occurs the handsome birch *Betula japonica* Sieb. var. *szechuanica* Schn., which here has a pinkish to flesh-colored bark while further south in the Min Shan its bark is bronce to copper-colored. *Sorbus thianschanica* Rupr., with white flowers and brilliant red fruits loves to be near spruces but never grows in their shade, more often it is confined to bluffs overlooking lower shrubs. Preferring plenty of light and higher altitude is *Rhododendron capitatum* Max., a shrub 3 feet high with bluish-purple flowers which grows in great masses above the spruce forests, in loess, often completely filling shallow grassy valleys. This plant the Tibetans call su-ru; enormous masses of it are cut and placed several feet thick on the top of the brick or stone walls of chanting halls in the larger lamaseries, immediately under the roofs of the buildings. The stems, of uniform thickness, are piled up to a depth of three feet or more, the ends flush with the wall are pressed tight and are then cut even, forming an ornamental panel into which are stuck the usual brass or bronce-guilt Buddhistic emblems. An Iris with yellow flowers grows among rocks, and in the actual spruce forest we find *Iris Potanini* Max., while *Primula flava* Max. loves open, exposed, loessy banks, and *Viola mongolica* Franch., (var. *floribus carneis*) adheres to grassy slopes as does *Hedysarum multijugum* Max. with purple flowers.

To rock walls and cliffs are partial *Paraquilegia anemonioides* (Willd.) Ulbr., with delicate mauve-lavender flowers and *Incarvillea compacta* Max., with rich red flowers and white stripes in its throat; it does however also occur on rocky or grassy bluffs with *Valeriana tangutica* Batal., with purplish fragrant flowers. Confined to the alpine meadows are the new variety *Trollius pumilus* Don. var. *alpinus* Ulbr. var. nov., with rich orange-yellow blossoms, *Orchis salina* Turcz., with deep purplish-red flowers, *Taraxacum mongolicum* Hand.-Mzt., *Iris gracilis* Max., displaying purplish lavender flowers with deep orange yellow stamens, *Aster heterochaeta* Bth., *Polygonatum bulbosum* Lev., with carmine flowers, and finally the stately *Rheum palmatum* L. forma *floribus rubris*, flounting its brilliant red flowers and growing to a height of 8 feet. Forming cushions over rocks and large boulders is *Sedum quadrifolium* Pall., with deep blackish-red flowers and yellow stamens. The deep red to purple Tibetan lady slipper *Cypripedium tibeticum* King, like in Yün-nan, is confined to open meadows as is the yellow flowered *Pedicularis versicolor* Wahlenb., at elevations of 11,000 feet, associated with the buttercup *Ranunculus pulchellus* C. A. Mey., *Anemone rupestris* Wall., with purple flowers, and the yellow *Meconopsis integrifolia* Franch., which often covers whole hillsides.

In the very wet alpine meadows we find the pink flowered, yellow-eyed *Primula gemmifera* Batal., which extends south to Hsi-k'ang and the deep pink-flowered *Primula sibirica* Jacq., a well known species, which extends from arctic Europe to arctic and central Asia, Alaska, and to the western Himalayas.

East of Ra-gya in the Lungmar or Red valley [plate **55-56**] so named on account of its red clayey soil, we collected *Potentilla fruticosa* L. var. *parvifolia* Wolf, a common plant on clayey slopes, with bright yellow flowers, first described from Sungaria, *Lonicera hispida* Pall., originally known from the Altai mountains, 3-4 feet high with pretty yellow flowers, *Sibiraea angustifolia* (Rehd.) Hao, 4-5 feet tall associated with willows, and *Rhododendron capitatum.* On the grassy slopes of Lung-mar valley grew *Stellera chamaejasme* L., but very different in its floral colors from the Yün-nan plant which has deep yellow and brownish purple flowers, while the northern plant has white flowers, pinkish purple in bud, and on the undersurface of the petals; these two color variations are constant, and never have we observed a yellow flowered one in the north nor a white and pink flowered one in the south, but pure white ones do occur in the north, but very rarely. *Viola biflora* L., its yellow flowers reminding of *Viola Delavayi* Franch., of the south, *Anemone rupestris* Wall., are also confined to the grassy slopes at 10,000 feet elevation and higher. Among rocks grow *Cotoneaster multiflorus* Bunge, a shrub 4-5 feet, first described from the Altai mountains, and *Caragana jubata* Poiret of variable habit, but with lovely pink flowers recalling sweet peas. On scree or talus slopes occur two species of Corydalis, as *Corydalis stricta* Steph., with reddish-purple flowers and a yellow-flowered unidentified one (no. 14018).

As spring advanced more and more herbaceous plants came into flower on the mountains around Ra-gya, and towards the end of May we collected on the grassy slopes at 10,000 feet, *Gentiana riparia* Karel and Kir., a pale-flowered species, *Viola bulbosa* Max., its flowers cream-colored and lower lip striped purplish, *Cardamine macrophylla* Willd., with lavender-pink, very fragrant flowers reminding of lilac, but much more intense *Microula sikkimensis* (Clarke) Hemsley, *Fragaria elatior* Chrh., and *Rheum pumilum* Max., while *Rosa bella* Rehd. & Wils., with red flowers, loved the shade of the spruce. Of ferns among rocks thrived at 10,000 feet *Polystichum molliculum* Christ., and *Polypodium clathratum* Clarke. Spring merged into the short summer with its daily downpours and electrical displays of great intensity, and although the thermometer did not rise much above freezing during the night, often showed seven degrees of frost at the 11,000 feet level where ice covered the streambed, and willow bushes encased in more than one foot of ice, flowered unconcernedly. From day to day there appeared new plants which had not flowered previously, such as *Fritillaria Roylei* Hook., a herbaceous plant with dull greyish blue bells, *Anemone rupestris* Wall., its flowers white but with petals bluish beneath, the composite *Scorzonera austriaca* Willd., first described from Europe with single flower-heads of a yellow color, and a species of Polygonatum (no 14106) with single greenish-yellow flowers; all but the last one which grew in spruce forest, occurred on the alpine meadows at an elevation of 12,000 feet. Some of these plants had flowered a little earlier at the 11,000 feet level, but as the snow melted the sun awakened them to bloom and hurry to produce seed to perpetuate themselves for soon they would again be covered with snow.

Open clearings were taken up by the grass *Torresia odorata* (L.) Hitch., among which grew various willows, while in ravines among rocks grew the blue-flowered labiate *Nepeta coerulescens* Max. f. *major*, the purple-flowered *Pedicularis kansuensis* Max., *Carum carvi* L., an umbellifer first known from Europe, the caryophyllaceous *Melandrium glandulosum* (Max.) Williams, and on conglomerate banks the pink flowered *Valeriana tangutica* Batal., endemic to this part of the world. Amongst rocks in general thrived the legume of prostrate habit *Gueldenstaedtia diversifolia* Max., which, with its dark purple flowers is endemic in this region. Covering Lonicera bushes among rocks on the banks of the Yellow River, flowered *Clematis tangutica* Korsh. var. *obtusiuscula* Rehd. & Wils., with the composites *Tanacetum falcatolobatum* H. Krash., a shrub 2 feet in height, and *Aster poliothamnus* Diels, both then newly described, the latter a handsome and distinct species with numerous lavender-colored flowerheads. Near them, the white-flowered *Anemone rivularis* Ham., hugged the gravel along the river. On the right alpine meadows at elevations between 12,000 and 13,000 feet thrived the rich purple flowered legume *Hedysarum obscurum* L., the lavender flowered *Phlomis rotata* Benth., *Rheum pumilum* Max. endemic here, and among rock outcroppings grew *Sedum Kirilowii* Reg., previously known only from the Celestial Mountains or T'ien Shan 天山; other plants found in the moist or wet high alpine meadows were *Corydalis dasyptera* Max., endemic here, also the new *Astragalus Peterae* Tsai et Yü, *Pedicularis cranolopha* Max., with yellow flowers and its congener *Pedicularis szechuanica* Max., proper with white flowers while the variety boasts purplish pink flowers. The labiate *Marrubium incisum* Benth., with pinkish flowers, first described from Siberia and later also found in North China, *Dracocephalum heterophyllum* Bth., flowers white, associated with a pure white *Stellera chamaejasme* L. fl. *alba*, of which only one plant was observed. Still higher at 13,500 feet grew *Ranunculus pulchellus* C. A. Mey. var. *sericeus* Hook. f. & Thoms., with *Anemone rupestris* Wall., which was at home at this height, as well as at the 10,000 feet level. Others belonging to the 12,000 feet level were *Astragalus Handelii* Tsai et Yü, but confined to wet meadows, the umbellifer *Pleurospermum linearilobum* W. W. Smith, *Ligularia virgaurea* (Max.) Mattf., first known from eastern Mongolia, then the deep yellow flowered *Corydalis stricta* Steph., confined to the rocky slopes, as was the white flowered *Morina chinensis* Bat. A thousand feet higher, in the alpine meadows, thrived the white flowered ground orchid *Aceratorchis tschiliensis* Schltr., first described from northeastern China. Several species of Allium, but not yet identified, grew on grassy slopes overlooking the Yellow River at 11,000 feet. Here also flourished *Brassica juncea* (L.) Czern. & Coss., an escape from cultivation, *Potentilla multicaulis* Bge., *Stachys baicalensis* Fischer, with purple flowers first known from Siberia, *Vicia* aff. *tibetica* Prain, *Pedicularis cheilanthifolia* Schrenk var. *isochila* Max., with yellow flowers, and finally the blue-flowered boraginaceous *Eritrichium strictum* (Dcne.) A. DC.

In the Lung-mar valley there appeared by end of June besides those already enumerated, the bluish-lavender flowered *Polygala sibirica* L., the purple flowered geraniaceous plant *Erodium stephanianum* Willd., *Heracleum millefolium* Diels, with white flowers, first found by W. Filchner in Tibet, and the Edelweiss *Leontopodium Dedekensi* (Bur. & Fr.) Bod. Near the mouth of the Lung-mar valley on sandbanks of the Yellow River occured the green flowered orchid *Habenaria spiranthiformis* Ames & Schltr., together with *Geranium Pylzowianum* Max., *Hypericum Przewalskii* Max.; in grassy places *Senecio thianschanicus* Reg. & Schmalt., *Pedicularis semitorta* Max., and *Pedicularis ingens* Max., found a foothold; most of the above are peculiar to the highlands of Northeast Tibet. Of as yet undetermined plants we collected a *Gentianella* (no. 14218) and an *Astragalus* (no 14219).

Common on the sandy banks of the Yellow River everywhere in the valley is the ordinary European thistle *Cirsium arvense* L., also *Oxytropis imbricata* Kom., and on the grassy slopes higher up *Oxytropis melanocalyx* Bge.

It will be seen that although the ligneous flora is very poor, the herbaceous one is well developed, and those that are endemic predominate. Gentians are rare in this region for we found only *Gentiana riparia* K. & K. apparently the earliest flowering species while all others flower in late summer or autumn.

We did not remain in the Yellow River valley at Ra-gya for the unfolding of the autumnal flora and therefore no Gentians, Aconites and Delphiniums are here recorded, but these were collected on journeys north to and from the Gyü-par (Jü-par) Range, and in the grasslands on the journey back to La-brang. Although certain species of the above mentioned genera may be found only in the Yellow River valley near Ra-gya and not elsewhere, yet those collected on the mountains north of Ra-gya, and the intervening grasslands, occur in all probability also on the mountains about Ra-gya. We have here a cross section of flora peculiar to the Yellow River gorges and that of its tributaries. As we go farther north the flora becomes poorer for the area north of the Gyü-par (Jü-par) Range and the Yellow River proper is adjoined by waterless desert areas, swept by winds.

The Gorges of the Yellow River North of Ra-gya and South of the (Gyü-par) Jü-par Range

On all modern maps, the great bend of the Yellow River from Sog-tshang dgon-pa the Chinese So-tsung-kung-pa north and west, as well as its course past the Am-nye Ma-chhen to where it cuts through the Gyü-par Range (brGyud-par) appears still dotted, indicating that it is still unexplored. Although we explored the gorges of the Yellow River from gTsang-sgar to north of the Gyü-par (Jü-par) range as far back as 1926, this is the first accurate account of the region and its plant covering. Only a popular article about the region had been published in the *National Geographic Magazine* of Washington D.C., entitled: Seeking the Mountain of Mystery.[38] For many years I had been planning to publish a scientific account of our explorations from copious notes, maps, etc., made at the time, but this is the first opportunity that has presented itself, my previous time having been occupied with other work such as the translations of Na-khi tribal literature and a historic and geographic account of their territory.[39] A historic-geographic account of the region I hope to publish in the not too distant future with translations of Chinese and Tibetan texts pertaining to the history etc., of the area in question. It remains here to give a geographic and botanical account of the hitherto unexplored region, which had not been visited either before or after our visits from 1925-1927.

---

[38] *National Geographic Magazine* 57.1930, 131-185, 54 pl., 1 map.

[39] *The ancient Na-khi kingdom of South-west China*, published by the Harvard-Yenching Institute, Memoir Series vols. VII & VIII, 1947. 2 vols., 257 plates, 4 maps, and *the Na-khi Nâga cult and related ceremonies* in Rome Oriental series, 2 vols, 60 plates, 2 in color, 823 pp. text, 1952. [J.F.R.]

It was expected that on account of the considerable drop in the Yellow River between its source and where it reaches Kuei-te hsien there would exist waterfalls, especially where it passes through the deep gorges east of the Am-nye Ma-chhen, but no such waterfalls exist anywhere only terrifically swift rapids, and cataracts follow each other whose velocity we could not ascertain. All the Tibetans we interrogated about the possibility of following the Yellow River down in its gorges were unanimous in their statements that it is absolutely unfeasable, especially with a yak caravan, even to follow on the upper edge of the gorge on account of the many tributaries, all of which flow in narrow valleys intersecting the loess plateau here, but one could not follow on the edge of the gorge on account of the steepness of the ravines. The only way possible was by crossing valleys higher up and then pass along the intersecting spurs and observe and photograph the river from some prominent point.

In order to ascertain the truth of these statements we thought at first to make a preliminary investigation of the gorge by arranging a short trip down the Yellow River valley on the grass-covered edge of the gorge. We were however soon convinced that to follow the gorge down along the edge on the top, was actually impossible especially with a yak caravan.

The right wall of the Yellow River valley is less high than the left side which becomes higher and higher the further west one procedes till the mountains finally merge into the Am-nye Ma-chhen range. The sides of the valleys are precipitous on both flanks in many places, especially on the west side where they are forested in parts with junipers facing south or southwest, and with spruces facing north or northeast. The entire valley is slate, schist and shale, with superimposed loess of tremendous thickness, the demarcation line is very prominently visibile in the photographs. Some parts of the gorge especially nearer Ra-gya, show broad horizontal bands of sandstone conglomerate under the loess, and schale, schist and quartz below the conglomerate.

From Ra-gya the Yellow River flows in a northwesterly direction making many short turns around projecting spurs, till the great tributary the Tshab Chhu, which descends from the center of the Am-nye Ma-chhen east, is reached. There the Yellow River turns north as far as Ta-ho-pa 大河垻 where it described a shallow arc, proceeds north-northeast, cutting through the western end of the Gyü-par (Jü-par) Range, and then flows in a northeasterly direction towards Kuei-te 貴德.

On May 27th we started with a small yak caravan of ten animals and several horses on this preliminary excursion. All the left valley walls were partly covered with *Picea asperata* Mast., while those on the right were partly clothed with *Juniperus Przewalskii* Kom. I climbed to a bluff whence I took photos up and down stream, the river flowing here from south to northwest, the latter being the general direction (See Plate **57-58**). The bluff or promontory was at the mouth of Hao-ba (Hao-wa) valley at an elevation of 10,902 feet, while the trail led at a height of 11,300 feet. From this spur we descended into Hao-wa valley or rather ravine full of *Juniperus Przewalskii* Kom., and many willow bushes which grow along the streambed as *Salix sibirica* Pall., *Salix Rockii* Görz, *Salix rehderiana* var. *brevisericea* Schn., *Salix wilhelmsiana* M. B., *Salix taoensis* Görz, all about 4-5 feet in height. These willows occured also in the Nya-rug Nang or Nya-rug valley which the Tibetans designate the willow valley, which precedes the Hao-wa valley (east). Associated with them are *Caragana jubata* Poir., *Rosa bella*

Rehd. & Wils., *Salix myrtillacea* And., and *Ribes Meyeri* Max.; on the grassy slopes grew a fleshy stemmed Euphorbia (no 13941), *Iris Potanini* Max., the latter prefering rocky situations. Above, on the alpine slopes, grew *Salix oritrepha* Schn. var. *tibetica* Görz with masses of *Rhododendron capitatum* Maxim., *Sedum quadrifidum* Pall. var. *fastigiatum* Fröderström (14042) (*venustum* Praeg.), formed large cushions on boulders and on the cliffs at 11,000 feet, its dark red flowers contrasting with the glaucous foliage.

With *Juniperus Przewalskii* Kom., there also grew *Juniperus tibetica* Kom., a tree 20 feet in height but only on the slopes of the intervening spur between Nya-rug Nang and Hao-wa valleys. Here in these juniper groves we shot the white-winged grosbeak *Perissospiza carnipes carnipes* (Hodgs.) with large thick, triangular beaks, they were feeding on the juniper forests. Partridges, *Alectoris graeca magna*, disported themselves in the willow bushes and on the rocky hillsides.

I descended the Hao-wa valley almost to the Yellow River, but half a mile from the mouth the valley becomes so narrow and steep that further progress was impossible. In the bushes of willows, Berberis, Caragana and Ribes we shot a pink finch, and on the trunks *of Juniperus Przewalskii*, which filled the valley floor, a woodpecker.

On May 29th we returned to Nya-rug Nang and from there climbed at a terrific angle to a pass 11,850 feet elevation (see Plate **59**); whence in the distance north, there was visible a cairn or Obo on a high promontory. From there, so our nomad yak driver told us, one could see the Am-nye Ma-chhen Range to great advantage. The Obo is directly north-northwest bearing 316° and is known as Mo-khur-ri-ser-ma. On the long spur immediately below us grew *Juniperus tibetica* Kom., scattered over the hillside, while on the right bank of the Yellow River we noticed groves of Junipers mixed with Betula (birches). In the distance there was visible a bare mountain extending from northeast to southwest called Ha-rlung (Ha-lung), at the foot of which the Yellow River turns west; our objective, the long and forested valley called sTag-so nang was some distance south east of Mount Ha-lung.

We decided to descend into Hao-wa valley and there pitch our camp at an elevation of 10,980 feet. It was a peaceful night, hidden away as we were in the fragrant juniper grove. A brilliant and cool morning found us again on our way; we sent the yak caravan over an easier trail while we climbed the steep forested slopes of the canyon over a zig-zag trail to the top. To ride was of course impossible. On reaching a pass 11,100 feet elevation we obtained a magnificent view over the gorge of the Yellow River and its tremendous rapids. The right valley slopes are absolutely bare, but the left one is forested with spruces, the trees extending to near to the top of the mountain (see Plate **60-62**). Below the pass is a bluff called rTa-ra-lung as is the valley near by. The trail was a difficult one as it led over knife-edge ridges from one canyon into another; descending from Ta-ra-lung we reached a very narrow canyon called Sa-khu-tu, its stream flowing at 10,420 feet elevation, the bushes and trees were the same as in the Hao-wa valley, with the exception that on the gravelly slopes we found two species of Corydalis, both yellow-flowered the *Corydalis adunca* Max., but the second species is as yet unidentified (no. 14047). Among the rocks near the streambed we shot a wren, *Nannus troglodytes idius* (Richm.), whose habit is to fly low along the water's edge.

From Sa-khu-tu the trail leads up a bluff elevation 10,910 feet and about 1,200 feet above the Yellow River. The region on the right (east) side of the Yellow River is one maze of narrow canyons while on the west side only two shallow, gentle sloping valleys debouch into the Yellow River, the Shog-chhung and Shog-chhen, these valleys are beyond the Sa-khu-tu ravine, while diagonally opposite is the long Shag-lung (valley) which has its source in a mountain or pass called the hBrug-dgu nye-ra (Drug-gu nye-ra), over which a trail leads to the Am-nye Ma-chhen.

Tsha-rgan-hor-sgo (Tsha-gen-hor-go) is another deep canyon separated from Sa-khu-tu by a ridge and pass 11,200 feet elevation whose crumpled, folded walls of schist and shale are nearly one thousand feet in height, its floor and slopes covered with *Juniperus Przewalskii* Kom., with here and there the common spruce. Along the streambed were the usual shrubs of *Caragana Maximovicziana* Kom., and *C. jubata* Poir., with *Potentilla fruticosa*, *Salix myrtillacea*, *Sibiraea angustifolia* (Rehd.) Hao, and others. Here in this valley we met with some mighty monarchs of junipers (*J. Przewalskii* Kom.), hundreds of years old and loaded with the dark purplish, glaucous rund dupes, the foliage of a pleasing green, very different from the dull brownish green *Juniperus zaidamensis* Kom., of the T'o-lai Range north of the Koko Nor. The trail crosses the streambed at 10,800 feet and leads up the valley among willows, etc., for a considerable distance, past some lonely nomad camps belonging to the sGar-rtse (Gar-tse) tribe or clan of whom our yak driver was a member. He knew the name of every valley, spur and bluff and as he was an old friend of W. E. Simpson, who also accompanied me, we could vouch for the correctness of all the names. Beyond the encampment the trail turns up a narrow rocky ravine, the rocks like those of the Yellow River valley being slate, schist and some white quartz, the ground was littered with the flat slabs of these rocks. In its lower part Tsha-gen-hor-go, like all the rest of the tributaries, is an appallingly steep canyon, while higher up it becomes a shallow valley, merging into the grassy plateau, but still showing the same rock outcroppings. The upper parts of these valleys are the camping grounds of Tibetan nomads, and as their mud-stoves are only capable of burning sheep manure, the Tibetans do not find it necessary to cut timber for fuel, thus the forests are intact, except that they suffer from the grazing of the sheep and yak. As the spruce forests have as sole ground covering moss (Mnium sp?) of over a foot in thickness, the sheep do not penetrate into these but confine their grazing to the bare hillsides or under the junipers which seem to prefer a dry or well-drained open slope. It is only where grass encroaches on the moss covering in the spruce forests, that both moss and spruce begin to die.

Short narrow valleys descend at right angles to the tributaries from the spurs which separate them, and one in Tsha-gen-hor-go leads to the pass called Mo-khur Nye-ra, 12,800 feet above sea-level. Mo-khur Nye-ra is on the ridge on which the Obo Mo-khur-ri-ser-ma is situated.

On the grassy slopes of the pass bloomed the dark mauve, to blackish-red *Anemone imbricata* Max., and the yellow flowered *Trollius pumilus* var. *alpinus* Ulbr., a new variety, and a little higher at 13,000 feet the pink flowered and very fragrant crucifer *Parrya villosa* Max., reminding on the scent of lilac.

On the Re-lung Nye-ra at 12,800 grew various shrubs as the prostrate *Lonicera thibetica* Bur. & Franch., its pink flowers exhaling a delicious fragrance, the 3 feet tall

Salix oritrepha Schn. var. *tibetica* Goerz, a new variety, *Potentilla fruticosa* L., and *Rhododendron capitatum* Max. From Mo-khur Nye-ra we had our first part-view of the Am-nye Ma-chhen Range, but only the northern massive dome was visible. In order to obtain a better view we climbed a higher peak to the right, northeast of the pass, to an elevation of 13,220 feet. I must confess that my first view of Am-nye Ma-chhen was rather diappointing, and this same disappointment was expressed by my Na-khi companions from Northwest Yün-nan, who dwell among snow mountains and who with me explored the snow ranges composing the so-called «hump» between India and China. We thought the Am-nye Ma-chhen less high than the Mekong-Salwin divide, but that may have been due to the many ranges, mostly flat topped, which intervened, while from the peaks of the Kha-wa-dkar-po there is a sheer drop of about 12,000 feet to the Mekong, which cause them to appear much higher. At any rate the Am-nye Ma-chhen is not nearly as sculptured as the ranges which form the Yün-nan-Tibet divide in the southwest.

From the pass we could see two outstanding features of the range, each at the opposite or extreme end of the range, the southern one a blunt pyramid, and the northern one a vast dome, the highest part of the range. Between these are lesser peaks, and the highest of these lesser is the Am-nye Ma-chhen after which the range derives its name. The highest peak could never bear the name Am-nye Ma-chhen for the latter is only a local sa-bdag or earth lord, while the higher pyramidal peak to the south of it bears the name of Spyan-ras-gzigs pronounced Chen-re-zig or the God of Mercy the patron deity of Lha-sa of which the Dalai Lama is considered the incarnation. Ma-chhen spom-ra is however the mountain god of the mGo-log tribes, who live in fear and dread of him who, as his classic reveals, controls the lightning, hail, etc.

The compass bearings of the range taken on the peak back of Mo-khur Nye-ra gave the following readings: extreme south promontory 268.5°, the pyramid Chen-re-zig 270°, intermediary peak 271°, Ma-chhen spom-ra (or Am-nye Ma-chhen) 272°, the summit of the dome 275°, and extreme northern end visible and all deeply in eternal snow, 277.5° For further details of this range see the chapter: The Am-nye Ma-chhen Range.

From Mo-khur Nye-ra the trail leads over a slope densely covered with the shrubby *Rhododendron capitatum* Maxim. which also extends into the next valley with a swampy floor called the Shangs-shub. This short valley extends northwest into the Ar-tsa Nang, a tributary of the Yellow River, the only one east of the great valley over 60 miles long, called sTag-so Nang (Tag-so Valley). Between Ar-tsa and Tag-so is a high spur elevation 11,700 feet forested mainly with *Juniperus Przewalskii* Kom. (see Plate **63**). On the left (southwest) valley slopes of the Yellow River gorge are extensive forests of *Picea asperata* Mast.

Tag-so Nang and Its Flora

After days of traveling over the bare grassy uplands this valley proved a botanist's delight. We pitched camp at the only available flat space near the streambed at an elevation of 10,146 feet (see Plate **64**). The tree which formed pure stands on the north-facing slopes was again the common spruces, it reached heights of 150 feet with trunks several feet in diameter [Plate **65**]; it clothed the steep walls of the canyon to the exclusion of everything else. [Plates **66-67**]

Birches *Betula japonica* Sieb. var. *szechuanica* Schneid., lovely to behold in their rich green spring foliage grew on the outskirts of these forests. The ground cover was thick moss of a species belonging to the genus Mnium, to which the leguminous *Thermopsis alpina* Ledeb., an herb, about one foot high, with yellow flowers was partial. The clear crystal stream was bordered by *Berberis Boschanii* Schneid., and *Ribes Meyeri* Max., with red, drooping inflorescences. Encroaching the mossy forests were *Lonicera hispida* Pall., and *Sibiraea angustifolia* (Rehd.) Hao (see Plate **68**), but confined to the streambed were *Caragana brevifolia* Kom., and the new shrubby willows *Salix pseudo-wallichiana* Goerz, *Salix juparica* Goerz, and *Lonicera syringantha* Max.; *Lonicera microphylla* Willd., formed compact shrubs 4-5 feet tall, perfectly globose, on rocky exposed bluffs as did *Cotoneaster acutifolius* Turcz., with pink flowers and reaching 4-5 feet in height. *Sorbus tianschanica* Rupr., overshadowed the willows, hugged the cliffs, or grew on top of fallen boulders in the streambed, its spreading crown bearing umbels of white flowers. The common *Ribes stenocarpum* Max., the Kan-su wild gooseberry, in company with *Hippophaë rhamnoides*, *Potentilla fruticosa*, and *Myricaria dahurica* Ehrenb., constituted the ligneous flora in the valley near the stream. On the grassy exposed slopes grew the tall *Rheum palmatum* L., with brilliant red flowers, *Stellera chamaejasme* L., *Iris gracilis* Max., *Viola biflora* L., and *Polygonatum bulbosum* Lev.

Near the upper edge of the canyon among rocks, *Berberis diaphana* Max., a shrub 4 feet high with single yellow flowers was associated with junipers which grew here and there on the south-facing slopes. On bluffs overlooking the Yellow River, partial to wet gravel, flourished the handsome *Incarvillea compacta* Max., displaying brilliant red flowers with white markings in its throat; higher still with Potentilla and *Caragana jubata* Poir., here a prostrate shrub, grew innumerable yellow poppies *Meconopsis integrifolia* Franch. On the wet, often swampy alpine meadows, on the flat tops of the spurs, *Primula Purdomi* Craib, displaying reddish-purple flowers with yellow eyes, and *Primula limbata* Balf. f. et Farr., had their being.

A considerable distance up the ravine is a small grassy area where huge spruces and junipers (*Juniperus tibetica* Kom.) form dense forests. (See Plate **69**)

From a bluff elevation 10,900 feet on the northern valley wall of Tag-so canyon called Ngar-khi gzhug-ma (Ngar-khi zhug-ma or the lower Ngar-khi) a view is obtained down the Yellow River gorge showing the many cataracts, the river flowing northwest. The north-facing slopes (left bank) are forested with spruces and birches, a species of poplar grew scattered among them but could not be collected as it grew west of the river. The highest slopes above the right bank were covered with masses of *Rhododendron capitatum* Max., with aromatic small leaves, and lavender-blue flowers.

On these lonely slopes, high above the Yellow River, two nomad families had pitched their tents, such isolation and seclusion from human contact must leave its imprint on the mind, and foster suspicion towards strangers; they could not have pitched their tent in a more inaccessible spot. From this vantage point a wonderful view could also be had up Tag-so valley showing the forest demarcation line.

Below the camp the streambed narrowed, and to where it debouches into the Yellow River is but a stone trough of great depth, the water roaring through the defile washing both walls. Rocks or boulders 20 feet high blocked the streambed like steps of a giant's stairway. A waterfall prevented all progress and forced us to climb up steeply the left rock wall but without obtaining a view down into the mouth of the ravine. Through the rocky gate the surging waters of the Yellow River could be seen; the rapids were gigantic and the water was thrown many feet into the air.

The highest promontory above the canyon is Ngar-khi gong-ma (or the Upper Ngar-khi), elevation of 11,150 feet; from this vantage ground the trend of the Yellow River gorges could be observed for quite a distance. Further north the valley became more arid, and forests ceased, the rock formation was the same, numerous talus slopes extended into the Yellow River Valley on which a few birches had taken a foothold. On a rocky bluff near the summit grew the lovely *Incarvillea compacta* Max., also *Paraquilegia anemonioides* (Willd.) Ulbr., with purplish pink Corydalis, and the pathetic-looking *Primula flava* Max.

The decorative, white-flowered *Androsace Mariae* Kom. var. *tibetica* (Max.) H.-M., adhered to the rocks while the rather unlovely, dull *Primula tangutica* Duthie, one of the most graceless and homely looking species, its doleful, somber red flowers drooping as if it were ashamed of itself or aware of its homeliness, vaunted its ugliness above it.

The forest and the bushes along the stream were alive with birds, of which the following frequented the willow bushes, *Phylloscopus magnirostris*, *Phylloscopus Humei praemium* (M. & I.), *Phylloscopus Armandi* (Milne-Ed.) and *Phylloscopus proregulus proregulus* (Pallas), all willow warblers.

About eight miles from the point where the Yellow River flows north and beyond the mouth of the Tshab Chhu, the Yellow River receives a fairly large tributary, the hJang-chhung or the small Jang River; the same distance beyond the latter it acquires a still larger tributary the hJang-chhen or the Great hJang River, both have their sources west of the mDzo-mo La or the Dzo-mo Pass (= Half-Breed Yak-Cow Pass), at an elevation of 13,290 feet. In the former valley live the Yir-chhung Tibetans, and in the latter the Wam-chhog Tibetan clan. Half way between these two tributaries, the hBrong-sde nang (Drong-de nang) or the wild Yak valley empties its waters into the Yellow River from the west.

There are no other large tributaries, till we come to the long Bâ Valley (hBah), and beyond it to the short tributary called the Mu-gyang (Mu-yang) which has its source in the Tho-thug Nye-ra (pass) 13,900 feet elevation. From brGyud-par mtshar-rgan (Gyü-par tshar-gen) the highest peak of the Gyü-par Range, 14,456 feet elevation another tributary, whose name we could not learn flows into the Yellow River or rMa Chhu (Ma Chhu).

Where the Ta-ho-pa 大河壩 joins the Yellow River, the latter flows in a broad valley at the foot of steep, and much eroded loess bluffs of considerable height, only to enter the gorge it has cut through the northwestern end of the Gyü-par (Jü-par) Range.

Upper Tag-so Valley and the Grassy Plateau and Passes East of It
Looking south from an elevation of 12,300 feet a snow-covered range between 16,000 and 17,000 feet in height, extended from east to west, it is probably a part of the Am-nye Ma-chhen system, and is the border of the Go-log country. Northeast of Tag-so canyon at an elevation of 12,780 feet, on the grassy spur we met with the long-spined *Caragana jubata* Poir., here only about 2 inches high and prostrate, among which grew masses of *Meconopsis integrifolia* (Max.) Franch., also the ultramarine colored *Corydalis curviflora* Max., *Potentilla fruticosa* L., etc. The grassy spur merged into broad meadows in which thrived *Primula sibirica* Jacqu., and *Primula limbata* Balf. f. et Farr., its flowers a deep lilac-blue with dark blue throat and leaves a dull green.

Other alpines, loving swampy situations, were *Draba oreodes* Schrenk. var. *racemosa* O. E. Schulz with yellow blooms, *Pedicularis oederi* var. *heteroglossa* Prain, also yellow flowered, a Euphorbia (no 14093) as yet unidentified with dark red flowers and bracts, and *Allium monadelphum* Turcz. var. *thibeticum* Regel. One of the handsomest plants found here was the new crucifer *Cheiranthes roseus* Max., forma *caespitosus* with deliciously fragrant pale pink flowers tinged purplish, associated with the similarly fragrant *Parrya villosa* Max., an exquisite herb with pinkish-mauve flowers, all happy at an elevation of over 13,000 feet, frozen and covered with snow and ice for the greater part of the year. The pink flowered *Caragana jubata* Poir., continued to an elevation of 13,200 feet as did the species *Cheiranthes roseus* Max., proper.

From this highest point of the ridge a marvellous view can be enjoyed of the Am-nye Ma-chhen Range on clear days for the mountains are lower here on the other side of the Yellow River, and due to the presence of a broad valley, the Tshab Chhu nang which extends directly from the Am-nye Ma-chhen east into the Yellow River. A small valley debouches here into the Tag-so canyon, called Kun-bde (Kün-de), on the inner side of which, on the gravelly slopes, the beautiful *Incarvillea compacta* Max., grew in profusion.

We followed the Kün-de valley into the Tag-so and were there overtaken by one of the fiercest electrical storms we had ever experienced, it made its way from the Am-nye Ma-chhen to our side of the Yellow River. The storm lasted for half an hour, a terrific downpour accompanied by hail and fierce lightning, a continuous discharge of electricity; the thunder rolled without interruption and sounded like a terrible battle at close range, peal followed peal in a terrifying continuity. There was nothing to be done but to continue down into the valley while the lightning crashed at 13,000 feet elevation on an open ridge. Near the mouth of Kün-de valley were many nomad tents, guarded by vicious Tibetan mastiffs the size of a small lion with huge manes. I hated the sight of nomad tents for it always meant fighting off the vicious brutes who have been trained to instinctively make for a man's throat. Whips are of little avail, they sometimes jump up the rear of a horse and bite the rider in the back, which happened once to one of my men. They came from all directions and I was forced to use my 45 colt automatic as one jumped up my horse trying to tear my limbs.

The Tag-so valley is here absolutely bare, the streambed is shallow and no one would surmise that in its lower part it harboured such lovely forests. Continuing for some distance we passed other encampments and to where two lateral valleys met we pitched camp on a grassy level spot. The valley extending southeast is called Wa-ru nang, the other extending northeast, is the U-su-to, a narrow rocky ravine with a few *Juniperus tibetica* Kom., trees on its upper slopes.

The junction of these ravines is known as Wa-ru mdo and is at 11,300 elevation. Here in the ravine of U-su-to we shot several birds among them *Prunella rubeculoides Beicki* Mayr.

Returning to Ra-gya we followed up the Wa-ru valley east and then southeast, the ascent being gradual. The valley wall facing southwest is composed of much folded and tilted schist, the cliffs contain also slate and shale especially near the valley floor; on the opposite side the hills slope gently, are of loess and grass-covered. Wa-ru valley is of considerable length and absolutely bare, a lonely forlorn place. In its upper third it is joined by several lateral valleys, two on the left, and it is in the last or upper lateral valley that the trail leads to a pass. In the streambed the ice was here two and a half feet thick covering the whole width of the stream, and this in June. Here grew *Rhododendron capitatum* Max., *Potentilla fruticosa* L. the latter not in flower and several willow bushes, *Meconopsis integrifolia* Franch., grew scattered over the grassy slopes with *Caragana jubata* Poir., *Cheiranthes roseus* f. *caespitosus* and a tiny Euphorbia. The ascent to the Wa-ru la or Wa-ru pass is quite steep especially the last few hundred feet, the ground was composed of muddy gravel in which grew the lovely *Primula limbata* Balf. f. et Farr., with the large single flowered *Oxigraphis glacialis* Bge., of the Ranunculaceae, first described from Siberia, and the fleshy creeping, blue-flowered scrophulariaceous herb *Lagotis glauca* Gaertn. Wa-ru La (pass) is 13,720 feet above the sea, to the left (north) of the pass is a high scree and as it promised to give us a view of the Am-nye Ma-chhen we climbed it. We expected Saussureas to grow on the top of the scree but were disappointed; that Saussureas do grow there is certain for we found old rosettes from the previous year, but no new ones, the time being still too early. The rocky summit was 14,350 feet elevation and had it not been cloudy we would have had an unobstructed view of the Am-nye Ma-chhen; as it was we saw the lower glaciers and parts of the highest peaks above the clouds. From this vantage point the Am-nye Ma-chhen looked higher than it appeared from Mo-khur Nye-ra.

Looking south there was outlined in clearest atmosphere, a long snow-covered range which I estimated to be 17,000 feet in height judging from the immediate mountains around, all of which were over 15,000 feet yet without a vestige of snow, while those in the south were snow-covered for about a thousand feet. Here it was made evident that the snow-covered range in the south (within the knee of the Yellow River) joined the Am-nye Ma-chhen indirectly, that is it formed a shallow curve of which the Am-nye Ma-chhen peaks were the highest of the entire range, but towards the northwest, beyond the great dome, the range dwindles rapidly.

On the grassy slopes of the hill south of the Wa-ru La we found another primrose, *Primula Purdomi* Craib with flowers of a deep lilac color with a yellow eye, and coriaceous leaves, it was the only plant in flower, but leaf-rosettes indicated that others would soon join it in a burst of color. There were also many rosettes of Meconopsis, Aconites, Delphiniums, Gentiana and others, but for these it was still too early.

Directly south lay another long valley bearing the same name; the Tibetans have the custom, as already remarked, to give two opposite valleys whose heads culminate in a pass, the same name, which name is also applied to the pass between them. From this pass the trail descends steeply into the Wa-ru nang over a swampy meadow and boggy slopes to the mouth of the valley, which branches here at 12, 870 feet; the left branch descends from the Wa-ru La, and the right one from the Wo-ti la, the two forming the main Wa-ru Nang. At the junction of these two valleys we shot three species of game birds, the partridge *Perdix Hodgsonii sifanica, Tetraogallus tibetanus Przewalskii* Bianchi and *Alectoris Graeca magna* and several hares.

The brGyud-par (Gyü-par) Range

From brGyud-par to the hBah (Bâ) Valley
As we had not yet been able to make arrangements for the journey to the Am-nye Ma-chhen we decided to explore the Gyü-par, actually pronounced Jü-par Range in the upper bend of the Yellow River south of Kuei-te. We would thus also traverse previously unexplored territory between Ra-gya and the Bâ valley and beyond, the Gyü-par Range, which had been crossed on its extreme eastern end by Karl Futterer, but had never been explored or its peaks climbed. The range is not marked on some Chinese maps and on others the Tibetan name is transcribed Chu-p'a-erh Shan 朱帕尔山 in Chinese. The Yellow River cuts through its western end and flows northeast of it. The range itself extends from northwewst to southeast and then east, describing a gentle arch. A long valley called the Gyü-par Nang stretches to the northeast, that is parallel to it, this valley has its main source in the Tho-thug Nye-ra, elevation 13,900 feet; a shorter branch joins it from the northeast with its source in the sGo-mang dgung-kha (Go-mang gung-kha). At the headwaters of the Gyü-par stream the mDah-tshang (Dâ-tshang) Tibetans have their encampments.

The very much feared robber tribe called the Shab-rang has its encampment both south and north of the Bâ Valley, their territory extending to the southern foot of the Gyü-par Range. At the mouth of the Gyü-par valley, south of the Yellow River dwell the dGah-hja (Gâ-ja) clan who are under the control of the lamasery of dGah-rang (Gâ-rang) situated west of the Gyü-par valley.

The Gyü-par Range has two high peaks, both located in the western fourth of the range, the highest being brGyud-par mtshar-rgan (Gyü-par-tshar-gen), elevation 14,546 feet, the second brGyud-par sher-snying (Gyü-par sher-nying), only slightly lower.

To the northeast of the Gyü-par Range is a waterless plateau known as the Ma-la-dge Thang (Ma-la-ge thang) which extends to the Yellow River, while further east and north of the range, actually an extension of the Ma-la-ge thang, are many sand dunes called Mang-ri bye-ma (Mang-ri je-ma). What the Gyü-par valley is to the northwestern end of the range, the sGo-mang nang (Go-mang nang) is to the other or eastern half of the range; it has its source in Go-mang gun-kha, but on the eastern slopes, and extends east the whole length of the range, it probably empties into a tributary of the Yellow River.

Owing to the feuds and internecine strife existing between the various Tibetan clans it was most difficult to hire yak to take us to the Gyü-par Range; it proved in fact impossible for any clansman to go into the territory of the other, for fear of being held as hostage. It was suggested that we hire yak from clan to clan. This I refused for one might be dropped in a neighboring tribal territory and left there, or be subjected to price exactions, if not exposed to robbery. We therefore hired yak from the western side of the Yellow River and engaged three lamas of Ra-gya monastery who belonged to an encampment west of the river, to take us to the Gyü-par Range and back again to Ra-gya. Lamas are more or less immune, but not always, and none prayed more fervently than our lamas, while we were in the territory of the Sha-brang tribe of unsavory reputation. That tribe was feared by all, as its members are ferocious and no respectors of persons; they are quickly disposed to robbing, even murder if profitable.

Having made arrangements with our lamas we left Ra-gya June 19th with fifteen yak and three well armed lamas, ourselves also being armed with modern rifles, and automatic weapons.

We followed the Kuei-te trail back of Ra-gya lamasery in and out lateral ravines which extend into Dreg-yang nang (Dre-yang Valley), the latter's outlet is east of Ra-gya and like all tributaries, flows through a vertical rock gate as it approaches the Yellow River, thus forming a gorge, here of conglomerate as are the cliffs in the vicinity of Ra-gya.

The flora of Dre-yang valley is much like that found in other valleys near Ra-gya, but owing to the advent of summer, additional, later flowering species had made their appearance. Among the bushes of willows we found two species of Thalictrum one with purple (no 14126), and one with white flowers (no 14125) these have not been included in B. Boivin's recent (1945) work: *Notes on some Chinese and Korean species of Thalictrum*, but other numbers of the same collection were described by him. In the wet grass grew *Pedicularis semitorta* Max., the white flowered *Anemone rivularis* Ham., *Ajuga lupulina* Max., a labiate with cream-colored flowers and purplish tinge, and the crucifer *Draba lanceolata* Rayle var. *leiocarpa* O. E. Schulz, also found in Mongolia, Turkestan and in the Himalayas; another but new variety of the same species, var. *latifolia* O. E. Schulz occured with it as did *Pedicularis scolopax* Max., with rather larger than usual, yellow flowers, first collected by Przewalski in Kan-su. Of crucifers besides the Draba we encountered *Torularia humilis* (C. A. Mey.) O. E. Schulz, and *Malcolmia africana* (L.) R. Br., the former a white and the latter a pink flowered species, and first described from Ethiopia. The primulaceous *Glaux maritima* L., and *Corydalis straminea* Max., the latter with yellow flowers, composed the herbaceous vegetation then in flower. The commonest shrub besides willows was the 2-3 feet high *Potentilla fruticosa* L. var. *parvifolia* Wolf., which occupied the rocky banks at the foot of the cliffs.

The grass flora was well represented and consisted of *Poa attenuata* Trin., *Poa flexuosa* Wahl., *Deschampsia caespitosa* (L.) Beauv., the new variety *brachystachys* of *Elymus sibiricus* L., *Koeleria argentea* Griseb., Trisetum sp? and others.

At 13,000 feet Primulas made their appearance, as *Primula limbata* Balf. f. et Farr., and *P. Purdomii* Craib, but were here less common than on the Wa-ru La. Yellow poppies *Meconopsis integrifolia* Fr., were in all their glory, their huge golden bells nodded all over the hillside. Less prolific grew the deep lavender poppy *Meconopsis quintuplinervia* Reg., and a bronze-colored *Fritillaria Roylei* Hook., with single nodding flowers. This is the renowned *Pei-mu*, a Chinese medicine, its small white bulbs forming an article of export. It is one of the most valuable wild products fetching high prices, as much as $ 20.00 U.S. per lb. Over the hillside here and there were scattered bushes of *Rhododendron capitatum* Max. Aconites and Delphiniums were plentiful but not yet in flower, and of Pedicularis only one species occurred, and that a new one *Pedicularis calosantha* Li sp. n., with purplish pink flowers, also found in the Min Shan to the south; it was associated with the new *Astragalus Peterae* Tsai et Yü, and *Carex striata* L. ssp. *pullata* (Boott) Kük. The pink *Caragana jubata* Poir., was in full bloom, but instead of growing erect kept close to the ground.

On the Wo-ti La, elevation 14,280 feet we collected among the rocks the small cushion plant *Arenaria kansuensis* Max., also *Potentilla Saundersiana* Royle, with yellow flowers, the purplish white *Oxytropis melanocalyx* Bge., the dark reddish-black flowered *Anemone imbricata* Max., and on grassy swampy spots *Coluria longifolia* Max., with *Draba oreodes* Schrenk. var. *Tafelii* O. E. Schulz, the yellow flowered *Cremanthodium Decaisnei* C. B. Clarke, *Saxifraga melanocentra* Fr. forma *Franchetiana* Engl. & Irmsch., while *Arenaria melanandra* (Max.) Mattf. an herb with pale pink flowers was partial to scree. On the swampy alpine meadows thrived the dull blue flowered, *Astragalus skythropos* Bge., the Edelweiss *Leontopodium linearifolium* H.-M., the crassulaceous *Sedum algidum* Led., whose flowers, when open, are of a pinkish-drab color, and the pale yellow flowered *Corydalis trachycarpa* Max. The only prostrate shrub found here was *Lonicera thibetica* Bur. & Fr., confined to the alpine meadows at over 14,000 feet elevation.

Looking south-southwest, there extended the Wa-ru nang (valley), and north-northwest Wo-ti nang. The rocks on the summit of the pass, and on the surrounding hills are slate. West of the Wo-ti La are two rocky hills one 14,350 feet, and the other 14,680 feet high, from the latter a fine view of the Am-nye Ma-chhen can be had. Here on the top, among the rocks we found *Anemone imbricata* Max., *Saxifraga Przewalskii* Engl., its flowers a yellowish red, *Oxytropis melanocalyx* Bge., the lovely *Primula Woodwardii* Balf. f., with its reddish-purple flowers displaying a yellow eye, *Corydalis dasyptera* Max., the crucifer *Cochlearia scapiflora* Hk. f. et Th., and *Pedicularis oederi* var. *heteroglossa* Prain, all were confined to the highest point west, up to an elevation of 14,680 feet.

On the loose slate grew one species of Saussurea, not then in flower, and a single Aster, *A. flaccidus* Bge., with purple disk and yellow ray florets.

The summit afforded an excellent view north of the Gyü-par Range whither we were bound, a long mountain range extending west across the Yellow River; its highest part in the west I judged to be 15,000 feet in height, but in the east it dwindles to low hills.

The descent into the Wo-ti nang (valley) is much more gradual that that from the Wa-ru la. The trail leads north from the pass; the whole hillside was one grand bog, and the mountains south of the Wo-ti la were snow covered. At the first grassy spot where it was feasable to pitch camp, we stopped. Our camp was one of the highest on the journey, 12,881 feet.

To the east of the Wo-ti valley, the mountains are higher in the immediate vicinity, while to the west a black range extends in front of the Am-nye Ma-chhen which I judged to be between 16,000 and 17,000 feet high. All around our camp were yellow poppies and willow bushes about 2 feet in height, *Salix oritrepha* Schneid., the same species which occurs also in the Wa-ru valley at about the same elevation.

The Wo-ti valley led into the upper Tag-so valley whose streams mingled at an elevation of 11,790 feet; the Tag-so stream comes from east-northeast and flows here west-southwest, the head of the valley is still ten miles east. We could not learn the name of the pass where it has its source. Crossing the valley diagonally to the mDzo-mo Nang (Dzo-mo Nang) or the Half-Breed Yak-Cow Valley, which extends from north-northeast to west-southwest, we entered it, and followed it upstream; the valley floor was very boggy and difficult for our horses to negotiate. Here we found cushions of the tiny *Primula fasciculata* Balf. f. et Ward, (synonym *Primula reginella* Balf. f.) or rather embedded in cushions or tussocks of grass, the entire individual plant being only one third of an inch or less high; the flowers are a loud pink with a yellow throat; with it grew the very pubescent, yellow flowered *Astragalus tatsiensis* Bur. et Franch., and a Euphorbia (no 14241) with reddish purple bracts.

Gradually the trail leads to the pass, the Dzo-mo La, at an elevation of 13,290 feet, a much easier one than the Wo-ti La. There were very few plants out on this pass, and none that had not been found on the higher Wo-ti La, except for *Leontopodium linearifolium* H.-M. here a rather small plant.

To the left or west of the pass is a rocky bluff 13420 feet in height and this we climbed; we were rewarded with a fine view of the Gyü-par Range, the bearing for its highest peak being 340°; the Bâ plain was visible for nearly its length parallel to the base of the mountain. To the extreme northeast of the range bearing 24°, could be seen the sand dunes, and many miles of waterless desert over which a trail leads to Kuei-te. In the northwest, the large plain of rNga-thang (Nga-thang) west of the Yellow River spread between the Chhu-sngon and Ta-ho-pa Rivers. Beyond the Gyü-par Range in the distance was visible the snow-capped range separating Hsi-ning from the Yellow River; Günther Köhler[40] calls this range the Ama surgu, while Hao marks it on his primitive map as Lagi Shan, half what? and half Chinese. Köhler has on his map opposite the Bâ valley west of the Yellow River, and north of the Chhu-sngon (Chhu-ngön) which he calls Tschürnong, a mountain chain he designates as the Ugutu range, but there is the great rNga-thang or Great Nga Plain. It is probably the high range to the northwest of the plain and identical with our Shar-gang.

The descent from the Dzo-mo La is rather steep; the valley extending north from the pass is also called Dzo-mo Nang, but this valley merges with a large valley of many branches with a stream designated as the sGar-rgan Chhu (Gar-gen Chhu) which debouches into the Bâ valley near (west) of a mountain named sGam-bu sum-na (Gam-bu sum-na). Some distance down, the valley becomes quite broad, and the eastern slopes and hillsides were dotted with the black tents of the nomads of the U-hjah clan (U-jâ).

Several small valleys open out from west into the Dzo-mo Nang which is enclosed by bare grassy hills ranging from 500-600 feet in height (above the valley floor). The last one of these valleys is called dGun-khai mar-kha (Gün-khai mar-kha) up which a trail leads to the monastery of gSer-lag dgon-pa (Ser-lag gom-pa); this lamasery is situated in a small valley which joins the hJang-chhen valley a tributary of the Yellow

[40] See Walter Fuchs: Günther Köhler in memoriam, 1901-1958. *Oriens Extremus* 5.1958, 246-251, portrait. Rock seems to refer here to Köhler's *Der Hwang-Ho: eine Physiogeographie.* Gotha: Perthes 1929. 104 pp., ill., maps (Petermanns Mitteilungen. Ergänzungsheft 203.). This is the author's 1927 Munich doctoral dissertation.

River. At the junction of Gün-khai Valley and the Dzo-mo nang we shot a finch *Leucostice nemoricula* (Hodgs.) among the herbaceous plants of the valley.

The Dzo-mo valley proved perfectly dry, so we were forced to continue till we came to a rocky cliff where we found a spring in the otherwise dry streambed, and as the valley floor was quite broad we decided to pitch our camp at an elevation of 11,890 feet. About 500 feet below the pass the valley slopes were wooded with *Rhododendron capitatum* Maxim., here two feet high and associated with one species of willow, *Salix oritrepha* Schneid.; over these bushes flew flocks of the rare and most curious finch, with its tail wine-red beneath. It is *Urocynchramus Pylzowi* Przew. named after Przewalski's companion M. A. Pylzoff[41]. For this finch a new genus was created. It flew close to the bushes and seemed to nest in them. We secured several specimens, also of another finch *Montifringilla nivalis adamsi* Adams. which also frequented these bushes.

While the U-jâ tribe or clan has its encampments in the Gar-gen valley, overflowing into the Dzo-mo Nang, the Wam-chhog inhabit the Jang-chhen valley, and the Nya-nag clan the region to the north of the latter valley. All these tribes are at feud with each other, and all fear the Sha-brang tribe of the Bâ Valley and plain.

Where the Dzo-mo Valley opens out into the Gar-gen valley, the latter is rather broad, and the junction of the two valleys is at 11,550 feet elevation. The Gar-gen valley extends from east-southwest to northwest and above the junction of the Dzo-mo nang, the Gar-gen receives a small affluent. The hills framing the valley are bare, with here and there rock outcroppings and are 500-700 feet in height in the vicinity of Dzo-mo nang. Gar-gen was dotted with the black tents of the U-jâ Tibetans and thousands of their sheep and yak roamed the hillsides.

Following the valley down along the left, west, hillside, we arrive at another branch of it coming from the southeast called Sha-la, it describes an arch towards northeast and has its source on the northwest side od the same spur whence the small valley Gün-khai-mar-kha descends, but in the opposite direction. The hills on both sides of the valley dwindle in height, and opposite the mouth of Sha-la valley is the small lamasery called sGo-chhen rdzong-sngon (Go-chhen dzong-ngön or the blue fort of the Great Gate or Outlet), above it on a hill is a large cairn or obo also known as Go-chhen.

The trail leads directly north, leaving the Kuei-te trail to the right, east, and follows a dry streambed up a steep valley bearing the Mongol name Wa-yan-sgol (Wa-yen gol, the last word meaning river in Mongolian) it is more a ravine than a valley, but half way up were nomad encampments; leaving these to our right the trail turns up a left branch of the valley to a pass, in the center of which is an obo, built up of rocks and stuck full of twigs, decorated with yak hair; the pass is 12,110 feet, no view can be had from it but from a grassy hill to the east, elevation 12,300 feet, a wonderful vista opened out over

---

[41] Second Lieutenant Mihail Aleksandrovič Pyl'cov accompanied Prževal'skij on his first journey, 1870-1873. See the travel report: *Reisen in der Mongolei, im Gebiet der Tanguten und den Wüsten Nordtibets in den Jahren 1870 bis 1873*, von R. von Prschewalski, Oberstleutnant im Russischen Generalstabe. Aus dem Russischen und mit Anmerkungen versehen von Albin Kohn. (2nd pr.) Jena: Costenoble 1881. XIV, 538 pp. (Bibliothek geographischer Reisen und Entdeckungen älterer und neuerer Zeit 12.) Original edition: Mongolija i strana Tangutov: Trehletnee putešestvie v vostočnoj nagornoj Azii N[ikolaja Michailoviča] Prževal'skago, Podpolkovnika Gen. Štaba. Sanktpeterburg: U. S. Balašev, 1875-76. 2 vols.

far-away mountain ranges, and deep valleys. Directly northwest there extended a high range which some shepherds told us was the Shar-sgang (Shar-gang), probably the Ugutu range mentioned by Köhler and found on the map of Przewalski; the name is probably Mongolian; it is northwest of the Nga-thang and not visible from further south, or it may have been hidden by clouds when we saw the Nga plain. The Gyü-par range as seen from here displayed absolute barren southern slopes, and some five miles to the south of it, flanking the northern rim of the Bâ Valley, rose a short chain, better termed mountain, called Lhab-bya (Lhab-ja). The Bâ plain is eight miles broad and is cut lengthwise by the Bâ valley; from its northern rim it slopes towards the foothills of the Gyü-par range. Part of the Yellow River gorge could also be seen, and though the Am-nye Ma-chhen was hidden in clouds, the valley of the Chhu-ngön which extends west from that mountain chain was plainly visible.

From the pass the trail descends into the northern Wa-yen-gol, a narrow valley with a few nomad tents. Here we were attacked by twenty of their fierce dogs of which we freed ourselves with the greatest difficulty and not ere we had fired a salve from a colt automatic into the air, which made them turn tail. This valley merges into the southern border of the Bâ plain and there we decided to camp at an elevation of 10,950 feet. We had hardly settled down when a terrific thunderstorm broke over the Bâ plain and deluged us, while a gale whipped our tents and almost snapped the tent poles.

Rain continued all night which turned into a drizzle in the morning. It was one of the most dismal camping places. Before us lay the huge Bâ plain enshrouded in mist and clouds nothing could be seen, it was as if every hill had vanished. Water stood everywhere in large pools, and the whole ground had been converted into a bog.

We cut straight across the vast plain with nothing to guide us, our lama yak drivers never having been here previously, and surrounded by the hostile Shab-rang tribe felt not altogether too comfortable. After a three mile ride we came upon the brink of the Bâ Valley.

The Bâ Valley
The entire Bâ plateau is loess, grass-covered, and in this plateau the Bâ stream has cut its bed. On the southern side of the valley the slopes are loess and coarse gravel, the northern side is deeply sculptured loess with bands of lighter colored gravel between deposits. (See Plate **70**) The southern valley rim is at an elevation of 10,400 feet, and is covered with bushes as are the slopes, but only in stretches, the most common are the tussock-forming *Caragana tibetica* Kom., *Lonicera microphylla* Willd., *Lonicera syringantha* Max., the former with small yellow, the latter with larger, fragrant pink flowers; *Rosa bella* Rehd. & Wils., is here a shrub 6-8 feet tall and displays large, rich red, single flowers; *Cotoneaster multiflorus* Bunge, the white flowered *Potentilla fruticosa* L. var. *veitchei* (Wils.) Bean, the yellow *P. fruticosa* var. *parvifolia* (Fisch.) Wolf., *Berberis caroli* Schneid., and *Stellera chamaejasme* L., grew mainly on the gravelly slopes but also on the valley floor in rubble and loess.

The Bâ stream flows at an elevation of 9,941 feet, or 459 feet below the valley rim and plain. Willows are here represented by two species, *Salix wilhelmsiana* M. v. B., a shrub reaching a height of 15 feet, and the new *Salix juparica* Goerz. of similar stature, also *Hippophaë rhamnoides* L., ever present where willows grow and their constant companion along streambeds. On the gravelly exposed slopes *Androsace Mariae* Kom. var. *tibetica* (Max.) H.-M., exhibited its whitish pink flowers, while *Primula sibirica* Jacq., flourished in swampy meadows on the banks of the stream. The most common herbaceous plant was a species of Corydalis as yet undescribed (no 14260). It was the largest I have ever come across, and formed bushes or large clumps three or even more feet in height, the plants being all of two feet in diameter, it was a very handsome showy species producing hundreds of brilliant flowers in large racemes.

Among the willows grew the new white flowered *Thalictrum Rockii* Boiv., and in their shade the very showy, purple flowered *Pedicularis muscicola* Max., with spreading branches formed cushions. The deep violet flowered *Solanum septemlobum* Bge., first described from Peking also preferred the shade of the shrubs, while the river bank was a profusion of pink *Primula sibirica* Jacq., the only species of that genus encountered. Here we shot *Passer montanus obscuratus* Jacobi, and in the bushes the stone pheasant *Phasianus colchicus Strauchii* Przew., whereas *Mergus merganser orientalis* Gould., fished in the waters of the Bâ stream. On the branches of the taller willows and on a rocky bluff perched *Milvus lineatus* (Gray), a species of kite. Botanically the region was rather poor, but the herbaceous flora proved richer than the ligneous one.

Along the streambed in loess and gravelly soil flourished various leguminous plants as the new *Astragalus Handelii* Tsai et Yü, *Astragalus adsurgens* Pall., *Oxytropis deflexa* (Pall.) DC., *Astragalus versicolor* Pall., *Astragalus* aff. *subumbellatus* Klotzsch., both blue-flowered species, as well as *Oxytropis Kansuensis* Bge., *O. falcata* Bge., and *Astragalus polycladus* B. & Fr. In the same locality occurred the new *Pedicularis bonatiana* Li, n. sp., *Pedicularis kansuensis* Max., the former with purple and the latter with pinkish-purple flowers, also *Astragalus tanguticus* Bt., with dark blue flowers which, like all the other Astragalus, formed prostrate rosettes. *Pedicularis cheilanthifolia* Schrenk var. *typica* Prain, its flowers a pale pink, and known also from south of Lhasa in Tibet proper occurred here with them.

In swampy meadows of the valley floor grew the white flowered *Gentiana leucomelaena* Max., also known from Mongolia and Tibet, a species of Gentianella of the section Crossopetalus, but not yet determined (no. 14247), *Juncus Thomsoni* Buch., and the ill-scented, dark reddish-black flowered *Scrophularia incisa* Weinm., first described from Siberia. In the long grass grew the orchid *Orchis salina* Turcz., with rich purple flowers, in the sand *Carum carvi* on the loess bluffs here and there a species of Euphorbia forming large clumps and as yet not determined (no 14259). *Hedysarum multijugum* Max., and *Triglochin maritimum* L. were also partial to the sandy streambed and first known from Europe. In the shade of the willows we found the boraginaceous *Microula sikkimensis* (Clarke) Hemsley. This comprised nearly the entire vegetation in this section of the Bâ valley.

The confluence of the Gar-gen and the Bâ stream is called sGam-bu-sum-na, and from here to the foot of the Gyü-par Range the plain is inhabited by the Sha-brang tribe, but east of the Gar-gen, and to the foot of the Gyü-par it is occupied by the Klu-tshang tribe (Lu-tshang). The Bâ receives various affluents from the south, and east of the Gar-gen, as the Nyin-shig nang, the Tsha-han-sgol, a Mongol name transcribed into Tibetan, and the Yer-gong-nang, all three having their sources in the territory of the sBa-bo-mar (Ba-wo-mar) tribe, and north of the Chha-shing Chhu which is the northern branch of the Tshe Chhu. The head waters of the Bâ are many miles east of here in the tribal territory of the dBon-hjah (Wön-jâ) Tibetans, but the Lu-tshang extends still to the north of them as fas as the great Wa-yin thang or Wa-yin plain. Here it receives three affluents from the northeast, that is from the mountains encircling the Wa-yin plain on the northeast which form the divide between the latter and a long valley and stream which flows from northwest to southeast and then south where it forms the Dar-smug Chhu (Dar-mug Chhu) in the tribal land of the sMad-shul (Me-shül); the southernmost of these affluents is called the Ngang-khung; one affluent coming from the west, the first nearest its source, is the Ngang-tshang. All that territory is composed of vast plains and grassy undulating plateaus without any tree growth.

This region of the Bâ valley would be of great interest to archaeologists, for here on both sides of the Bâ valley, on the plain, we came across hoary ruins over one thousand years old; the southern ruin was that of an ancient village site once occupied by the Hor. It was according to Tibetan oral tradition, the seat of the ancient Hor kingdom and the traditional site of the wars fought by King Ke-sar against the Hor, and of whose exploits Tibetan bards sing endless epics. To the north of the Bâ valley not far from its rim are the square ruins of ancient forts and villages situated on a promontory above the plain; they were said to have been erected against Ke-sar. We climbed to the top of one of the ruins and found it surrounded by a trench, but this had been recently dug and was used in warfare by either robbers or nomads, or vice versa, both terms being here synonymous. Remnants of Hor who were formerly Mongols, and are probably the descendants of the ancient Mongols of Jenghiz Khan, still dwell to the southwest of these ruins, but they are now indistiguishable from Tibetans, except by name. Unlike the Sog-wo A-rig they dwell however in Tibetan black yak hair tents and not in yurts.

The Bâ plain on the north side, is 10,575 feet elevation or about 650 feet above the stream. To the right, east of the trail stood the square ruins of the ancient Hor fort, previously mentioned. The direction of the trail across the Bâ plain is northerly or 7° west of north and the distance to the mouth of the mKhas-rabs nang (Khe-rab valley) at the foot of the Gyü-par Range, five miles. It traverses the Bâ plain in a direct southerly course and enters the Bâ valley between the two Hor ruins, and East of the trail. On the plain we met two young nomads who were sitting in the grass twisting yak hair into rope. We asked them about the trail across the Gyü-par mountains, and although we were almost at the foothills, and they belonged to the Sha-brang tribe, answered in a surly manner that they did not know. They were exceedingly unfriendly but said that they were Sha-brang Tibetans. They then asked us where we were going whereupon we replied «here», and this ended the interview. A short distance beyond we came to an encampment of eight tents and were as usually attacked by their fierce dogs. The whole atmosphere, the sullen temper of the nomads, their savage dogs, the dismal dreary plain with a black curtain of storm clouds about to discharge their contents, a bleak cold north wind driving the rain into our faces, all this, with the possibility of being attacked by the churlish nomads made me feel very lonely, and for once I wished myself elsewhere.

Arrived at the mouth of the Khe-rab valley where it emerges on the plain I took the altitude which showed 11,330 feet making an incline of 755 feet from the rim of the Bâ valley to the foot of the range.

The Gyü-par Range

According to the Tibetans of Ra-gya, the Gyü-par Range was said to be densely covered with forest on its northern slopes, but as much as I tried scanning the range with binoculars, I found no sign of a single tree, let alone forest. However the Tibetans were right for later we did find forests, if not vast ones. To nomads used to grasslands, forests need not be large, to be termed vast. Two forests were confined mainly to the extreme northwestern end of the range. (Plates **71-73**)

Entering the Khe-rab Valley and hoping that it would bring us to a pass, and down to the northern slopes of the range, we followed the right branch where a trail indicated access to the northern face of the mountain. Here were the first signs of woody plants, in the yellow flowered *Caragana tibetica* Kom., bushes which clung to the much broken shale and schist with underlying slate of which the mountain is here composed; a white flowered Corydalis as yet undetermined (no 14408) grew on the rocky cliffs of Khe-rab valley together with the beautiful pale pink, to mauve-colored fragrant crucifer *Cheiranthus roseus* Max., while *Salix oritrepha* Schneid. covered the southern slopes of the valley. Other plants observed in Khe-rab valley were *Caragana jubata* (Pall.) Poir., the large brilliant yellow Corydalis (14260), *Androsace Mariae* var. *tibetica* H.-M., *Potentilla fruticosa*, *Lonicera hispida* Pall., the yellow poppy, and many other herbs.

The ravine is exceedingly rocky and narrow but soon opens out into a wide grassy valley into which several valleys merge. The narrow rocky ravine we had traversed leads only through the foothills of the Gyü-par Range which itself is quite bare on its southern slope. Directly ahead of us lay a gentle sloping valley and this we entered; travelling with a caravan of yak proved difficult in this region and applied also to our horses for the valley floor was either a rocky bed or a quagmire into which our animals sank a foot deep or even more. Splash, splash, we made our way through the morass, while the water ran down our coats from the incessant rains, we could see nothing for the driving rain, mist, and cold wind. We tried to wait for the caravan but waiting in the mud and rain became so unpleasant that we decided to continue. We passed a valley on our left and followed the main branch to a pass which proved to be exactly 13,000 feet in height. The pass called mKhas-rabs Nye-ra was boggy to a degree and the loaded yak could not ascend it by the main branch. The lamas sent one of our men ahead to tell us that they would take the smaller left valley, and for us, when arrived on the pass not to descend it, but to turn down a left valley, the trail to Kuei-te continuing straight down. This left valley and the one the caravan took would meet in a larger valley, and in the latter we were to wait for the yak. The hill tops were enshrouded in mist and clouds and arrived in the broad valley we waited for the caravan. The mist lifted over the spur enclosing the head of the valley and I saw with my field glasses our caravan on the top of the hill. They did not continue down but remained for a considerable time on the mountain. When looking again I saw them unpack and arrange our boxes on the ground, behind which the lamas and some of our Tibetans stationed themselves using our baggages as a barricade. I thought that they were in trouble, so sent one of my men, we were all mounted waiting for them, to see what had happened. It was fortunate that we did not all ride up to the head of the valley; the lamas had taken us for bandits and would have opened fire on us had we approached together, to defend themselves from behind our loads. Seeing only one man approach they waited, the lamas holding the loaded rifles ready to shoot, while one of them went within calling distance. When they found out who we were they reloaded the yaks and descended; they were a chagrinned bunch when they joined us, but at any rate they proved their alertness. The valley was called the sTong-chhags (Tong-chhag) which emptied into the sGo-mang Nang. The Tong-chhag valley was nothing but a bog, water was standing everywhere and it was difficult to find a fairly decent camping ground. Wherever the water was stagnant the grass was yellow, but where it could run off the grass was green. We pitched camp behind a rocky promontory in the center of the valley fairly sheltered from the wind. It kept on pouring with rain, the temperature at 3 p.m. was 42° F., while not cold, it was the penetrating dampness which made one feel cold and miserable.

Our camping place was at 12,230 feet elevation; the rain continued all night, but by 5 a.m. next morning, June 25th, blue sky was visible directly above us, while the mountains around us were still enshrouded in mist. When we left our camping place the sun had conquered, the mist was dispersed and the sky blue and cloudless, the atmosphere was so clear that every ridge was sharply outlined.

We descended the Tong-chhag valley to where it joins to long Go-mang valley, at the mouth of which we found an encampment of the kLu-tshang (Lu-tshang) tribe from whom we obtained a guide in payment of four squares of cotton drilling (blue dungaree) to show us the way to the forested region of the Gyü-par mountain.

Entering the Go-mang valley we followed it up on the right flank crossing spur after spur ranging from 12,500 to 12,600 feet elevation till we reached a pass at the head of the valley at 12,850 feet. From this pass we descended into the long Gyü-par valley which extends the whole northern flank of the Gyü-par range, and joins that of the Yellow River in a northwesterly direction. Actually the northern foothills of the Gyü-par chain slope gradually towards the Yellow River when they drop steeply into the Yellow River valley.

We now followed the Gyü-par Valley west, the mountain slopes being still bare of any tree. At 10 a.m. the thermometer registered 48° F., and by then heavy cumulus clouds had gathered over the hills all around us permitting only here and there a blue open space to be seen. The elevation of the Gyü-par valley floor was 12,400 feet, and the vegetation consisted here mainly of small willow bushes 2 ½ feet, with *Caragana jubata*, but instead of growing prostrate or forming a branching shrub, it grew perfectly erect, single stemmed as we had found it also on the northern slope of the Min Shan 岷山, in the south, resembling the habit of a Carnegia cactus. Other associates were *Potentilla fruticosa* L. var. *Veitchei* Dean, the new *Salix juparica* Goerz, and *Salix tibetica* Goerz. In the gravel here and there grew *Incarvillea principis* B. et Franch., with its lovely white tubular flowers, and in the wet meadows hundreds of *Meconopsis integrifolia* (Max.) Franch. whose petals the Tibetan lamas collect to extract a yellow dye.

The valley becomes constricted and further down impassable, forcing us to ascend a steep lateral valley to the right, to an elevation of 11,800 feet, thence to the left across a spur and down the main Gyü-par valley on the grassy slopes 600 feet above the stream. The yaks had great difficulty continuously crossing these high spurs and were nearly all completely played out. After crossing a 12,400 feet high pass we descend into the mGrin-gong gong-ma (Drin-gong gong-ma or Upper Drin-gong) valley; so far there was as yet not a sign of forest but still there was another spur and pass to be negotiated. As the yak caravan was nowhere in sight we decided to camp at 4 p.m. at an altitude of 12,100 feet. The sky was again overcast and gloomy and rain began to fall and continued all night until 7 a.m. the next morning when we broke camp. We crossed the middle and lower Drin-gong valleys and in the latter, on the northwestern slopes, we found the first spruce forest, but many of the trees were dead undoubtedly due to grazing. Further down the valley nomads were encamped; turning into and continuing down a large ravine where there was lovely forest of spruce, but alas again of the species *Picea asperata* Mast., we pitched our camp several hundred feet above the streambed which was also lined with spruces and birches *Betula japonica* Sieb. var. *szechuanica* Schneid., its constant companion. Poplars were also present.

Below our camp I spied an interesting plant which turned out to be *Potentilla salesoviana* Steph., rare in this region, and only found besides here in the Pien-tu-kʻou 扁都口 gorge across the eastern end of the Richthofen Range or Nan Shan 南山; it was first described from Siberia. It is a lovely shrub 3-4 feet tall with large white flowers and leaves silvery beneath. With it grew *Potentilla fruticosa* L. var. *Veitchei* Bean., *Caragana brevifolia* Kom., *Lonicera syringantha* Max., and many willows as the three new species *Salix juparica* Goerz, *Salix tibetica* Goerz, and *Salix pseudo-wallichiana* Goerz, and *Salix oritrepha* Schneid., the latter a small shrub 2-3 feet, while the former reached over 15 feet in height, also *Salix rehderiana* Schn. var. *brevisericea* Schn., a small tree of similar stature as the last one. All the above lined the streambed where, in the sandy margins, we collected *Astragalus chrysopterus* Bunge, growing under or on the outskirts of the willows as well as *Aster tongolensis* Franch., with purplish lavender flowerheads. In fact nearly all the shrubby vegetation was confined to or congregated near the streambed, with other moisture loving shrubs as *Cotoneaster multiflorus* Bge, and *Lonicera hispida* Pall. *Sorbus tianschanica* Rupr. was confined to the spruce forest, some of the spruces of which reached a height of 150 feet, with trunks three and four feet in diameter.

The rocks along the streambed are slate, crushed by large blocks of superimposed schist. We explored the valley in its upper part which narrows into a gorge where a cross section of the rock walls forming the gorge showed the lower strata to be composed of thin slate crushed by a wall of schist, large blocks of which were lying on the top of the slate. At the foot of these walls grew *Caragana jubata* Poir., willows, Loniceras, Rubus sp?, *Ribes Meyeri* Max., etc. As can be seen from the enumeration of the woody plants, the ligneous flora is certainly not rich, and the herbaceous plants proved also to be poor in species. This undoubtedly is due, like in the northern Nan Shan, to the proximity of the desert and prevailing northern or northwestern winds from the barren, waterless wastes. Yet precipitation is here abundant, however the short summer and cold temperatures do not permit the development of a varied flora, and allow only such species to become established which are hardy in such a climate.

Many of the plants found here occur also in Siberia; genera with more than one species are mainly Salix and Astragalus, while of Sorbus only one species reaches this far north. Of primula only *Primula sibirica* Jacq., of such wide distribution as from arctic and central Asia to Alaska, Tibet, and the northwestern Himalaya occurs here. The underlying rock formation has also a great deal to do with a plant cover, and it has been my observation that in these latitudes, mountains of schist, shale and slate support a much poorer flora than limestone mountains, vide Lien-hua Shan, only one degree latitude further south, a limestone mountain rising in a country of practically pure loess with underlying sandstone, schist and shale. It is a typical boreal flora with some endemic elements. Of conifers only one species, *Picea asperata* Mast., is widely spread from the Min Shan to the Nan Shan facing Inner Mongolia, now known as Ning-hsia; north of La-brang (Hsia-ho Hsien) this species forms pure stands, and where one could find Abies above the spruces as in the south, here its place is taken by the willows and junipers, the latter genus being rich in endemic species.

On the alpine meadows of the Gyü-par Range at elevations of 13,500 feet occur *Corydalis dasyptera* Max., *Rheum palmatum* L., *Astragalus acythropus* Bge, and *Cremanthodium plantagineum* Max. Other herbaceous plants found on meadows but on the western part of Gyü-par Range were a species of Pleurospermum (no 14307) related to *P. Candollei* C. B. Clarke, *Meconopsis quintuplinervia* reg., and *Primula tangutica* Duthie.

The spruces extend to an elevation of a little over 11,000 feet, and growing in the moss of the spruce forest we encountered again *Rubus idaeus* L. var. *strigosus* Max. [Plate **74**] The valley slopes in the Gyü-par Range are very steep and the spruces cling to them tenaciously, but many landslides have taken place and in general it appeared that the spruce forest of this range was doomed.

Several thunderstorms, as many as four a day occurred while camping on this mountain range, with intervals of clear skies. On June 28th we decided to explore the dryer slopes of the range down to the valley of the Yellow River. This necessitated first to cross the Gyü-par stream, and the climbing of a steep rocky hillside west of the stream, to an elevation of 11,300 feet, after which we followed the grassy slopes, skirting lateral valleys, some forested with spruces and willows, others bare. The trail led to 11,600 feet with the Gyü-par stream deep below in a veritable canyon which becomes narrower, and the walls steeper, as it approaches the Yellow River. About half way between our camp and the Yellow River we began to descend into steep ravines until we arrived at bare, yellowish red, gravelly bluffs covered here and there with grass. To our left was a narrow, spruce-forested, ravine and to our right the Gyü-par canyon. We followed a central spur on to a higher bluff which we found covered with large tussocks of *Caragana tibetica* Kom. (see Plate **75**).

From here we obtained an extensive view of the Yellow river and the surrounding country. The land towards the Yellow River valley is very much broken up into ravines and canyons. The Yellow River valley consists mainly of loess with strata of sandstone and gravel, while at the bank of the river is deep lead-blue slate, above that is gravel followed by red sandstone and a thick covering of loess (see Plate **76**).

The river, after cutting through the extreme western end of the Gyü-par Range, takes a complete turn to north-northeast. From a bluff 10,480 feet elevation I took several photos up stream looking southwest; the river described many sharp bends, and in the south-southwest there was visible a snow mountain, probably the Ugutu Range, while the Ta-ho-pa stream joins the Yellow River near the great bend, around the western end of the Gyü-par Range.

Directly northwest the Yellow River has the configuration of a trough, and beyond these bends the Gyü-par stream debouches into the Yellow River. From a lower bluff 10,380 feet, looking north-northeast, downstream, there were visible large poplar trees growing on its banks. A long rocky spur, east of the Gyü-par valley, stretches for a considerable distance into the Yellow River and forces it to make a sharp turn around it.

Northeast on the northern side of the Yellow River are extensive sand dunes, and beyond them, a large grassy plain, a vast loess plateau which is deeply eroded into innumerable steep and short canyons. They are back of the terrace on which is situated a lamasery called A-tshogs dgon-pa (A-tshog Gom-pa) consisting of only a few houses and one chanting hall. Looking north there stretches a long sand-spur in the centre of which rises a conical mountain called Am-nye Wa-yin, a landmark in the region.

The vegetation on the bluffs overlooking the Yellow River consists mainly of the tussock-forming *Caragana tibetica* Kom., the white-flowered *Cotoneaster tenuipes* Rehd. & Wils., a shrub 3-4 feet tall, the low shrub (1-2 feet high) *Caragana Roborowskyi* Kom., the herb *Peganum Harmala* L., with pale yellow flowers confined to the gravelly slopes, and a purple flowered Iris (no 14322) which grows in the Caragana bushes, and also in clumps of *Berberis Caroli* Schneid.

As regards the forests of the Gyü-par Range, they are doomed; spruce forests are now found only in patches and on the more inaccessible places on the steep valley slopes and ridges. the nomads with their sheep and yak have ruined this region. From all appearances the forests covered once all the valley slopes, at least those facing north, and wherever in this region forests occur, they are in a dying condition. Their undergrowth like that of the forests of Tag-so is a species of Mnium moss, and this has here almost entirely disappeared, and only where there are small groups of healthy trees to be found, this moss is also present, often over a foot in thickness and covering the ground completely. Thousands of dead trees are witness to the evil work of the yak and sheep. Where the forests have died willows have sprung up everywhere. When grass once encroaches on the spruce forest the moss is killed off, and when the latter is gone the spruces follow. A really healthy spruce (*Picea asperata* Mast.)[Plate **77**] forest such as I found on Niu-hsin Shan in the T'o-lai Range south of the Nan Shan, has no undergrowth whatever only moss, except *Rubus idaeus strigosus* Max. and *Thermopsis alpina* Ledeb. which penetrate a few feet deep only or remain on the very edge of the mossy carpet. All other shrubs such as roses, Berberis, willows, Lonicera, etc., are confined to the banks of the streams and long brooks, but never enter the somber spruce forest where no light penetrates, on account of the density of these forests.

In the Yellow River valley itself, looking west, spruce trees covered the upper valley slopes in patches. The actual limit of the spruce is here 12,000 feet, these are followed by willows but mainly by the new variety of *Salix oritrepha* Schneid. var. *tibetica* Goerz. (a shrub 2-3 feet high), while the others remain in the valleys along the streams.

The species *Salix oritrepha* Schneid. is also confined to high alpine meadows and rarely occurs along streambeds, unless it is in the high alpine regions.

One characteristic of the tributaries of the Yellow River which have their source in the high alpine grasslands is, that at their head they are wide and shallow, and as they approach the Yellow River become exceedingly narrow, being mere rock gates a few feet wide but of great depth.

Gyü-par Tshar-gen the Highest Peak of the Gyü-par Range

Having exhausted the botanical possibilities of the Gyü-par forests we broke camp and proceeded to the upper part of the Gyü-par valley to near the foot of the higehst peak of the range Gyü-par tshar-gen where we pitched our camp on a small meadow near the Gyü-par stream at an elevation of 11,660 feet (boil. pt.)

The most common plant in this region is the pink-flowered *Caragana jubata* Poir., next common the yellow flowered variety of *Potentilla fruticosa* L. var. *parvifolia* Wolf, which is confined to the higher levels, then *Salix oritrepha* Schneid., *Salix myrtillacea* And., *S. brevisericea* Schneid. and *Salix Rehderiana* Schneid. var. *brevisericea* Schneid., but the two latter confined to the streambed.

Lonicera hispida Pall., with hirsute branchlets and yellow flowers and the red flowered *Rheum palmatum* L., grow on the grassy slopes of the valley, while *Juniperus tibetica* Kom., although it extends high above the spruces, descends here to 11,400 feet associated with *Berberis caroli* Schneid. On the cliffs in the upper part of the Gyü-par valley one met with the reddish flowered *Sedum algidum* Led. var. *tanguticum* Max., *Cheiranthus roseus* Max. forma *elatior*, a *Hedysarum* (no 14347), an *Astragalus* (no 14340), a yellow flowered *Corydalis* (no 14344), and *Geranium pratense* L., first described from Europe. The rocks are all slate and schist and the streambed is here and there hemmed in by cliffs.

On June 30th we ascended Gyü-par tshar-gen. At 5 a.m. the temperature had dropped to 32° F., and ice had formed on the margins of the stream and our tents were frozen. We left camp at 5.30 a.m. with a nomad guide who had his encampment above the grassy head of the valley. We ascended the ridge to the left of our camp and then descended into the valley of Tshar-gen Khog which opens into the Gyü-par valley from the west. It is so named because it leads to the foot of the peak. We followed the lonely Tshar-gen valley up stream about three miles, and then turned up a lateral valley to the right to the foot of Gyü-par Tshar-gen.

The vegetation in these valleys is composed mainly of grass with here and there a yellow poppy still in bloom and a few stunted plants of *Caragana jubata*, nothing else was visible. There was still another grassy valley or steep gully to ascend, and then the zigzag trail up the main ridge of the mountain, still grass covered. The last stretch was over gravel to the top of the peak, a small flat area with a cairn or obo. We made the height by boiling point 14,546 feet above sealevel.

We had a race with the clouds which had gathered, although the sky was a deep blue and free of clouds when we had started. An easterly wind had brought clouds in its train behind us up the peak. Had we arrived ten minutes later we would have had no view at all, but as luck favored us there before us lay the whole Am-nye Ma-chhen Range.

West-southwest of the Yellow River is the Nga thang or Nga plain, the river flows here at the foot of the eroded loess cliffs, and in the immediate foreground were the bare slopes of the western end of the Gyü-par Range.

The Amnye Ma-chhen suggests very much the Gangs-chhen mdzod-lnga (Gang-chhen-dzönga) known to foreigners as Kanchenchunga, of the Himalayas. The lower slopes of the Amnye Ma-chhen are not visible from here on account of a black, then snow streaked, range which extends parallel to the Am-nye Ma-chhen. The eastern rim of the Nga plain drops vertically into the Yellow River and is eroded into thousands of pinnacles and turrets, similar to those at Hsün-hua to the east. To the right, 2° south of west, the Ta-ho-pa stream which the Tibetans call the Hang Chhu, enters the Yellow River coming from west-northwest; on terraces near the streambed, forest could be observed as well as at the foot of the loess bluff forming the bank of the Yellow River. Northeast there was visible a huge sandy plain or rather desert of sand dunes, flanked by high mountains beyond.

The summit of the Gyü-par tshar-gen consists of slate and schist with here and there quartz outcroppings. The plants found on the summit represented no great variety, and as the ligneous flora, so proved the herbaceous one, poor in species. Partial to grassy areas was the yellow *Trollius pumilus* D. Don, first described from Nepal but found throughout west China; in the gravel grew the very lovely *Pedicularis pilostachya* Max., first collected by Przewalski, with deep carmine flowers and thickly greyish-pubescent spikes, *Lagotis brevituba* Max., with bluish-white flowers which extends to the eastern Himalaya, the intense yellow flowered *Corydalis melanochlora* Max., and a still undescribed species with rich, deep blue flowers, exuding a fragrance like roses, *Corydalis* sp? (no 14328); a rhubarb, *Rheum spiciforme* Royle, a prostrate fleshy plant with broad leaves appressed to the ground, was partial to the muddy gravelly slopes, while *Meconopsis racemosa* Max., with deep steel blue flowers grew among rocks protruding from the meadows. On the scree occurred a *Saussurea* (no 14337), *Polygonum sphaerostachyum* Meisn., with a white globose infloresence, *Chryso-splenium nudicaule* Bge, a greenish-flowered saxifrage first known from the Altai mountains, the yellow flowered *Iris Potanini* Max., the white flowered crucifer *Eutrema compactum* O. E. Schulz, first described from Turkestan, but occurring in Tibet, North China and Mongolia, the Edelweiss *Leontopodium linearifolium* Hand.-Mz., and *Saxifraga Przewalskii* Engl.; on alpine meadows adjoining the scree, but not on the latter, *Corydalis dasyptera*, was at home.

Here on the rocky summit of the Gyü-par mountains we shot *Prunella collaris tibetanus* (Bianchi), these birds are found at very high altitudes, and only on rocks; on the much higher Li-chiang Snow Range we collected a related species *Prunella collaris ripponi* Hart., also at high elevation, nearer 16,000 feet, among limestone crags.

As a matter of record I give here the bearings of the Amnye Ma-chhen as viewed from the highest point of the Gyü-par Range. The southernmost snow-covered part was 219°, the second peak 222.5°, the third 226.5° and the highest dome-shaped peak 228°, the last or northern snow-covered peak was 231°; a glacier descended between bearings 222.5° and 226.5°. The atmosphere was too hazy to make out other glaciers.

From the Gyü-par Range to Ragya via the Tho-thug Nye-ra and the Ho-tog Nye-ra
From the Gyü-par range the trail follows the Gyü-par valley its entire length, the valley being forked near the source of the stream, to a pass called the Tho-thug Nye-ra, elevation 13,900 feet. A rocky bluff to the left (southeast), elevation 14,100 feet, and a higher peak on the right (northwest) proved to be 14,570 feet elevation, or 34 feet higher than Gyü-par tshar-gen the highest peak of Gyü-par Range. However, Tho-thug Nye-ra and the hills flanking it were part of the Gyü-par Range, in about the center of it with the Gyü-par Nang to the left, and the Go-mang Nang to the right. Looking northeast the eye met a vast expanse of sand dunes the Mang-ri bye-ma (Mang-ri je-ma) which drop abruptly on to a grassy plain. Directly east there stretched a high mountain range, the highest peak of which is known as Am-nye Brag-dkar (Am-nye Drag-kar) or the White Rock (peak) which I judged to be 16,000 or more feet in height, the range being situated south of Kuei-te. A peak in this range is marked on some maps as Mt. Djakhar with a height of 3,779 m, or about 12,400 feet. The name Djakhar corresponds to Drag-kar which is the correct spelling, the «dr» is sometimes pronounced like «dj», but khar is wrong as kh represents an aspirated k which it is not. Am-nye being only an honorific prefix. Apparently the range itself has no particular name. The Am-nye Ma-chhen Range was however hidden in clouds.

Northeast of the Gyü-par Valley a stream which has probably its source north of Go-mang gung-kha flows at first parallel to the Gyü-par valley, but then turns west-northwest to enter the Yellow River; the valley is called rMang-ra (Mang-ra) and the junction with the Yellow River rMang-ra mdo (Mang-ra mdo) or simply Mang-mdo.

The flora of the higher of the peaks at the Tho-thug Nye-ra proved quite interesting. In the scree occurred three Saussurea, one *Saussurea medusa* Max., resembled very much the Li-chiang species *S. laniceps* H.-M., with an oblong large, hood-like infloresence densely covered with a white flocculent substance with deep purple flowerheads embedded in the latter, *Saussurea hypsipeta* Diels, and Saussurea sp? (no 14368). *Astragalus mattam* Tsai & Yü, formed mats with bluish-purple flowers, adhering flat to the ground, while *Draba lichiangensis* W. W. Sm., preferred gravelly mud as did *Pleurospermum thalictrifolium* Wolff.; also partial to gravel and scree proved *Anemone imbricata* Max., *Sedum juparense* Fröd. sp. n. (14365), *Corydalis melanochlora* Max., and in muddy gravel *Rheum spiciforme* Royle, anchored with a big taproot. On the scree which covered the slopes of the hill we shot four *Tetraogallus tibetanus Przewalskii* Bianchi (see Pl. **78**), these birds seem to live on medicinal alpine plants, causing its flesh to taste bitter; they can only be found on the highest screes where they are hardly distinguishable from the rocks, the color of their feathers being mottled, even when shot we had often difficulty finding them.

From the pass the trail descends south-southwest into the Tho-thug Nang which becomes rather narrow, very rocky, and most trying for both man and beast. The Tho-thug Nang leads out on to the Bâ plain and Bâ valley which here is a veritable bog where flies and mosquitoes were most annoying. The altitude of the valley floor is 9,593 feet boiling point, aneroid 9,600 feet. We followed down the Bâ valley towards its junction with the Yellow River, it was however impossible to reach its mouth.

In the Bâ valley are several small hamlets, the only villages in the entire grassland area. [Plate **79**] The inhabitants are Tibetans who had migrated from Rong-bo (Rong-wo), and as they first lived in caves in the loess walls of the Bâ Valley the nomads called them Sa-og rong-ba (Sa-og Rong-wa) or the Rong-wa (living) under soil or ground. Another village further down the river is called Gad-mo-chhe (Ge-mo-chhe) elevation 9,500 feet and inhabited by five families, these sedentary Tibetans had settled here only since 1919. The houses of the Tibetans are built of loess and bricks, with flat roofs and dirt floors. In their fields they cultivated mainly barley and wheat; wild grew *Convolvulus arvensis* L. var. *sagittifolius* Fish., *Convolvulus Ammanii* Desr., and *Geranium pratense* L. In the swampy meadows of the Bâ valley grew *Orchis salina* Turcz., *Cremanthodium plantagineum* Max., and the yellow flowered *Pedicularis longiflora* typica Rud., while above on the dry embankment *Nitraria Schoberi* L., and *Peganum harmala* L., were common. The streambed was filled with willows mentioned previously, also Hippophaë.

The arid upper loess slopes were covered with the cushions of *Caragana tibetica* Kom., *Oxytropis kansuensis* Bge. and *Allium polyrhyzum* Turcz. Mosquitoes, despite the elevation, proved here a veritable plague. The trail ascends the gravelly slopes of the Bâ valley where all the shrubs, as Lonicera, Cotoneaster, Berberis, and Caragana were already past flowering, except *Stellera chamaejasme* L., which scented the air with its fragrance, not being dependant on seed for reproduction it could continue to flower till winter came.

The Bâ plain is here at an elevation of 10,400 feet, whence a way leads across it to the mouth of the Ho-tog Nang (valley) up to a pass, the Ho-tog Nye-ra, elevation of 12,300 feet. Skirting the head of a valley on the southern side of the pass and going southeast, the trail descends into the eastern gSer-lag valley (Ser-lag), leaving the western one in which gSer-lag gon-pa (Ser-lag Gom-pa) is situated on the right; the latter having its source in a pass to the west of the Ho-tog nye-ra. The lamasery houses about 300 monks all sons of nomad families of the encampments around here, each monk being supported by his family. As the lamaseries rule over the nomad clans of a given area, the families who have sons represented in the lamasery have thus an influential voice in any decisions arrived at.

The Ser-lag valley joins the Sha-la valley at an elevation of 11,542 feet. The Ser-lag and Sha-la valleys are usually dry, except at their junction with the Gar-gen valley where there are springs. In the Gar-gen and Dzo-mo valleys and on the Dzo-mo pass, a number of plants had come into flower not previously seen as *Polygonum Hookeri* Meisn., *Corydalis glycyphyllos* Fedde, *Corydalis hannae* Kanitz, *Arenaria melanandra* (Max.) Mattf., *Astragalus Fenzelianus* Pet.-Stieb., *Cremanthodium plantagineum* Max., *Crepis Hookeriana* C. B. Clarke, *Saussurea medusa* Max., *Primula optata* Farr., *Microula Rockii* Johnston, *Hedysarum tanguticum* Fedt., *Meconopsis racemosa* Max., and *Potentilla Saundersiana* Royle all growing at an altitude of 13300 feet.

After having crossed the Dzo-mo La or Dzo-mo pass, our lama yak drivers yelled «the gods are victorious» for there was now nothing more to fear from Tibetan robbers; we could now sleep peacefully and unconcerned about possible night attacks, for we had come again into the territory of the Gar-tse under Ra-gya lamasery.

On the Wo-ti La we found a few plants which we had not encountered on our way up, these were *Potentilla Forrestii* W. W. Sm., the new *Parrya* var. *albiflora* O. E. Schulz, *Cremanthodium decaisnii* C. B. Clarke, *Astragalus Licentianus* Hand.-Mzt., *Pedicularis oederi heteroglossa* Prain, a native also of central Himalaya to Tibet, with deep yellow flowers, the white flowered *Cochlearia scapiflora* Hook. f. et Thom., the pale blue *Microula tangutica* Max., the crucifer *Dilophia fontana* Max., with white flowers, the prostrate, rosette-forming *Dontostemon glandulosus* (Kar. & Kir.) O. E. Schulz, with pale pink flowers and the yellow flowered *Corydalis hannae* Kanitz.

Many of the alpine Compositae, Delphinium, Aconitum, etc., showed as yet no flowers, but three Meconopsis bloomed, the yellow *M. integrifolia* (Max.) Franch., the lavender M. *quintuplinervia* Reg., and the blue *M. racemosa* Max., the yellow flowered one being always first, followed by the second, and the last named the latest to open its flowers. It was strange that we did not observe a single plant of *Meconopsis punicea* Max., although this is a later flowering species and for that reason did not come under our observation, yet further south in the Min Shan where its station is less high it flowers in June.

South of the Wo-ti La in the Wa-ru valley we encountered the white flowered *Incarvillea principis* Bur. et Franch., but it was rare; with it grew *Ligularia plantaginifolia* Franch., with yellow flowers, at an altitude of 13,000 feet on July 6th. In the swampy meadows south of the Wo-ti la at 12,800 feet occurred *Pedicularis rhinanthoides labellata* (Jacq.) Prain., and the dark purple flowered *Cremanthodium discoideum* Max. In the streambed of the Wa-ru valley grew *Ranunculus affinis* R. Br., and the lavender and white flowered *Phlomis rotata* Benth., at an elevation of 13,000 feet. Had we been able to visit the region at a later time, in late summer or autumn, undoubtedly the number of species would have been considerably augmented. But as it was we did cover part of that ground on our return journey to La-brang, later in summer, and the plants collected then will give a fair picture of the plants covering of the region.

The valley slopes of the Dzo-mo (nang) and upper Tag-so were one mass of the lavender-flowered *Rhododendron capitatum* Maxim., which forms impenetrable thickets of over 12,000 feet, the only shrub in this region.

Ice covered the streambed in places to a thickness of a foot or even more, indicating that in these regions it never melts.

The Am-nye Ma-chhen Pom-ra (Am-nye rMa-chhen sPom-ra)

The Go-log tribes
There are three main tribes of the mGo-log (Go-log) who look upon Ma-chhen pom-ra also called rMa-rgyal-spom-ra (Ma-gyal-pom-ra) as their mountain god. Ma-gyal Pom-ra, legend relates, was the tea boiler of Chen-re-zig the god of mercy. (The Yellow River also bears his name rMa, pronounced Ma, chhu = river). The mountain god or *sa-bdag* = earth owner, is known as the great Ma, Pom-ra, or the King Pom-ra.

The Am-nye Ma-chhen always had attractions for foreigners, ever since General George Pereira a British Brigadier General expressed the idea or belief that the range exceeded Mount Everest in height.

Pereira was preceeded by many years by the Russian travelers Przewalski and Roborowski, the latter of whom approached nearer the Am-nye Ma-chhen than any traveler either before of after him.

The greatest obstacle to the exploration of that famous mountain range have been the Go-log tribes who jealously guard that mountain, and who are strenuously opposed to any one not Tibetan to visit that mountain. Should any one wish to tarry at the range or still more photograph it, they will put every obstacle in his way, even threaten murder if he should have the audacity to come near their encampments or their mountain. They are a most suspicious, superstitious, and unfriendly lot, unapproachable, perfidious and crafty, whose mountain fastness guarantees them complete isolation. The difficult struggle for existence and the stern nature of their environment have left their mark on them, their features are coarse and never a smile lights their feace. Accustomed as they are to being surrounded by adversaries they are always ready for battle and woe to the outsider who penetrates to their mountain fastness. Nominally they were under the rule of the Moslems of Hsi-ning who controlled the Koko Nor or Ch'ing-hai but who could never subdue them or make them pay taxes.

In 1921 General Ma Chi-fu of Hsi-ning who was then in control of Ch'ing-hai 青海 sent letters addressed to the chiefs of the main three Go-log tribes named the Ri-mang, Khang-gsar, and Khang-rgan; these letters he sent to the high incarnation A-rig Ra-gya dgar de'i bla-ma Shing bzah paṇ-ḍi-ta (A-rig Ra-gya gar-dei la-ma Shing-za Pandita) to be forwarded to the afore mentioned Go-log chiefs. This the incarnation was obliged to do, as he was under the control of the Moslems of Hsi-ning. The Go-logs having had word that such letters were being sent, and who had also a knowledge of their purport, refused to accept the letters and promptly returned them to Ra-gya. They declared that all the Go-logs will be the enemy for life of the Ra-gya incarnation, and the whole lama fraternity for forwarding the letters. They had been ordered to pay a grass and water tax, one silver dollar per yak per year.

Thereupon Ma Chi-fu organised a punitive expedition consisting of 5,000 soldiers with machine guns, etc., against the Go-logs; he first attacked the most powerful Ri-mang tribe, drove off their sheep and yak and impoverished them completely. This forced the Go-logs to declare their submission and to agree to the payment of the tax. When Ma Chi-fu sent his tax collectors the following year to the Go-logs they killed them all. Ma Chi-fu himself told me that he found it impracticable to enforce the tax payment, for it would cost more to collect it than the tax was worth.

At our visit the Ri-mang chief Trul-ku Dan-drag [lDan-brag] was absent in Lha-sa to declare his submission to the Dalai Lama of Lhasa. What the communist have been able to achieve with the Go-logs is not known, but it is very doubtful if their teachings will have much influence on these war-like people, and one thing is certain, they will never be able to disarm them. The Go-logs are ruled by their own chiefs who are crafty and whose word cannot be trusted.

The Go-logs do not differ from the other Tibetans in their mode of living, but there is something about a Go-log which makes him stand out from other Tibetans, yet it is not his mode of dress. Their heads are as round as a bullet and once one has seen a Go-log he can never mistake a Go-log for another Tibetan nomad. They have their own dialects which also distinguish them from the other Tibetans.

These marauding tribes bring terror to the hearts of their neighbours and passing caravans. Even lamaseries are not immune to their depradations. The year before our arrival at Ra-gya Gompa, a group of Go-logs crossed the Yellow River, occupied the house and compound where we were put up by the lama of Ra-gya, and there robbed all and sundry; they drove off the yak and sheep of the high incarnation who ruled Ra-gya and for days they robbed anyone who came to the lamasery whose inmates were powerless against them.

A foreigner can only travel here if he is adequately armed with modern weapons and in a larger group, but the larger the party the more cumbersome his caravan. When traveling in this area one must be mobile, have no slow yak, but only fast horses, plenty of ammunition and superior rifles. The low mentality and their most primitive way of life give great scope to monstruous superstitions, and we were often told that with our field glasses we could see through mountains, and that the powder of our guns was sufficient to kill animals without any bullet hitting them. At that time they had not seen aeroplanes, but they said they had heard that we could get into eagles and fly.

They are arrogant and rude, and their actions are unpredictable. They are at home on these bleak mountain fastnesses and seeing a caravan of a foreigner incites wild curiosity and the pleasant anticipation of robbing him to find out what treasures he carries and to come into their possession.

When traveling in the land of the Go-log or in Tibetan nomad country in general one must be prepared for all eventualities; the Go-log not acquainted with high powered rifles that carry much further than his flint lock guns or his guns of 1870 from the Franco-Prussian war, relies on his numerical strength and is surprised when he meets unexpected resistance by a small party and due to superior arms can stave him off. To travel in that area is hazardous to say the least, and one must be constantly on the look out and prepared for one might be treacherously attacked at any moment. They have been the bane of caravans who cross from Hsi-ning on the main caravan road to Tibet, for Go-logs often go in large bands, at times of six hundred or more, to loot and rob.

After the first world war rifles of American make have come into the hands of Go-log and Tibetan nomads in general. These rifles had very primitive stocks and and were made for the Russians at a time when they had practically nothing to fight with towards the end of the first world war. I was told that they could be bought for nine US dollars a piece in the United States. The Russians, not having had any further use for these rifles, sold them to Mongol and Moslem traders who again sold them at a tremendous profit to the Go-logs and Tibetans, smuggling them from Mongolia into the Kokonor or via Ordos where Chinese had little or no control.

The Go-log are the proudest and most independent of all Tibetans, they call themselves a free people and declare that King Ke-sar of Ling who fought wars with the Hor, the Chinese and many other people, had his tents at the Am-nye Ma-chhen and that his miraculous sword, still hidden somewhere in the Am-nye Ma-chhen has lost nothing of its power. Their courage is derived from Ke-sar and his miraculous sword, and from Am-nye Ma-chhen Pom-ra their protector. There are few tribes who dwell around the

Am-nye Ma-chhen who have not been conquered by the Go-logs, the exception being the Ngu-ra and the Ong-thag tribes, the first living within the knee of the Yellow River, and the second northwest of the Am-nye Ma-chhen. It was the latter tribe which attacked Roborowski and his caravan. He speaks of later meeting what was then a former incarnation of Shing-za Pandita the highest incarnation of Ra-gya Gom-pa; the last one whom we met was a brother of the chief of the Ong-thag tribe. Although a young man of 20 years, he has since died and whether his incarnation has been found or not is now not known, it is very likely that the present regime will prohibit incarnations from re-appearing.

It can thus be seen that to work peacefully and to explore as one might elsewhere in China, in the wild and inhospitable land of the Go-logs is next to an impossibility. There was no necessity on the part of Leonard Clark[42] to make a rush-dash, hurried trip, and then run. With such an outfit, or such large escort, with machine guns and what not, one could have stayed a month or longer; they could have had nothing whatever to fear, and it is a pity, that having had such an opportunity, no better use was made of it, except to produce a phantastic story.

As regards K. S. Hao's journey[43], any one who knows the country can see at a glance that the line on the map which is to indicate his itinerary is not reconciled with the terrain as it actually exists, and as he figures it on his primitive map. No one who claims to have made the long detour within the loop of the Yellow River could have ignored the large rivers as Chhu-ngön and the Tshab Chhu. Had he gone to the east of the Am-nye Ma-chhen as indicated on his map, he would have had to cross the Chhu-ngön River which is unfordable in that region; he never even indicates the existence of such a river. He states that he spent three days at the Amne Matchin, as he calls the mountain, which he puts in an entirely different location from the one it actually is. He gives a French name, from an old French atlas to a range, which is near where the real Am-nye Ma-chhen is, but draws it as extending from east to west, when it really stretches from northwest to southeast. Where he has been is probably at one end of the range which is the divide between the To-su Nor, the Chinese T'o-szu Hu 托絲湖, and the Tibetan sTong-ri mtsho-nag, or the Black Lake of the Thousand Mountains; To-su Nor is the Mongol name, and the Chhu-ngön which flows from west of the Am-nye Ma-chhen, makes a big bend north, and debouches into the Yellow River a little above the Bâ River. He marks no range southeast of the To-su nor (which he calls Tso-go) and the Chhu-ngön, this alone is sufficient to show that he could not have been further than to the northwestern side of the To-su nor and the Chhu-ngön divide, for the latter river, the largest and most difficult to cross on account of its many rapids and cataracts, except in its northern bend, is not at all on his map. Then the most impossible of all, he shows his route as crossing practically the center of the range. No more need be said. Had he actually been even at the western foothills of the Range he would not have drawn or published the map he did. The map shows a total ignorance of the country.

During our absence from Ra-gya robbers had come to our quarters and stole some bags of grain, but nothing of importance, my men, a few of whom I had left behind,

---

[42] *The marching wind.* By Leonard Clark. New York: Fung & Wagnalls 1954. XVI, 368 pp.

[43] K. S. Hao: Pflanzengeographische Studien über den Kokonor-See und über das angrenzende Gebiet. *Botanische Jahrbücher* Bd 68.1938, 515-668.

chased them off with rifle shorts; the lamas begged them not to kill any robbers, they could catch them, but must not shoot at them. A rather peculiar system, the robbers are permitted to shoot and kill, but one must not shoot *them*. A peculiar incident happened during our absence. Moslem traders had come from Rong-wo in the north with grain, mostly barley, to exchange for wool from the nomads. They had gone across the Yellow River where Go-logs were encamped, to buy yak from them to transport the wool back to Rong-wo. They had bought 88 yak, but it was dark before the bargain had been struck and the yaks paid for. As it was known to the ferry men at Ra-gya that the traders were buying yak, and they notified their encampment which was not far from the Yellow River that there was an opportunity to steal some yak during the night, for it had been too late to swim the yak across the river to Ra-gya monastery. During the night robbers came and tried to drive off some yak, the third time the robbers came, they had been chased away twice, the Moslem traders opened fire and unfortunately they killed one of the would-be bandits. The remainder fled to their encampments, and the Moslems being scared, crossed the Yellow River to Ra-gya and left their yak on the oother side.

Early in the morning the whole tribe had come down to revenge the would-be robber's death, and the first act of theirs was to drive off the 88 yak. Not satisfied, they threatened to cross over to Ra-gya to fetch the traders. The latter were scared, and as the Tibetans demanded life money for the man they had shot, they sent them 50 taels of silver about $30.00 U.S., one rifle and ammunition and three horses to appease them and prevent them from coming over to the Ra-gya side. The members of the tribe to whom the dead robber belonged thereupon demanded 4,000 taels for the life of their confrere which equalled about 2,960 dollar U. S. However, they were satisfied with the horses, one rifle and 40 taels of silver plus 88 yak and stayed on the other side. This gives a fair idea of conditions under which people carry on trade with these haughty devils.

As to the different Tibetan clans or tribes who dwell in the region of the knee of the Yellow River and the Am-nye Ma-chhen, the following notes will be of interest as they give for the first time the names of clans and the territory each owns. They are nomads only to the extent that they move from their winter encampments which are at lower elevations to their summer encampments on the higher alpine meadows up to 13,000 feet and more. Each clan jealously guards its tribal lands, for should other clans herd their sheep in either the winter or summer encampments during the clan's absence, their animals would starve to death if others should have allowed their sheep or yak to overgraze the land.

To the right, north of the Am-nye Ma-chhen is a red rocky range, mostly scree called Mang-dgun Ula (Mang-gün) the last is a Mongol word meaning mountain. A pass leads over the range called the Man gün La reached by Roborowski in the winter of 1895. He was there attacked by what he calls Tanguts. It is the Ong-thag tribe who live at the Mang-gün Ula; this range is visible in the picture of the Am-nye Ma-chhen taken from Am-nye Drug-gu. It is the dark range which seems to extend from the northern end of the Am-nye Ma-chhen, but is actually east of it, the Mang-gün pass is also visible in the photograph, it is the deep gap in the range. It seems that the incarnation of Shing-za of Ra-gya has been found several times among the family of the chief of the Ong-thag tribe, for when Roborowski met the Shing-za incarnation it was in the territory of the Ong-thag where he had been attacked. He told him that if he had only met him previously the attack would not have taken place and he invited him to come to Ra-gya, but Roborowski left for the Koko Nor.

At the headwaters of the Dom Khog the Lür-di [Lus-rde] tribe has its encampment. This tribe was ruled over by a so-called queen. She became known as the queen of the Go-log, but the way she acquired the title was rather a distasteful one. During the war between the Go-log and the Moslems of Hsi-ning she was captured and brought to La-brang and held there; General Ma conceived the idea to cohabit with her, and when she had produced a sone he declared her queen of the Go-log, thinking that by so doing he would gain the favor of the Go-log. As it was the Go-log rejected her for she was the first to submit. At the time when she was captured there was also taken prisoner the mother of the powerful chief of the Ri-mang Go-log, who, to save his own life fled across the Yellow river on a raft and sent the raft down with the current, leaving his mother on the bank of the river.

Later on the so-called queen of the Go-log ransomed the mother of the Ri-mang chief with 500 heads of yak, whereupon she was permitted to return to her tribe, but she is not now, nor ever was queen of the Go-log, it was a Mohammedan invention. She only ruled over 600 tents.

The Tshang-rgur tribe, whose chief is called Tshang-ba rku-chhung (Tshang-wa ku-chhung) has its encampments at the head of Gur-zhung Valley which extends from south to north and debauches into the Tshab Chhu. [Plate **80-81**]

The hBu-tshang mGo-log live to the south of the Ri-mang tribe, that is south of the Sha-ri yang-ra range (see Plate **82**). To the southwest of the Am-nye Ma-chhen is a red scree range called gLang-me-btsag-dmar (Lang-me tsa-mar) which is the mountain god of the Me-tsang Ta-wo; in reality they are not a Go-log tribe, but one of the nomad Tibetan clans who suffered constantly at the hands of the Go-log robbers, and in order to gain the protection of the powerful Khang-sar chief they joined his tribe and thus became naturalized Go-log.

The largest of all the Go-log, the Ri-mang, have their encampments south of the Yellow River and west of the Nga-ba tribe. The Khang-sar live to the west of Ri-mang and the Khang-gen to the west of the Khang-sar. Their mountain god is the beautiful limestone range gNyan-po-gyu-rtse-rdza-ra (Nyen-po-yu-tse-dza-ra).

We were still trying to make arrangements for yak and guides to take us to the Amnye Ma-chhen, when the high incarnation Shing-za sent word to us that he would like to see us. I sent Mr. Simpson to his dwelling and he informed us that there was an opportunity to leave for the Am-nye Ma-chhen early next morning and under the following circumstances: A few days previously an incarnation from Ra-gya had returned from the Am-nye with his steward and some lamas, where they had been camping among the Go-log to bless their herds etc. On their return journey they were waylaid by some Go-logs who shot the steward and robbed the lamas. Ra-gya lamasery had decided to send sixty lamas to the territory of the tribe to curse them and we were to join the lamas on their cursing expedition to the tribe whose encampment was at the foot of the Am-nye Ma-chhen.

The news only reached us in the evening and it was impossible to get yak and arrange for such a trip at a few hours notice, and we saw the lamas leave the next morning unable to join them as our yak which we had sent for had not shown up.

In the meantime word was sent to us from the tribes living at the Am-nye Ma-chhen that should we decide to come they would lie in wait for us and would kill us. They also threatened tribes living near Ra-gya that should any of one take us across they would be their enemy for life.

Everybody was scared to rent us yak or take us across the river to their sacred mountain, until on July 14th we actually left having made arrangement with the rGya-bzah (Gya-zâ) clan, a Tibetan tribe which had its encampment west of the Yellow River in the Shag-lung Valley, to take us as far as it was wise and possible. We engaged all the male members of the clan; only the chief came to our quarters the morning of our departure while the other Tibetans were waiting for us at their encampment where we were to spend the first night. The ferry people who took us across on their skin rafts shook their heads and chided the Gya-zâ chief for taking us to the mountain. [Plate **83-84**]

The cook had packed the necessary food for a fortnight, I took my small tent and the men theirs, the Tibetans had no need for a tent, they pulled their sheepskin garment over their heads and slept on their saddle blankets. In fact we took just the absolutely necessary, we were all armed and hoped to shoot for the pot, for game on the other side of the Yellow River is plentiful.

Start for the Am-nye Ma-chhen
Early on July 14th we crossed the Yellow River on the flimsy goatskin raft, the horses swimming. Through the carelessness of the ferry men, or on purpose, one horse went adrift and was lost in the current. It was not a propitious beginning. We had hardly been all across with our riding and packing animals when a terrific thunderstorm overtook us.

The blackest clouds I ever saw discharged their contents over the Ra-gya valley, so that I took shelter in a nomad tent near the river. The lone woman in the tent had nothing but a scowl for us, such hostile and unfriendly people I have never met anywhere in the world, it seems that a smile never crosses their coarse features. As soon as the storm had somewhat abated, it was then 3.30 p.m., we followed the sandy streambed to the little U-lan Valley at the mouth of the Yellow River defile. The U-lan brook had become a brown torrent which we had to cross and recross many times, the steep hillside had become an impassable, red clay, mud slide. Riding was impossible and the ascent from the Yellow River proved most difficult. The storm had passed over Ra-gya and the hills beyond, while towards the west blue sky became visible. We reached a pass 11,150 feet with a few tents to the left (south), then came to a higher one 11,650 feet, and from there we could look west into a clear sky. From this last pass our trail led down a small gully which brought us into the Shar-lung valley, an affluent of the Shag-lung Chhu, which has its source in the hBrug-dgu Nye-ra (Drug-gu Nye-ra). It flows first east and then north in a canyon into the Yellow River, opposite the Kha-khi pass. The water was a deep cinnabar red, but the fording did not prove difficult. Continuing up a lateral valley and skirting small gullies, we crossed a spur and arrived at the encampment of the Gya-zâ (rGya-bzah) clan at 7:30 p.m., just before the darkness set in. This was the first time that any white man hat pitched camp west of the Yellow River, and east of the Am-nye Ma-chhen.

We rose at 5 a.m. and after a frugal breakfast we assembled our Gya-zâ nomad escort and sallied forth into the unknown. There seemed to be no trail, but we went in a southwesterly direction over bare grassy hillsides without a tree visible anywhere, till we came to cliffs of red conglomerate, in a line with those back of Ra-gya but east-southeast. These cliffs culminate into high rocky red bluffs which were crowned by *Juniperus tibetica* Kom. These red conglomerate crags represent the mountain god of the Gya-zâ clan who is called Am-nye dGe-tho (ge-tho); at the base of the cliffs, on the left of the trail (going southwest) is an obo or cairn of rocks, sticks and rags where the clan burns juniper boughs as offering to Am-nye Ge-tho. The elevation at the Obo is 11,800 feet. The trail continued to lead west over grassy hills, up and down at an elevation of 12,100 feet, turns northwest leaving a deep valley to our left with sunken terraces, one above the other called rDo-btseg (Do-tse), this valley the trail descends, it is the left branch of the Shag-lung, here called the Dragon Valley or hBrug-nang (Drug-nang) which led to a pass, the Drug-gi Nye-ra [hBrug-gi nye-ra] or the Pass of the Dragon, at an elevation of 14,250 feet.

The vegetation on this pass is very similar to that found on the Wo-ti La, with the exception that here we found an abundance of the red poppy *Meconopsis punicea* Max., which is absent east of the Yellow River. On the schist scree we found the dark purplish blue *Saussurea hypsipeta* Max. Another still undescribed species (Saussurea sp? no 14413) with dark, blackish to purple flowerheads we collected on a high rocky range to the east of the pass at an elevation of 15,000 feet, where it forms clumps. A crucifer *Dilophia macrosperma* O. E. Schulz, with white flowers flourished on the scree on the hills around the pass, together with *Anemone rupestris* Wall., here with dirty-straw-colored to white flowers, forming rosettes, while the rest of the vegetation was the same as found east of the Yellow River on the high passes leading north.

The ascent as well as descent of the Drug-gi Nye-ra were very difficult as the pass was one huge bog. A short distance below the pass we spied three huge sheep, they were of enormous size, with horn which must have been a foot or more in diameter at the base, they are called rNyan (Nyen) by the Tibetans. Their horns did not spread laterally but extended forward in close spirals as I could see with my glasses. I fired but missed, the bullet kicking up the rocks under their feet. They differed considerably from the big horns I have seen in the New York Natural History Museum, as *Ovis Ammon*, and *Ovis Poli*, and I am firmly convinced that the Am-nye Ma-chhen big horn or Nyen represents an undescribed species. The Tibetans told us that old ram often die of starvation, as owing to their huge horns whose spirals extend forward, beyond their snouts, they cannot reach the grass in the winter or dig it out of the snow.

This pass is the divide between the Gya-zâ and the gYon-gzhi (Yön-zhi) tribe, and led into the Tsha-chhen Valley; their territory extends to the Tshab Chhu and Yellow River, and the number of tents which make up their encampments do not exceed two hundred. In order not to meet with the Yön-zhi tribe so as not to give them an opportunity to spread the news of our presence, we camped in a small valley called mTshan lung (Tshen-lung) above juniper forests (*Juniperus tibetica* Kom.) at an elevation of 12,890 feet.

After a tranquil night in our secluded valley we made our way down into the valley of Tsha-chhen and there encountered lovely groves of *Juniperus tibetica* Kom., huge trees with large trunks of two and more feet in diameter and forty to fifty feet in height (see Plate **85**). On the gravelly slopes of the high spurs enclosing the Tsha-chhen we found *Hedysarum pseudastragalus* Ulbrich, a prostrate plant at an elevation of 14,200 feet, in company with *Primula Purdomi* Craib, and among rocks a species of *Saussurea* (no 14425) as yet not described, with flowerheads which exuded a sour-sweetish, sickly odor resembling that of decayed bananas. In the shade of the junipers grew the deep bluish-purple *Salvia Prattii* Hemsl., and on the moist meadows occurred *Primula sikkimensis* Hook., the only place where we found it in the entire region. We stopped for lunch in a meadow in the Tsha-chhen Valley which was one mass of the blue poppy *Meconopsis racemosa* Max., now called *M. horridula* Prain; it is possible that the two plants are identical, but I prefer to keep this plant of the highlands of the far northwest separate from the southern plant, first found in the west Tibetan Himalaya, or retain it at least as a variety of *horridula*.

While we were collecting plants and taking photographs of the Tsha-chhen Valley and part of our Nomad escort, Mr. Simpson and the chief of the Gya-zâ clan went up a small valley to see the chief of the Yön-zhi clan; he had however shifted his camp to the mouth of the Tsha-chhen Valley near the Yellow River.

The scenery increased in beauty as we ascended the valley both sides of which were forested with junipers. The blue poppy and the yellow *Primula sikkimensis* Hook., grew everywhere in the wet meadows. Willows were common as *Salix oritrepha* Schneid., and its variety *tibetica* Goerz var. nov. and *Salix Rockii* Goerz nov. sp., *Sibiraea angustata* (Rehd.) Hao, *Potentilla fruticosa* L., with varieties, *parvifolia* with yellow flowers, and var. *Veitchii* with cream colored flowers, also small flowered *Spiraea alpina* Pallas, and the yellow flowered *Caragana brevifolia* Kom., all of which grew along the stream or in moist meadows at the foot and valley slopes.

Climbing the left valley slopes, as the valley itself became too narrow, we descended at the mouth of the Ta-rang Valley where we encountered the new *Pedicularis calosantha* Li sp. n., with pink flowers spotted purple, it grew in meadows of the Ta-rang Valley with *Pedicularis szechuanica* Max., *typica* Li, at 13,000 feet elevation, also *Leontopodium linearifolium* H.-M., *Leontopodium Souliei* Bod., and the new *Pedicularis paiana* Li.

The trail led through absolute virgin forest of *Juniperus tibetica* Kom., and here suspended over the trail on yak hair string from a juniper branch were a row of mutton and yak shoulder blades, one below the other, incscribed with the sacred formula Om mani padme hum. In order to pass along the trail it became necessary to push the string of bones aside, for they hung very low, thereby saying the prayers written on the bones, for the benefit of the person who so suspended them.

Our aim was to reach the foot of Am-nye Ma-chhen Drug-gu a high mountain in the territory of the Yön-zhi clan whence a fine view could be obtained of the Am-nye Ma-chhen.

After crossing the wooded spur over an execrable and slippery trail we descended into a lateral valley, the hBrug-dgu Nang (Drug-gu Nang) or Nine Dragon Valley forested with *Picea asperata* Mast., on the northern slopes and *Juniperus tibetica* Kom., on the southern. In this valley we encountered the tents of the Yön-zhi tribe or clan; they seemed quite friendly to us, although they are absolutely a law unto themselves, and acknowledge no authority. Yet they seemed perplexed about our cavalcade, strange tents, and the escort which consisted practically of all the male members of the Gya-zâ clan [plate **86**], for we had appeared as out of nowhere. They were so suspicious that during the following night they packed up, and in the morning they had vanished and there was not a vestige of a tent or a nomad to be seen.

Some infectious disease had broken out among them, probably relapsing fever, carried by lice, of which the Tibetans are never free. We saw some dying outside their tents covered with rags, and chief Gomba of the Gya-zâ clan who was with me, held his nose and gave them a wide berth and motioned to me to do likewise.

We followed up to the head of the valley past all the nomads and pitched camp at the foot of the Am-nye Ma-chhen Drug-gu at an elevation of 12,500 feet. [Plate **87-88**] As we had arrived quite early, I decided to climb to the top of the mountain to see if I could obtain a view of the Am-nye Ma-chhen; it was a stiff climb after a hard day's ride, but we reached the summit in due time and made the altitude of 14,450 feet, or nearly 2,000 feet above our camp. On the top were a few prayer flags an indication that the Yön-zhi Tibetans burnt juniper boughs as offerings to the mountain god Am-nye Ma-chhen. Am-nye Drug-gu is the protector and mountain god of the Yön-zhi tribe. We had occasional glimpses of the mountain mostly hidden in clouds. The scenery was superb; below, in front of us flowed the mGur-zhung (Gur-zhung) in a deep valley, the river not visible, and debouching into the Tshab Chhu, and the latter into Yellow River about 15 miles below the Tsha-chhen. To the right of the Am-nye Ma-chhen we could see the Ye Khog (valley), and to the left the Yön Khog (valley) which together with the hDom Khog form the Tshab Chhu.

The vegetation on the summit of Am-nye Drug-gu consisted mainly of scree plants as *Saussurea hypsipeta* Max., *Dilophia macrosperma* O. E. Schulz, an endemic crucifer with white flowers; on muddy gravel flourished *Cremanthodium Decaisnii* G. B. Clarke, with yellow flowerheads. In large rubble of schist thrived the peculiar *Saussurea medusa* Max., *Arenaria Przewalskii* Max., a very ornamental white flowered caryophyllaceous plant, the umbelliferous *Pleurospermum thalictrifolium* Wolff, a prostrate plant with grey flowers, a purple? flowered Saussurea, undetermined as the flowers were not fully developed, *Meconopsis racemosa* Max., and *Pedicularis chenocephala* Diels, its flowers a rich red; not on the scree, but on the grassy slopes below the summit, at 13,000 feet, and five hundred feet lower, grew the lovely *Codonopsis bulleyana* Franch.; but it may be a related species.

On the summit in grassy areas occurred also the pink flowered *Caragana jubata* Poir., here a prostrate shrub quite stunted and very pubescent, and a thousand feet lower *Meconopsis punicea* Max., whose red drooping flowers swayed in the wind like little bells. It is strange that it should be found west of the Yellow River and not again east until one approaches La-brang. Preferring the moist alpine slopes at 15,000 feet, that is half way up the mountain from the camp, we found many individuals of Pedicularis as

Pedicularis Przewalskii Max., with deep red flowers and its yellow-flowered congener *Pedicularis lasiophris* Max., also a monkshood *Aconitum rotundifolium* K. & K. var. *tanguticum* Max., with purplish blue flowers. Scattered on the grassy slopes grew close to the ground in the form of a hollow rosette with yellow flowers, *Crepis Hookeriana* G. B. Clarke, and on the scree near the summit the saxifrage *Saxifraga melanocentra* Franch., var. *pluriflora* Engl. & Irmsch., a curious plant with white corolla and deep purple calyx, but no Corydalis. These with the here common *Anemone imbricata* Max., and *A. rupestris* Wall., formed the plant growth on this mountain.

We spent some time on the summit, enjoying the glory of a setting sun over the huge massive of the Am-nye Ma-chhen.

The following morning was absolutely cloudless and we made haste to reach the summit of Am-nye Drug-gu; it was a perfect morning and the range lay before us in all its whiteness and purity, the sun shining on its dazzling glaciers directly to the west of us, and thus making it appear somewhat flat. I had no means to measure its height but judging from the elevation from which we beheld the range, I thought then that it probably might be in the neighborhood of 28,000 feet. but after having seen Mi-nyags Gangs-dkar from a much higher elevation, and knowing the height of it, I cannot help but come to the conclusion that the Am-nye Ma-chhen is between 20,000 and 21,000 feet, and especially since having seen Mt. Everest so recently (1950-51) and so close, from a plane which flew at 12,500 feet, there is no comparison as height is concerned between these two mountain ranges. As to Clark's recent measurements trying to prove the mountain higher than Everest, and reckoning the height from a base elevation which he found on a Chinese map, needs no further comment.

We took many photographs of the range (see Plate **71-73**) also of the bare ranges to the north with the Gyü-par mountains in the distance and the gorges of the Yellow River indicated by the opposite converging slopes of the much disected plateau.

We descended from Am-nye Drug-gu a spur which led directly north-north-east to a bluff whence we could overlook the Yellow River in its gorges, as there was no trail it was more or less difficult as willow bushes and masses of *Rhododendron capitatum* Max. barred our way. The bluff was about 1,500 feet above the Yellow River whence we took photos down and up stream. *Picea asperata* Mast., and *Juniperus tibetica* Kom, and *Juniperus Przewalskii* Kom., grew below the bluff with willows, and up to the grass-covered spur *Rhododendron capitatum* Max. Looking north the Yellow River flows in bare canyons making sharp zig-zags which are visible in the photograph.

Hardly had we returned to our camp when a thunderstorm deluged it. As the place was not a delightful one but was selected only so as to make the ascent of Mount Drug-gu more easy, we struck camp and descended the Tsha-chhen Valley where we pitched camp in the juniper forest at an elevation of 10,950 feet. Our camp was in the midst of *Primula sikkimensis* Hook., and *Meconopsis racemosa* Max.; here was also a small cave where our Tibetans could sleep more or less protected from rain.

A few words about the Tshab Chhu Valley: The valley is forested on its northern slopes with spruces, *Picea asperata* Mast., and with junipers on the southern slopes, the former occur only near the mouth where it opens into the Yellow River, while the latter extend much further further up the valley. The region is poor botanically. Although the summer was well advanced there were few herbaceous plants about, gentians were absent, Pedicularis and Corydalis were scarce, Delphiniums and Aconites had not yet flowered. The region is apparently too high and the summer too short for the development of an alpine flora such as is found on the Min Shan and the mountains further south.

Everywhere one looked there was game, near the stream in a bend of the valley we met a huge stag but did not get a shot at it. Musk deer were abundant, but the great sheep, the Tibetan Nyen, were restricted to the very high crags at 15,000-16,000 feet; unlike the blue sheep who go in large groups, the Nyen are only to be found in pairs or with one young. They are very wary and difficult to approach.

Camp was pitched below the Drug-gi Nye-ra at 12,700 feet opposite a small valley called Ti-nag near the head waters of the Tsha-chhen valley. Back of our camp was a high rocky spur and this I climbed to get another view of the Am-nye Ma-chhen Range. The summit of the spur was 14,900 feet, the rocks were schist and shale and some with quartz, but there were very few plants to be seen and none we had not already collected. As it proved we could see only the great pyramid Chen-re-zig, the eastern peak of the range. To the west I saw a high rocky range of scree which formed the wall of the Ta-rang Valley, and I decided to climb that range to the highest point to get a closer view of the Am-nye Ma-chhen.

It rained all night and in the early morning the rain had changed to snow, as I looked out of my tent the whole landscape was covered with a mantle of pure white. Several inches of snow had fallen, and as I could not see our Tibetans, I called, when the snow moved and out looked the Tibetans from under their snow covered felt rain coats which had served as their bedding, they laughed, the first time I saw nomads laugh; they were very cheerful and seemed to enjoy the situation. Clouds hung low over the passes and the mountains which enclosed our valley. The thermometer registered 32° F. at 7 a.m. Inspite of all the snow, the petals of the blue and red Meconopsis which were common here in the grass, were as bright and fresh and unharmed, each flower wore a cap of snow, the leaves buried in snow, it was a beautiful picture and showed the hardiness of these alpines, as hardy as the nomads who inhabit these mountains.

When the Nomads had boiled their tea, a man would take a large ladle, dip it into the large pot, and amidst the chattering of prayers would throw the tea into the air as offering to the mountain gods; only then would they sit down and eat their frugal meal of buttered tea and tsamba (roasted barley flour).

We left our camp opposite the Ti-nag valley and climbed the opposite valley wall deeply covered with snow, and this on July 19th; the snow increased in depth as we ascended and the ground became boggy. We reached a pass at 14,100 feet and saw the clouds lifting and the sun peeped faintly through the mist, necessitating snow glasses against snow blindness. From the pass we descended into the Ta-rang valley which we followed down stream for a short distance and left our men to pitch camp at the mouth of a small lateral valley which led to the summit of Sha-chhui-yim-khar (Sha-chhui-yim-khar). Chief Gomba of the Gya-zâ clan and some of my men had been hunting and brought back a male musk deer with large tusks, I was glad for it replenished our provisions. In the evening the sky had cleared and we hoped for fine weather in the morning; the elevation of our camp in the Ta-rang valley was 12,300 feet. The vegetation consisted of willow shrubs, of *Salix oritrepha* Schneid., and its variety *tibetica* Goerz, *Salix Rockii* Goerz, *Spiraea alpina* Pall., and *Potentilla fruticosa* L. var. *parvifolia* (Fisch.) Wolf., while junipers grew in the mouth of the valley.

The morning of July 20th dawned brightly without a cloud in the sky. It was a difficult climb to the summit of the Sha-chhui-yim-khar as there was no trail. We left our horses at the foot of the scree and climbed over the frozen scree and slate; the peak we saw from our camp was only a preliminary one, a rocky eminence, the main peak being still quite a distance beyond it. On the slaty slopes grew blue-flowered Saussureas, tiny rosettes with wooly heads on the top of which protruded the blue flowerheads; the plant has not yet been determined (no 14412) and should it prove to be new, I would like to propose the name *Saussurea Simpsoni*, in honor of William E. Simpson Jr. an intrepid traveller and an expert in Tibetan; he had an extensive knowledge of the grasslands and of the nomads. He fell victim to the bullets of Moslem bandits while on an errand for his father in Shensi.

From the slopes of the mountain I gained a peep of the large pyramid of Chen-re-zig of the Am-nye Ma-chhen Range and we shouted for joy, being certain of a glorious view from the summit for there was not a cloud in the sky.

A cold wind blew at the top which we made 15,200 feet, and as the temperature was 25° F., we wrapped ourselves in our fur coats. It felt bitterly cold for the 20th of July. Soon the wind ceased and the sun appeared and we soon forgot about the cold for before us lay one of the grandest mountain ranges of Asia. The dome in the north is the highest part, but it is not so imposing as the large pyramid at the southern end. There was no haze and we secured some good photos. In front of us lay the Gur-zhung Valley extending from South to North, and the Tshab Chhu from East to West, the Gur-zhung bearing junipers on its upper slopes.

It was difficult for me to tear away from this sublime view, especially as I knew I would never see it again. The range was covered for about 4,000 feet with eternal snow and was indeed a grand spectacle, the pyramid of snow Chen-re-zig being especially beautiful. I could have remained for hours on that summit never tiring of this grand view, Simpson and I being the first white men privileged to view the range from west of the Yellow River. The Go-logs so far had been ignorant of our presence, we were above their camps. While we were photographing the range there appeared a Go-log from the Gur-zhung valley to burn juniper boughs to the Am-nye Ma-chhen. We were astonished and so was he. He did not tarry long after he had set his junipers on fire and mumbled his prayer. He was certainly to spread the news of our presence.

After all had descended I still remained on the summit of Sha-chhui-yim-khar drinking in the the glorious view and collecting the queer Saussureas on the slaty summit. With a heavy heart I tore myself away, and with one last glance back at the great Am-nye Ma-chhen I descended the steep slopes to our camp.

After a repast of musk deer and rice we struck camp and followed up the Ta-rang Valley to near its head, the valley is shallow, its sides gently sloping, its head an amphitheater of slaty scree, and boggy. We turn left up a pass to 14,200 feet whence we had one more view of the snowy range against a grey sky for clouds were beginning to gather. We descended a narrow valley composed of slate in its upper part and red conglomerate in its lower. We made haste to leave the region before the Go-logs should be aroused and perhaps pursue us, for the chief of the Bu-tshang Go-logs, who was the steward of the incarnation in whose house we lived in Ra-gya, swore before crossing the Yellow River that should we come anywhere near his encampment he would kill us. The reason being that I refused to give him one of our colt automatics.

On the Brag-nag Nye-ra (Drag-nag nye-ra) or Black Rock pass, elevation 14,520 feet, we found a lone Corydalis which proved new and was named by the late Dr. Fedde[44] *Corydalis Rheinbabeniana* Fedde, also the *Astragalus tongolensis* var. *glaber* Peter-Stib.; other plants found were *Pedicularis szechuanica* Max. *typica* Li; *Pedicularis calosantha* Li sp. n. *Leontopodium linearifolium* Hand.-Maz., and *Leontopodium Souliei* Beauvd.

We now returned to the Gya-zâ encampment the same way we had come and finally to Ra-gya Gom-pa without delay.

The Summer Flora of the Grasslands Between Ra-gya and La-brang

In the Wa-ru valley grasses did not form regular meadows but among them grew willows, and the familiar bushes of Potentilla, Sibiraea, Caragana, Ribes, etc. The most common grasses were *Poa attenuata* Trin., *Koeleria argentea* Grieseb., *Poa arctica* R. Br., *Deschampsia cespitosa* (L.) Beauv., *Elymus sibiricus* L., and its variety *brachystachys* Keng, a species of Trisetum, *Festuca ovina* L., *Stipa mongolica* Turcz., and the new *Koeleria enodis* Keng, among them grew *Leontopodium linearifolium* H.-M., a species of Allium (no 14449) but this extended also to the conglomerate cliffs of the Wa-ru Valley, and a yellow-flowered Pedicularis as yet not determined (no 14450).

On Wa-ru Khang-mdun pass (Wa-ru Khang-dün) elevation 13,840 feet, the pass being opposite the Wa-ru La, up to an elevation of 14,150 feet I found the first *Delphinium albocoeruleum* Max., well named for its pale blue flowers, it was restricted to the gravelly slopes as was its congener *Delphinium Souliei* Franch., which contrasted by its deep blue to dark purplish flowers. Associated with them were *Leontopodium linearifolium* H.-M., the lavender flowered *Meconopsis quintuplinervia* Reg., *Cremanthodium decaisnei* G. B. Clarke, the cyperaceous *Kobresia Prattii* C. B. Cl., the new *Pleurospermum Dielsianum* Fedde, and *Pedicularis cheilanthifolia* Schrenk *typica* L.; the latter white-flowered, a rather uncommon color for a Pedicularis, which turn however yellow on drying; it extends from Mongolia to the East Himalaya range and the Koko Nor. All these plants grew on loose scree from a elevation of 13,800 feet up to the summit of the bluff, Pedicularis invading also the grassy slopes.

In the Ser-chhen Valley up stream we found in the grass the rosette-forming *Microula tibetica* Benth., with blue flowers like a forget-me-not and belonging to the same family.

At 13,000 feet on the moist slopes among rocks thrived the pink flowered *Primula stenocalyx* Max., and on the grassy slopes the *Saussurea pygmaea* Spreng., or perhaps a variety of it, with purple flowers. This species is also known from the Altai and T'ien Shan 天山, and seems to reach here its southern limit. At 13,200 feet in swampy

---

[44] Friedrich Fedde, editor of *Repertorium novarum specierum regni vegetabilis*. Berlin 1905 ff.

meadows with stagnant water were masses of the yellow flowered *Cremanthodium plantagineum* Max., one of the commonest composites in swampy areas.

One episode occurred on our return journey worth recording. The whole of the sGar-rtse (Gar-tse) tribe had assembled in the upper Ser-chhen Valley and hundreds of their black tents were spread over the flat grassy slopes, but the scene did not have a peaceful aspect for mounted armed men rushed about excitedly. Four men rode up to us and asked us if we had engaged the U-jâ tribe in a fight and how many we had killed. We looked at them in astonishment whereupon they informed us that while they were celebrating a sort of thanksgiving for the increase of their herds, their neighbors the U-jâ people, who have their encampment across a spur to the north in the Gar-gen valley, had come over and robbed the Gar-tse clan of a hundred yak which they had driven off. It seemed particularly unkind of the U-jâ people to come and rob when all the members of the Gar-tse tribe were attending their annual summer prayers, but I suppose the Gar-tse people had played similar tricks to the U-jâ, and the latter were probably only retaliating.

We came to a circle of white lama tents, one with windows painted on the canvas, this was occupied by rNying-ma-pa or red sect lama-sorcerers who were performing a religious ceremony. Seated on the ground within the tent were twenty-five lama sorcerers with their long hair (this sect never cuts the hair but lets it grow to enormous length) wrapped around their heads like a turban. From the ceiling of the tent were suspended ten flat, circular drums which the sorcerers were beating rhythmically, while others were blowing brass trumpets and clashing cymbals. They all chanted in a peculiar voice very different from Yellow sect lamas. In front of the tent stood three sorcerers with exceedingly long strands of dirty hair, one in particular had let his hair fall over his shoulder on his arm and then again thrown over his shoulder from which it hung down to the hem of his robes near his feet. The strands of hair were the thickness of a rope and filthy beyond description. Each of the three wore a large round hat from the brim of which hung long black fringes which reached to their mouth. On the top of the hat was the hideous head of some demon, while small skulls were fastened around the rim of the hats. One wore a red garment, the center lama an imitation tiger garment, and the right hand one a purple garment. To the left, further back of the three chanting sorcerers stood a tall old lama with long strands of grey hair wound around his neck and falling over his shoulders. He held a spear in his hand to which was tied a square flag, black in the middle with a red border. In the center was the syllable hum. This banner he dipped while he chanted Om-a-hum, a mystic formula in many repetitions. His cheeks were sunken and so was his mouth for lack of teeth, and in all he had a very peculiar physiognomy. The chanting continued alternatingly and then a single voice would drone forth to be joined by the chorus of the assembled sorcerers. From the black hats of the sorcerers flowed silk scarfs of red, yellow, green, white and blue. In the rear and center of the tent hung a painted scroll covered with scarfs. The left sorcerer of the three in front of the tent was given a brass bowl in which reposed a gtor-ma (tor-ma) an offering, pyramidal in shape, and made of barley flour and butter, with this he walked about 200 yards followed by the other two and threw the tor-ma out on the grass, all this was carried out in slow motion.

We left the sorcerers and followed up the valley where I found that our tents had been pitched near the stream in the middle of the valley. We met more excited and armed men on horseback who asked us if we had engaged in the fight. They all were after the U-jâ nomads in the next valley, and that fighting was going on between the two clans. I was about to retire when I heard shooting and people running about, five had already been killed on the Gar-tse side, also six horses, and many had been wounded. Not to be in the midst of the fray we moved our camp up a small ravine farther up the valley, at 12,700 feet elevation, so as to be out of the line of fire should the fight spread into the Ser-chhen valley.

All the male members of the Gar-tse tribe were being gathered together and everyone who owned a horse was obliged to go and fight the U-jâ clan. This, I learned to my consternation, included the man from whom we rented the yaks for he informed us that he could not take us any further for he had to join the other Gar-tse men and fight the U-jâ.

To this I demurred for he had agreed to take us to La-brang; I said we could not remain in the midst of tribal warfare and expose ourselves and our belongings to the danger of being either killed or looted. He thereupon suggested that we go with him to his chief's tent to arrange for him to hire a substitute, that is a man who did not own horses, and who was thus not obliged to participate in this war. The Tibetan is a poor infantry man, and in fact foot soldiers are non existant.

Fortunately we were able to buy a substitute for our yak owner and so were glad to get away the next morning from the scene of war. When we left camp, all the remaining Gar-tse men were also leaving to pursue the U-jâ tribe and try to recover their yak.

I was very glad to get away from the Ser-chhen valley and into other tribal territory, that adjoining the Gar-tse in the east, the Rong-wo clan. The grass was very long and many flowers were out as the pink *Primula stenocalyx*, Pedicularis, Cremanthodium, Delphinium and all the grasses previously enumerated. Here we also shot a new owl *Athene noctua imposta* B. & P. which was sitting on rocks in the bare valley. The valley is all of forty miles long and at its head is a bog whence a trail leads to a pass called Tshe-bde-ra (Tshe-de-ra) which is the divide between the Gar-tse and Rong-wo encampments. The name of the pass being derived from the Tshe Chhu River. At this pass, and below on the eastern slopes, were large herds of gazelles or huang yang [黃羊], but no blue sheep who are partial to rocks and cliffs.

On the pass in the scree at an elevation of 13,550 feet we encountered *Saussurea apus* Max., with purple flowerheads forming mats 1-2 feet in diameter first discovered by Przewalski; in its company grew *Melandrium apetalum* (L.) Fenzl., which enjoys a wide distribution from the Himalayas westwards of Afghanistan and Turkestan. These were the only plants we had not met with on the other passes to the west, but those encountered previously grew here also. This included also most of the grasses.

The trail leads from the pass to a plain with the tents of the Rong-wo nomads (see Pl.), the plain is called Na-mo-ri-on-chhung-ba (Na-mo-ri-ön-chhung-wa), this we follow east along the edge of the swampy hummocky expanse. The head waters of the Tshe Chhu are directly west only about six miles distant. The Tshe Chhu consists of two parallel flowing branches about 18 miles long, of which the lower or southern is considered the main Tshe Chhu, which has its source in the Tshe-de-ra (pass), while the northern one is called the Chha-shing Chhu; these two unite and flow around a mountain called Sa-ri mkhar-sgo in a southwesterly direction, and then straight south to the Sog-wo A-rig encampment where we crossed it on our way to Ra-gya.

The only plant of interest and not collected previously grew on the gravelly banks of the Tshe Chhu, namely *Senecio thienschanicus* Reg. & Schmalh., with yellow flowerheads.

The trail traverses the Na-mo-ri-on-chhe-ba (Na-mo-ri-ön-chhe-wa) in the center of which is a small hill which the nomads designate as Na-mo-ri on-rdza-sde (Na-mo-ri-ön-dza-de). Southwest of Rong-wo, a lamasery situated northwest on the dGu Chhu (Gu-chhu) or Gu River, is a rocky range with a great Y-shaped gap, called sGam-chhen and the peak to the right of it Sha-dar which I estimated to be 16,000 feet in height. There are so many ranges and peaks which to explore thoroughly would take years, the country is so vast and diverse, although the high elevation and the northern latitude and short summer, flanked in the north by arid regions and actual deserts, prohibit the development of a rich flora and I doubt if many more species would be added to those found traversing the region from east to west; the northern end undoubtedly becomes poorer and the southern one richer in ratio.

South of the Rong-wo encampments, i.e. in the lower half of the plain are the summer grounds of the Hor tribe now a Tibetan nomad clan, probably the descendant of Turco-Mongol tribes with whom the famous Ke-sar, King of Ling, fought many battles. Their winter encampments we had encountered in the Bâ valley south of the Gyü-par range where the ruins of ancient Hor forts are still to be observed. Beyond the Hor nomad tents, at the extreme eastern end of the plain, are the pastoral ground of the So-nag clan. The camps of all these various clans are always arranged in a large circle to enable them to accomodate their herds of sheep and yak in the centre and prevent their being driven off by marauding nomads.

The Na-mo-ri-ön-chhe-wa plain has an average elevation of 12,000 feet, swampy in places which necessitates travelling along the enclosing foothills, for the central part is often covered with a tussock formation, as is the very end of the plain.

East of a low spur elevation 11,937 feet is a smaller plain southeasterly of which are the tents of the Sog-wo A-rig tribe. Directly east, and extending from north to south is a long rocky range apparently without a name, which is crossed by three passes the dBang-chhen Nye-ra (Wang-chhen Nye-ra) in the north, the mKhas-chhags Nye-ra (Khe-chhag Nye-ra) in the center, and the mKhas-thung Nye-ra (Khe-thung Nye-ra) to the southeast.

A mountain called Sa-ri mkhar-sgo (Sa-ri khar-go) separates the plain from the Na-mo-ri-ön-chhe-wa. In the far distance south-southeast is visible the kLui-chhab-rag Range or the Bathing place of the Nagas, a beautiful limestone mountain mass similar to the Min Shan of which undoubtedly it is a part, although separated by many miles of grassland; both lie in the same direction but the latter is somewhat to the south of the former. As has already been remarked the Lui-chhab-rag, which is the Chinese Hsi-ch'ing Shan is placed too far north on foreign maps, there is no perpetual snow on that range, nor on the Min Shan.

The ford across the Tshe Chhu, not deeper than a foot, is near an old ruin of a former monastery, of which only part of one mud house was still standing. This place is called

Seng-ge khang-chhags (Seng-ge khang-chhag), the altitude at the ford is 11,600 feet, or 350 feet higher than the ford 16 miles further south. There are quite large fish in the Tshe Chhu but none could be preserved for identification; of birds, sheldrake, *Casarca ferruginea*, snipes and cranes are common here.

On the rocky slopes flourished *Pedicularis ingens* Max., a yellow flowered species, first known from northern Ssu-ch'uan 四川, where it was discovered by Potanin, while the dull bluish-purple, new *Delphinium labrangense* Ulbrich, grew on the banks of the stream associated with the pinkish-mauve, and fragrant, *Nardostachys Jatamansi* D.C., an undetermined Saussurea sp? (no 14485) with purple flowerheads more confined to the meadows at an elevation of 11,600 feet, and with it the common *Potentilla anserina* L., a prostrate herb with pale yellow flowers and nodulose roots which are collected by the Tibetans who call them Gro-ma, pronounced like Jo-ma or dro-ma. These form an article of diet of the nomads; the plants are especially common near encampments. The pale lavender *Aster Bowerii* Hemsl., the blue flowered, prostrate *Oxytropis dichroantha* C. A. Mey., the common Edelweiss *Leontopodium linearifolium* H.-M., and the umbelliferous, prostrate *Heracleum millefolium* Diels, first found by Filchner in Ch'ing-hai province, were all confined to the banks of the stream.

The Tshe Chhu forms the boundary between the territory of the Sog-wo A-rig tribe which numbers about 2,000 families and the dGon-shul (Gön-shül) tribe; that of the former adjoins the Ngu-ra in the south within the knee of the Yellow River, and that of the latter extends east of the Tshe Chhu. The trail follows up a valley with a small stream which debouches into the Tshe Chhu and whose source is on the slopes of the Wang-chhen Nye-ra. Here on the eastern bank of this valley on grassy slopes grew the bluish flowered *Delphinium densiflorum* Duthie, the yellow-flowered *Aconitum anthora* L., the prostrate pink flowered thistle *Cirsium Souliei* Franch., and the pink *Saussurea arenaria* Max., first discovered by Przewalski in Tshai-dam, the salt swamps west of the Koko nor (Lake), endemic to the region.

Beyond the plain a round hill at an altitude of 11,600 feet is called Am-nye sGar-dang (Am-nye Gar-dang) near an affluent of the Tshe Chhu. Northeast stretches the hJo-bu Thang (Jo-wu Thang) or Jo-wu plain to the foot of the range which is crossed by the Wang-chhen Nye-ra. Here on the banks of the stream we shot two terns of *Sterna hirundo tibetana* Saund.

The trailless landscape is vast and marshy and covered with grassy hummocks which necessitates keeping to the foot of the hillsides. The Jo-wu Thang (plain) averages here a height of 12,000 feet and is the home of large herds of gazelles or antelopes. South of this plain there is visible a long range composed of old grey limestone which I judged 15,000 feet in height; it is called Sho-mdo-tsha-hkhor-ri and extends from east to west connecting with the higher craggy limestone range, the Lui-chhab-rag Range, which, as already remarked, is undoubtedly a part of the Min Shan system, being of the same geological formation. The T'ao River or Lu Chhu or Nâga River, after cutting through it, flows north and then east at the foot of the Min Shan. To the south of the source of the T'ao River and the Lui-chhab-rag Range, a lower range extends from northeast to southwest which can be crossed by a pass called dBu-ru-a-si (U-ru-a-si). The geological formation of this range is not known. North of the T'ao River a pass leads in a line with U-ru-a-si, over the Lui-chhab-rag, called the Dar-rdzong-dkar-hjah-la (Dar-dzong-kar-jâ

La) leaving the T'ao River to the right (east), flowing through a limestone gorge north. It receives two southern affluents the western one called the Am-nye nang, and the eastern one near its bend north, the Mir-rdzang nang (Mir-dzang nang), between these two streams is a mountain called the Am-nye Mir-dzang, each stream being designated by one half of the name of the mountain they enclose.

The U-ru-a-si Range forms the Yellow River and the T'ao River divide, and several tributaries which have their source on the southern slopes flow into the Yellow River, while smaller ones flow from the northern slopes into the lower or third branch of the Lu Chhu or T'ao River.

The Jo-wu Thang (plain) was one mass of yellow from the thousands upon thousands of *Senecio thianschanicus* Reg. & Schmalh., which extended to the very foothills; associated with them were the deep blue flowered endemic *Gentiana siphonantha* Max., a species of Gentianella (no 14497), a prostrate, pink flowered thistle *Cirsium Souliei* Franch., and many of the grasses previously enumerated.

From the plain a small valley extends to the northern pass over the range, called the Wang-chhen Nye-ra, whence the trail becomes distinct, and leads steeply over slate and schist and gravel. Delphinium were in their glory and five species, *Delphinium albo-coeruleum* Max.*, D. Souliei* Franch., the new *D. labrangense* Ulbr., *Delphinium Henryi* Franch., *D. Forrestii* Diels, as well as another as yet undetermined species (no 14505) grew on the gravelly slopes and meadows of the pass at an elevation of 13,400 feet. The moist grassy slopes were also dotted with the brilliant red-flowered *Meconopsis punicea* Max., the rich deep blue, greenish striped *Gentiana algida* Pall., or perhaps a form of it, the species is spread from Siberia to west China; *Arenaria Przewalskii* Max., white flowered, and forming thick patches, the boraginaceous *Microula tangutica* Max., a pale blue flowered species, and the new *Microula Rockii* Johnston, also with pale blue flowers occupied the moist grassy slopes. On the summit in the grassy patches occurred the yellow flowered *Cremanthodium bupleurifolium* W. W. Smith, *Aconitum tanguticum* (Max.) Stapf, with deep blue flowers, and among rocks the yellow flowered *Saxifraga unguicula* Engl., with the purple flowered *Saussurea phaeantha* Max., only known from these grasslands. *Crepis Hookeriana* C. B. Clarke with hollow stems, entire leaves, and yellow flowerheads, preferred the eastern slopes not far below the pass; here we also met with another species of Delphinium (no 14520) as yet undetermined and probably new.

The larkspurs were of special interest as they displayed all shades of blue, from smoky grey to almost white, and then again deep purple blue; they formed stocky, bushy plants, a foot or more in height.

From Wang-chhen Nye-ra the trail descends to the head waters of the dBang-chhen Nang (Wang-chhen Nang) which was lined with Delphiniums and most of the other alpines previously enumerated. Directly ahead there loomed up a large, perfect cone of a mountain the dBang-chhen Shar-snying (Wang-chhen Shar-nying), commonly called the Bullock Heart Mountain of Wang-chhen, past which the trail leads up a valley to a small pass 12,250 feet, and down another called dBang-ra-rgan Nang (Wang-ra-gen Nang) the head of which is at an elevation of 12,200 feet.

At the foot of the grassy cone as well as on the slopes there basked in the sunshine the grayish blue flowered *Codonopsis ovata* Benth., the yellow flowered *Pedicularis lasiophrys* Max., *Crepis trichocarpa* Franch., or perhaps a related species, with lyrate leaves and yellow flowers; *Saxifraga pseudohirculus* Engl., an herb with yellow flowers, *Ligularia virgaurea* (Max.) Mattf., resembling a Senecio in habit, were common in the tall grass with the white flowered *Pedicularis cheilanthifolia* Schrenk. Of grasses *Deschampsia cespitosa* (L.) Beauv., as well as *Stipa mongolica* Turcz., with Poa, Elymus, Koeleria and Trisetum formed over 60% of the vegetation.

The plain from which Bullock Heart Mountain rose is called Dar-chhog Thang, it gradually sloped towards the streamlet which formed the Wang-ra-gen Valley, an affluent or branch of the larger Chhu-nag Nang or Black River Valley. This joins the Wang-chhen further south and united they flow into T'ao River south, under the name «The Great Wang River».

The trail continues east, then north of east, across marshy undulating country, past many nomad encampments of the Sang-khog clan, traverses several very boggy places and thence at the foot of grassy hills with the Sang Khog (Valley) below it. It crosses over the Sang chhu and T'ao river watershed, merely a low marsh, gently sloping on either side, but so inconspicuous, that one would never suspect the marsh to be the divide between the two rivers. Our route was well marked across the grassland here called the rTa-brag-gi (Ta-wrag-gi pronounced so by the nomads of this region) ranging from 11,500 to 11,800 feet in height.

Here grew *Ligularia sagitta* (Max.) Mattf., *Delphinium Forrestii* Diels, with flowers greyish purple and yellowish tinge, and the new *Pedicularis decorissima* Diels, with exceedingly long corolla tubes, large petals, and twisted keel, it is of a beautiful rose pink, and one of the most striking Pedicularis of the genus.

Directly south the huge limestone range Lui-chhab-rag could still be seen, phantastically served into fluted columns and pyramids, certainly one of the most extraordinary of mountain ranges, rivalling the Min Shan, further southeast. The trail skirts many valleys, crossed many streams up and down, till the last pass is reached called Yob-sha Nye-ra sGar-dang (Yob-sha Nye-ra Gar-dang). On the grassy slopes at an elevation of 11,600 feet grew *Gentiana dahurica* Fisch., first described from Dahuria, white flowered and of rather prostrate or spreading habit. It is a veritable species as far as color of flowers is concerned for deep purple shades occur also, but here only the white form flourished. Here and there occurred *Aster Vilmorini* Franch., with deep lavender ray florets and orange colored disc florets, *Leontopodium linearifolium* Han.-Mazt., and *Pedicularis decorissima* Diels.

Yob-sha Nye-ra Gar-dang is 11,700 feet above the sea and hardly worthy of the name of pass as one saunters gradually from one valley into the other, the valley east of the pass being called Yob-sha Nang. From the latter the Yob-gzhung Nang (Yob-zhung Nang) is reached, with its stream at an elevation of 10,740 feet. From this valley it is only half a day's journey to La-brang.

Index of Personal and Ethnic Names

Index of Zoological Names

Index of Botanical Names

Aster trinervius Roxb. (Asteraceae) 64
Aster vilmorini Franch. (Asteraceae) 49, 104, 220
Astragalus adsurgens Pall. (Leguminosae) 190
Astragalus aff. *subumbellatus* Klotzsch. (Leguminosae) 190
Astragalus chrysopterus Bunge (Leguminosae) 194
Astragalus fenzelianus Pet.-Stib. (Leguminosae) 200
Astragalus handelii Tsai et Yü (Leguminosae) 174, 190
Astragalus licentianus Hand.-Mzt. (Leguminosae) 201
Astragalus longilobus Peter-Stib. (Leguminosae) 58, 59
Astragalus mattam Tsai & Yü (Leguminosae) 199
Astragalus melilotoides Pall. (Leguminosae) 67
Astragalus moellendorffii var. *kansuensis* Pet.-Stib. (Leguminosae) 105
Astragalus monadelphus Bge. (Leguminosae) 64
Astragalus peterae Tsai et Yü (Leguminosae) 174, 186
Astragalus polycladus Bureau & fr. (Leguminosae) 190
Astragalus scaberrimus Bge. (Leguminosae) 38
Astragalus skythropos Bge. (Leguminosae) 186, 195
Astragalus tanguticus Bat. (Leguminosae) 190
Astragalus tatsienensis Bureau & Fr. (Leguminosae) 187
Astragalus tongolensis var. *glaber* Peter-Stib. (Leguminosae) 214
Astragalus versicolor Pall. (Leguminosae) 190
Astragalus yünnanensis f. *elongatus* Simps. (Leguminosae) 71
Athyrium acrostichoides (Sw.) Diels (Woodsiaceae) 87, 100
Athyrium filix-femina (L.) Roth (Woodsiaceae) 55, 78, 86
Athyrium spinulosum Milde (Woodsiaceae) 97
Beckmannia erucaeformis (L.) Host. (Poaceae) 104
Benzoin umbellatum (Thbg.) Rehd. (Lauraceae) 86
Berberis boschanii Schneid. (Berberidaceae) 117, 122, 126, 137, 172, 180
Berberis caroli Schneid. (Berberidaceae) 137, 189, 196, 197
Berberis dasystachya Max. (Berberidaceae) 34, 36, 54, 122, 124
Berberis diaphana Max. (Berberidaceae) 44, 69, 71, 103, 116, 122, 126, 180
Berberis kansuensis Schn. (Berberidaceae) 51, 69, 87, 94, 124
Berberis mouillacana Schneid. (Berberidaceae) 39, 42, 100
Berberis parvifolia Sprague (Berberidaceae) 34, 36, 64
Berberis silvataroucana Schn. (Berberidaceae) 36, 54, 66
Berberis tibetensis Laferr. (Berberidaceae) 167
Berberis vernae Schn. (Berberidaceae) 26, 66, 83, 123, 126, 133, 137
Berchemia pycnantha Schn. (Rhamnaceae) 86
Betula albo-sinensis Burk. (Betulaceae) 51, 54, 58, 70, 73, 95, 99, 103, 104
Betula albo-sinensis Burk. var. *septentrionalis* Schn. (Betulaceae) 40, 80, 87, 96, 97
Betula delavayi Franchet (Betulaceae) 95
Betula japonica var. *Rockii* Rehd. (Betulaceae) 122
Betula japonica Sieb. var. *szechuanica* Schneid. (Betulaceae) 36, 66, 86, 98, 123, 166, 167, 172, 180, 194
Biebersteinia heterostemon Max. (Geraniaceae) 68
Brachypodium durum Keng (Poaceae) 121, 132

Brassica juncea (L.) Czern. & Coss. (Brassicaceae) 174
Buddleja albiflora Hemsl. (Buddlejaceae) 86
Buddleja alternifolia Maxim. (Buddlejaceae) 103
Bupleurum longeradiatum Turcz. (Apiaceae) 106
Cacalia deltophylla (Max.) Mattf. (Asteraceae) 55
Cacalia potaninii C. Winkl. (Asteracae) 58
Caltha palustris L. (Ranunculaceae) 36
Caltha scaposa Hook. f. & Thoms. (Ranunculaceae) 36
Campylotropis macrocarpa (Bge.) Rehd. (Leguminosae) 85
Cannabis sativa L. (Cannabaceae) 115, 125
Caragana brevifolia Kom. (Leguminosae) 44, 122, 166, 180, 194, 209
Caragana densa Kom. (Leguminosae) 34, 81, 95
Caragana frutex K. Koch (Leguminosae) 106
Caragana jubata Poir. (Leguminosae) 28, 43, 46, 47, 54, 61, 65, 71, 93, 116, 122, 126, 132, 133, 136, 138, 173, 176, 178, 180, 182, 183, 186, 192, 193, 194, 197, 210
Caragana maximovicziana Kom. (Leguminosae) 34, 36, 53, 172, 178
Caragana pygmaea (L.) DC. (Leguminosae) 133
Caragana roborovskyi Kom. (Leguminosae) 196
Caragana tangutica (Asteracae) 80, 86, 95
Caragana tibetica (Leguminosae) 79, 156, 189, 192, 195, 196, 200
Cardamine macrophylla Willd. (Brassicaceae) 173
Cardamine macrophylla ssp. *polyphylla* (Don) O. E. Schulz (Brassicaceae) 51
Carduus euosmos W. W. Sm. & Forr. (Asteraceae) 118, 127
Carex atrofusca Schkuhr (Cyperaceae) 36
Carex striata L. ssp. *pullata* (Boott) Kük. (Cyperaceae) 186
Carum carvi L. (Apiaceae) 173, 190
Carpesium lipskyi C. Winkl. 55
Caryopteris tangutica Max. (Lamiaceae) 100, 106
Celtis Bungeana Blume (Ulmaceae) 103, 105
Cheilanthes argentea (Gmel.) Kze (Adiantaceae) 166
Cheiranthus roseus Max. (Brassicaceae) 182, 192
Cheiranthus roseus Max. forma *elatior* (Brassicaceae) 197
Cheiranthus roseus Max., forma *caespitosus* (Brassicaceae) 182, 183
Chrysosplenium nudicaule Bge (Saxifragaceae) 198
Chrysosplenium sphaerospermum Max. (Saxifragaceae) 38
Cirsium arvense L. (Asteraceae) 175
Cirsium euosmos W. H. Sm. & Forr. [vgl. Carduus euosmos W. W. Sm.) Forrest] 55
Cirsium setosum M. Bieb. (Asteraceae) 106
Cirsium souliei Franch. (Asteraceae) 218, 219
Cladonia gracilis (L.) Willd. var. *elongata* (Jacq.) Floerke (Cladoniaceae) 50
Clematis aethusifolia Turcz. (Ranunculaceae) 40, 44, 84, 106
Clematis brevicaudata DC (Ranunculaceae) 78, 82, 103, 123
Clematis fargesii Franch. (Ranunculaceae) 40, 63, 64
Clematis fruticosa Turcz. (Ranunculaceae) 82, 84
Clematis glauca Willd. var. *abeloides* (Max.) Rehd. & Wils. f. *phaeantha* Rehd. (Ranunculaceae) 64, 106

Clematis gouriana Roxb. var. *Finetii* Rehd. & Wils. (Ranunculaceae) 84
Clematis gracilifolia Rehd. & Wils. (Ranunculaceae) 38, 43
Clematis lasiandra Max. (Ranunculaceae) 86
Clematis macropetala Ledeb. (Ranunculaceae) 43, 64
Clematis tangutica Korsh. var. *obtusiuscula* Rehd. & Wilson (Ranunculaceae) 103, 118, 119, 174
Clematoclethra integrifolia Max. (Actinidiaceae) 81, 95
Clematoclethra lasioclada Max. (Actinidiaceae) 80, 86, 103
Clintonia udensis Trautv. & Mey. (Convallariaceae) 87
Cochlearia scapiflora Hk. f. et Th. (Brassicaceae) 186, 201
Codonopsis bulleyana Franch. (Campanulaceae) 210
Codonopsis ovata Benth. (Campanulaceae) 219
Codonopsis viridiflora Max. (Campanulaceae) 57, 105
Coluria longifolia Max. (Rosaceae) 186
Convolvulus ammanii Desr. (Convolvulaceae) 200
Convolvulus arvensis L. var. *sagittifolius* Fisch. (Convolvulaceae) 200
Cornus macrophylla Wall. (Cornaceae) 44, 80, 85
Cortusa matthioli L. (Primulaceae) 102
Corydalis adunca Max. (Papaveraceae) 177
Corydalis curviflora Max. (Papaveraceae) 182
Corydalis curviflora var. *pseudo-Smithii* Fedde (Papaveraceae) 39, 45, 46
Corydalis dasyptera Max. (Papaveraceae) 48, 71, 174, 186, 195, 198
Corydalis glycyphyllos Fedde (Papaveraceae) 200
Corydalis hannae Kanitz (Papaveraceae) 200, 201
Corydalis melanochlora Max. (Papaveraceae) 59, 198, 199
Corydalis rheinbabeniana Fedde (Papaveraceae) 214
Corydalis straminea Max. (Papaveraceae) 43, 185
Corydalis stricta Steph. (Papaveraceae) 173, 174
Corydalis trachycarpa Max. (Papaveraceae) 186
Corylus sieboldiana Bl. var. *mandschurica* Schneid. (Corylaceae) 35, 102
Cotoneaster acutifolius Turcz. (Rosaceae) 34, 69, 87, 125, 180
Cotoneaster acutifolius Turcz. var. *villosulus* Rehd. & Wils. (Rosaceae) 34, 87
Cotoneaster adpressus Bois (Rosaceae) 34, 73, 97, 100, 166
Cotoneaster horizontalis Decne (Rosaceae) 105
Cotoneaster lucidus Schlecht. (Rosaceae) 81
Cotoneaster multiflorus Bunge (Rosaceae) 34, 42, 66, 78, 84, 88, 102, 125, 173, 189, 194
Cotoneaster multiflorus Bunge var. *calocarpus* Rehd. & Wils. (Rosaceae) 103
Cotoneaster nitens Rehd. & Wils. (Rosaceae) 81
Cotoneaster obscurus var. *cornifolius* Rehd. & Wils. (Rosaceae) 86
Cotoneaster racemiflorus K. Koch var. *soongaricus* Schn. (Rosaceae) 68, 69
Cotoneaster tenuipes Rehd. & Wilson (Rosaceae) 122, 196
Crataegus kansuensis Wils. (Rosaceae) 37, 78, 82, 102
Crataegus kansuensis forma *aurantiaca* Wils. (Rosaceae) 66, 83
Cremanthodium bupleurifolium W. W. Sm. (Asteraceae) 71, 219
Cremanthodium decaisnii C. B. Clarke (Asteraceae) 186, 201, 210, 214

Cremanthodium discoideum Max. (Asteraceae) 201
Cremanthodium humile Max. (Asteraceae) 59
Cremanthodium lineare Max. (Asteraceae) 118
Cremanthodium limprichtii Diels (Asteraceae) 76
Cremanthodium plantagineum Max. (Asteraceae) 58, 195, 200, 215
Crepis hookeriana C. B. Clarke (Asteraceae) 71, 200, 211, 219
Crepis paleacea Diels aff. (Asteraceae) 58
Crepis rosularis Diels aff. (Asteraceae) 59
Crepis trichocarpa Franch. (Asteraceae) 220
Cupressus funebris (Cupressaceae) 57
Cyclophorus sticticus (Kze) C. Chr. (Polypodiaceae) 86
Cynanchum inamoenum (Max.) Loesn. (Asclepiadaceae) 37
Cypripedium luteum Franchet (Orchidaceae) 56
Cypripedium nutans Schl. (Orchidaceae) 51
Cypripedium tibeticum King (Orchidaceae) 51, 71, 172
Cystopteris moupinensis Fr. (Woodsiaceae) 80
Daphne giraldii Nitsche (Thymelaeaceae) 40, 63
Daphne tangutica Max. (Thymelaeaceae) 43, 54, 56, 63
Daphne wilsonii Rehd. (Thymelaeaceae) 54
Delphinium albo-coeruleum Max. (Ranunculaceae) 62, 155, 214, 219
Delphinium coelestinum Franch. (Ranunculaceae) 49
Delphinium densiflorum Duthie (Ranunculaceae) 218
Delphinium forrestii Diels (Ranunculaceae) 49, 219, 220
Delphinium grandiflorum L. (Ranunculaceae) 55, 61
Delphinium henryi Franch. (Ranunculaceae) 44, 62, 155, 219
Delphinium labrangense Ulbrich (Ranunculaceae) 218, 219
Delphinium maximowiczii Franch. (Ranunculaceae) 48, 49, 60
Delphinium pylzowi Max. (Ranunculaceae) 60
Delphinium souliei Franch. (Ranunculaceae) 48, 127, 214, 219
Delphinium sparsiflorum Max. (Ranunculaceae) 62
Delphinium tanguticum Huth (Ranunculaceae) 49, 59
Delphinium tatsienense Franch. (Ranunculaceae) 37, 116, 155
Delphinium tongolense Franch. (Ranunculaceae) 45, 50, 54, 57
Deschampsia cespitosa (L.) Beauv. (Poaceae) 163, 185, 214, 220
Deutzia albida Batalin (Hydrangeaceae) 94
Dianthus sinensis L. (Caryophyllaceae) 101
Dianthus superbus L. (Caryophyllaceae) 44, 104, 155
Dilophia fontana Max. (Brassicaceae) 201
Dilophia macrosperma O. E. Schulz (Brassicaceae) 208, 210
Dioscorea nipponica Makino (Dioscoreaceae) 95
Dioscorea quinqueloba Thunb. (Dioscoreaceae) 95
Dontostemon glandulosus (Kar. & Kir.) O. E. Schulz (Brassicaceae) 201
Doronicum stenoglossum Max. (Asteraceae) 55
Doronicum thibetanum Cavill. (Asteraceae) 47
Draba lanceolata Royle var. *latifolia* O. E. Schulz (Brassicaceae) 185
Draba lanceolata Royle var. *leiocarpa* O. E. Schulz (Brassicaceae) 185

Gentiana piasezkii Max. (Gentianaceae) 60
Gentiana przewalskii Max. (Gentianaceae) 49, 59
Gentiana quinquinervia Turr. (Gentianaceae) 61
Gentiana riparia Karel and Kir. (Gentianaceae) 173, 175
Gentiana sino-ornata Balf. f. (Gentianaceae) 116, 127
Gentiana siphonantha Max. (Gentianaceae) 116, 219
Gentiana spathulifolia Kusnez. (Gentianaceae) 55
Gentiana straminea Max. (Gentianaceae) 117
Gentiana striata Max. (Gentianaceae) 69
Gentiana szechenyii Kan. (Gentianaceae) 75
Gentiana tetraphylla Kusnezow (Gentianaceae) 49
Geranium aff. *pratense* L. (Geraniaceae) 37
Geranium eriostemon Fisch. (Geraniaceae) 39, 68
Geranium pratense L. (Geraniaceae) 197, 200
Geranium pylzowianum Max. (Geraniaceae) 40, 42, 174
Glaux maritima L. (Primulaceae) 36, 185
Gueldenstaedtia diversifolia Max. (Leguminosae) 173
Habenaria conopsea Benth. (Orchidaceae) 55, 64
Habenaria cucullata (L.) Hoefft. (Orchidaceae) 86
Habenaria spiranthiformis Ames & Schltr. (Orchidaceae) 174
Hedysarum esculentum Ledeb. (Leguminosae) 44, 55
Hedysarum multijugum Maxim. (Leguminosae) 172, 190
Hedysarum obscurum L. (Leguminosae) 174
Hedysarum pseudastragalus Ulbrich (Leguminosae) 209
Hedysarum tanguticum Fedt. (Leguminosae) 200
Hemerocallis dumortieri Morren (Hemerocallidaceae) 106
Heracleum millefolium Diels (Apiaceae) 174, 218
Heracleum millefolium Diels var. *longilobum* Norm. (Apiaceae) 105
Herminium tanguticum Rolfe (Orchidaceae) 104
Hippophaë rhamnoides L. (Elaeagnaceae) 34, 95, 116, 122, 127, 132, 135, 157, 168,
   180, 190
Humulus lupulus L. (Cannabaceae) 100, 104
Hydrangea bretschneideri Dipp. (Hydrangeaceae) 37, 81, 86
Hydrangea bretschneideri var. *glabrescens* Rehd. (Hydrangeaceae) 87
Hydrangea longipes Franch. (Hydrangeaceae) 86, 96
Hypecoum erectum L. var. *lectiflorum* (Kar. & Kir.) Max. (Papaveraceae) 30
Hypericum przewalskii Max. (Clusiaceae) 68, 174
Incarvillea compacta Max. (Bignoniaceae) 44, 136, 172, 180-182
Incarvillea principis B. et Franch. (Bignoniaceae) 193, 201
Indigofera bungeana Walp. (Leguminosae) 84, 94
Indigofera bungeana Walp. forma *spinescens* Kob. (Leguminosae) 88
Iris dichotoma Pall. (Iridaceae) 94
Iris ensata Thunb. (Iridaceae) 35, 63, 124, 125, 127, 138
Iris gracilis Max. (Iridaceae) 36, 37, 172, 180
Iris potanini Max. (Iridaceae) 172, 177, 198
Iris tenuifolia Pall. (Iridaceae) 38, 116, 118, 119, 121, 124, 125, 127, 172

Parnassia setchuenensis Franch. (Parnassiaceae) 44, 55
Parrya var. *albiflora* O. E. Schulz (Brassicaceae) 201
Parrya villosa Max. (Brassicaceae) 178
Pedicularis affinis *P. plicatae* Max. (Scrophulariaceae) 105
Pedicularis alaschanica Max., *typica* Li (Scrophulariaceae) 38, 118
Pedicularis armata Max. (Scrophulariaceae) 44, 48
Pedicularis bonatiana Li (Scrophulariaceae) 36
Pedicularis bonatiana Li, n. sp. (Scrophulariaceae) 190
Pedicularis calosantha Li (Scrophulariaceae) 71, 186, 209, 214
Pedicularis cheilanthifolia Schrenk (Scrophulariaceae) 220
Pedicularis cheilanthifolia Schrenk var. *typica* Prain (Scrophulariaceae) 190, 214
Pedicularis cheilanthifolia Schrenk var. *isochila* Max. (Scrophulariaceae) 36, 174
Pedicularis chenocephala Diels (Scrophulariaceae) 71, 210
Pedicularis chinensis Max. (Scrophulariaceae) 54, 104
Pedicularis cranolopha Max. (Scrophulariaceae) 174
Pedicularis cranolopha Max. var. *longicornuta* Prain (Scrophulariaceae) 64
Pedicularis cristatella Penn. & Li (Scrophulariaceae) 68
Pedicularis davidi Fr. (Scrophulariaceae) 60, 76
Pedicularis decorissima Diels (Scrophulariaceae) 155, 220
Pedicularis ingens Max. (Scrophulariaceae) 174, 218
Pedicularis kansuensis Max. (Scrophulariaceae) 173, 190
Pedicularis labellata Jacq. (Scrophulariaceae) 45, 50, 57
Pedicularis lasiophrys Max. (Scrophulariaceae) 211, 219
Pedicularis lasiophrys var. *sinica* Max. (Scrophulariaceae) 73
Pedicularis longiflora typica Rud. (Scrophulariaceae) 200
Pedicularis merrilliana Li (Scrophulariaceae) 48
Pedicularis muscicola Max. (Scrophulariaceae) 36, 42, 190
Pedicularis oederi var. *heteroglossa* Prain (Scrophulariaceae) 45, 46, 71, 182, 186, 201
Pedicularis paiana Li (Scrophulariaceae) 73, 209
Pedicularis pilostachya Max. (Scrophulariaceae) 198
Pedicularis przewalskii Max. (Scrophulariaceae) 211
Pedicularis recurva Max. (Scrophulariaceae) 50
Pedicularis rhinanthoides labellata (Jacq.) Prain. (Scrophulariaceae) 201
Pedicularis rudis Max. (Scrophulariaceae) 37, 54
Pedicularis scolopax Max. (Scrophulariaceae) 185
Pedicularis semitorta Max. (Scrophulariaceae) 104, 155, 174, 185
Pedicularis striata Pall. var. *poliocalyx* Diels (Scrophulariaceae) 68
Pedicularis szechuanica Max. (Scrophulariaceae) 76, 174, 209, 214
Pedicularis torta Max. (Scrophulariaceae) 42, 68
Pedicularis tristis L. var. *macrantha* Max. (Scrophulariaceae) 104
Pedicularis versicolor Wahlenb. (Scrophulariaceae) 172
Peganum harmala L. (Zygophyllaceae) 196, 200
Peltigera aphthosa (L.) Willd. (Peltigeraceae) 50
Pertya sinensis Oliv. (Asteraceae) 54, 85
Philadelphus pekinensis Rupr. var. *kansuensis* Rehd. (Hydrangeaceae) 35, 40, 54, 67,
   86, 95, 104

Polystichum molliculum Sieb. var. *szechuanica* Schneid. (Dryopteridae) 43
Populus balsamifera L. (Salicaceae) 115, 125
Populus cathayana Rehd. (Salicaceae) 80, 84
Populus nigra L. (Salicaceae) 137
Populus nigra L. var. *italica* Duroi (Salicaceae) 85
Populus przewalskii Max. (Salicaceae) 38
Populus simonii Carrière (Salicaceae) 34, 35, 38, 42, 66, 115, 124, 125, 134
Populus suaveolens Fisch. (Salicaceae) 133, 134
Populus szechuanica Schn. var. *Rockii* Rehd. (Salicaceae) 87
Potentilla anserina L. (Rosaceae) 34, 218
Potentilla biflora Willd. (Rosaceae) 59
Potentilla bifurca L. (Rosaceae) 38
Potentilla bifurcata L. (Rosaceae) 118
Potentilla forrestii W. W. Sm. (Rosaceae) 201
Potentilla fruticosa L. (Rosaceae) 65, 125, 126, 132, 133, 136-138, 140, 155, 157, 158,
    166, 167, 178-180, 182, 183, 192
Potentilla fruticosa L. var. Purdomi Rehd. (Rosaceae) 70, 75
Potentilla fruticosa L. var. *dahurica* Ser. (Rosaceae) 35, 57, 64, 70, 71, 75, 99
Potentilla fruticosa var. *parviflora* Wolf. (Rosaceae) 39, 48, 49, 57, 61, 102, 116, 119,
    122, 125, 172, 185, 189, 197, 209, 213
Potentilla fruticosa var. *veitchii* Bean (Rosaceae) 121, 189, 193, 194, 209
Potentilla multicaulis Bge. (Rosaceae) 174
Potentilla potaninii Wolf. (Rosaceae) 35
Potentilla salesoviana Steph. (Rosaceae) 138, 194
Potentilla saundersiana Royle (Rosaceae) 186, 200
Potentilla sericea L. (Rosaceae) 35, 37
Prenanthes tatarinowii Maxim. subsp. *necrantha* Stebb. (Primulaceae) 55
Primula aerinantha Balf. f. et Purd. (Primulaceae) 104, 105
Primula alsophila Balf. f. & Farrer (Primulaceae) 74
Primula chionantha Balf. f. & Forrest (Primulaceae) 44, 46
Primula conspersa Balf. f. & Purdom (Primulaceae) 38, 44, 69, 104
Primula fasciculata Balf. f. et Ward (Primulaceae) 159, 184
Primula flava Max. (Primulaceae) 46, 51, 166, 172, 181
Primula gemmifera Batal. (Primulaceae) 44, 49, 57-59, 74, 172
Primula graminifolia Pax & Hoffm. (Primulaceae) 47, 59
Primula lichiangensis Forr. (Primulaceae) 41
Primula limbata Balf. f. et Forr. (Primulaceae) 47, 58, 180, 182, 183, 185
Primula optata Farr. (Primulaceae) 76, 200
Primula pinnafitida Fr. (Primulaceae) 104
Primula polyneura Franch. (Primulaceae) 41, 42
Primula pumilio Max. (Primulaceae) 48
Primula purdomii Craib (Primulaceae) 47, 63, 180, 183, 185, 209
Primula reginella Balf. f. (Primulaceae) 160, 187
Primula sibirica Jacq. (Primulaceae) 36, 172, 182, 190, 195
Primula sikkimensis Hook. (Primulaceae) 209, 211
Primula stenocalyx Max. (Primulaceae) 38, 43, 51, 105, 214, 216

Ribes moupinense Fr. var. *tripartitum* Jancz. (Grossulariaceae) 83, 87
Ribes stenocarpum Max. (Grossulariaceae) 155, 171, 180
Ribes vilmorini Jancz.(Grossulariaceae) 56, 86, 167, 171
Rodgersia pinnata (Saxifragaceae) 86
Rosa bella Rehd. & Wils. (Rosaceae) 63, 122, 125, 173, 176, 189
Rosa biondii Crep. (Rosaceae) 57, 58, 102
Rosa omeiensis Rolfe (Rosaceae) 34, 40, 45, 57
Rosa sertata Rolfe (Rosaceae) 86
Rosa sweginzowii Koehne (Rosaceae) 44, 103, 122
Rosa wilmottiae Hemsl. (Rosaceae) 36, 83, 133
Rosa xanthina Lindl. (Rosaceae) 34
Rubus amabilis Focke (Rosaceae) 42, 64, 68
Rubus idaeus L. var. *strigosus* Max. (Rosaceae) 49, 171, 195, 196
Rubus pileatus Focke (Rosaceae) 64, 75, 81, 87
Rubus przewalskii Ilj. (Rosaceae) 125
Rubus xanthocarpus Bur. & Fr. (Rosaceae) 106
Sageretia theezans Brongn. (Rhamnaceae) 82
Salix alfredi Goerz (Salicaceae) 40, 51, 58
Salix brevisericea Schneid. (Salicaceae) 197
Salix cereifolia Goerz (Salicaceae) 38, 65, 70
Salix denticulata Anderss. (Salicaceae) 58
Salix ernesti Schneid. (Salicaceae) 71
Salix flabellaris Anderss. (Salicaceae) 58
Salix flabellaris Anderss. forma *spathulata* And. (Salicaceae) 58
Salix hypoleuca Seem. var. *kansuensis* Görz (Salicaceae) 51, 54, 58
Salix juparica Görz (Salicaceae) 180, 189, 193, 194
Salix juparica Görz x *S. sibirica* Pall. (Salicaceae) 171
Salix myrtillacea (Salicaceae) 46, 51, 54, 70, 166, 171, 177, 178, 197
Salix myrtillacea x Rockii Görz (Salicaceae) 171
Salix oritrepha Schneid. (Salicaceae) 186, 188, 192, 197, 209, 212
Salix oritrepha var. *tibetica* Görz (Salicaceae) 71, 105, 171, 177, 179, 196, 209, 213
Salix paraplesia C. Schneid. (Salicaceae) 102, 171
Salix plocotricha Schneid. (Salicaceae) 44, 51, 64, 70, 104
Salix pseudospissa Goerz (Salicaceae) 47, 50, 51, 105
Salix pseudowallichiana Görz (Salicaceae) 167, 180, 194
Salix rehderiana var. *brevisericea* Schn. (Salicaceae) 51, 54, 74, 95, 96, 166, 171, 176,
     194, 197
Salix rockii Görz (Salicaceae) 171, 176, 213
Salix sibirica Pallas (Salicaceae) 34, 42, 46, 58, 70, 104, 176
Salix taoensis Goerz (Salicaceae) 34, 42, 44, 166, 167, 171, 176
Salix tibetica Görz (Salicaceae) 193, 194
Salix wallichiana Anderss. (Salicaceae) 34, 42, 65
Salix wilhelmsiana M.B. (Salicaceae) 34, 166, 176, 189
Salvia prattii Hemsl. (Lamiaceae) 209
Salvia przewalski Wolf. (Lamiaceae) 45
Salvia roborowskii Max. (Lamiaceae) 39

Sedum kirilowii Reg. (Crassulaceae) 174
Sedum progressum Diels (Crassulaceae) 73
Sedum purdomii W. W. Sm. (Crassulaceae) 105
Sedum quadrifidum Pall. var. *fastigiatum* Fröderström (Crassulaceae) 177
Sedum quadrifolium Pall. (Crassulaceae) 172
Sedum venustum Praeg. (Crassulaceae) 51, 177
Selaginella involvens (Sw.) Spring. (Selaginellaceae) 88
Senecio acerifolius C. Winkl. (Asteraceae) 55, 64, 104
Senecio argunensis Turcz. (Asteraceae) 39, 64
Senecio campestris D.C. (Asteraceae) 40
Senecio nemorensis L. (Asteraceae) 87, 97
Senecio thianschanicus Reg. & Schmalh. (Asteraceae) 174, 217, 219
Serratula centauroides L. (Asteraceae) 37, 55
Sibiraea angustata (Rehd.) Hao (Rosaceae) 132, 209
Sibiraea angustifolia (Rehd.) Hao (Rosaceae) 168, 173, 178, 180
Sibiraea laevigata var. *angustata* Rehd. (Rosaceae) 37, 61, 73, 132
Silene fortunei Vis. (Caryophyllaceae) 96
Smilax oldhami Mig. (Smilacaceae) 104
Smilax rubriflora Rehd. (Smilacaceae) 86
Smilax trachypoda Norton (Smilacaceae) 86
Solanum septemlobum Bunge (Solanaceae) 190
Sorbaria arborea Schn. var. *glabrata* Rehd. (Rosaceae) 44
Sorbus hupehensis Schneid. var. *aperta* (Koehne) Schn. (Rosaceae) 82, 86, 103
Sorbus koehneana Schneid. (Rosaceae) 38, 44, 51, 67, 70, 81, 103, 106, 122, 125
Sorbus prattii Koehne (Rosaceae) 44, 64, 67, 103
Sorbus tapashana Schn. (Rosaceae) 41, 48, 50, 54, 58, 71, 74, 96
Sorbus tianschanica Rupr. (Rosaceae) 117, 122, 125, 137, 172, 180, 194
Souliea vaginata (Max.) Franch. (Ranunculaceae) 41, 46
Spiraea alpina Pall. (Rosaceae) 41, 73, 172, 209, 213
Spiraea blumei G. Don. var. *microphylla* Rehd. (Rosaceae) 84
Spiraea canescens G. Don var. *glaucophylla* Fr. (Rosaceae) 35, 44
Spiraea longigemmis Max. (Rosaceae) 35, 37, 38, 58, 73
Spiraea uratensis Fr. (Rosaceae) 84
Spiraea wilsonii Duthie (Rosaceae) 81
Stachys baicalensis Fischer (Lamiaceae) 174
Stellera chamaejasme L. (Thymelaeaceae) 39, 120, 173, 174, 180, 189, 200
Stereosanthus Souliei Franch. (Asteraceae) 37
Sticta henryana Muell.-Arg. (Lobariaceae) 86
Stipa conferta Poir. (Gramineae) 118, 121
Stipa mongolica Turcz. (Gramineae) 214, 220
Stipa splendens Trin. (Gramineae) 121
Syringa microphylla Diels (Oleaceae) 39
Syringa oblata Lindl. var. *affinis* Lingelsh. (Oleaceae) 39
Syringa oblata Lindl. var. *Giraldii* Rehd. (Oleaceae) 84, 91, 95
Syringa pekinensis Rupr. (Oleaceae) 94, 103
Syringa potanini Schn. (Oleaceae) 95

Woodsia macrospora C. Chr. & Maxon (Woodsiaceae) 97
Zanthoxylum setosum Hemsl. (Rutaceae) 85
Zygophyllum xanthoxylum Maxim. (Zygophyllaceae) 137

List of Rock Numbers
(mostly undetermined species at the time of writing)

12138 Iris
12314 Corydalis
12337 Anisodus
12407 Primula
12470 *Primula chionantha* Balf. f. & Forr. var.
12476 Incarvillea
12495 Corydalis
12567 Thalictrum
12587 *Philadelphus pekinensis* Rupr. var. *kansuensis* Rehd.
12675 Hedysarum
12691 Gentianella
12704 Astragalus
12719 Adenophora
12734 *Heracleum millefolium* Diels var. *longilobum* Norm, type
12740 Saussurea
12772 Hypericum
12784 *Pedicularis* affinis *P. plicatae* Max.
12786 Ruellia
12789 Allium
12851 Iris
12997 Senecio
13010 Hedysarum
13013 Artemisia
13031 Gentianella
13037 Allium
13049 Cremanthodium
13063 Hedysarum
13065 Oxytropis
13105 Adenophora
13110 Ligularia
13116 Adenophora
13118 Adenophora
13121 Saussurea
13122 Cirsium
13144 Dicranostigma
13156 Saussurea
13184 Geranium
13196 Codonopsis
13284 Sorbus
13306 Meconopsis
13347 Gentianella
13626 *Anemone Rockii* Ulbr.

13662 Gentianella
13700 Saussurea
13941 Euphorbia
14042 *Sedum quadrifidum* Pall. var. *fastigiatum* Fröderström
14093 Euphorbia
14106 Polygonatum
14125 Thalictrum
14126 Thalictrum
14218 Gentianella
14219 Astragalus
14241 Euphorbia
14259 Euphorbia
14260 Corydalis
14307 Pleurospermum
14322 Iris
14328 Corydalis
14337 Saussurea
14340 Astragalus
14344 Corydalis
14347 Hedysarum
14365 *Sedum juparense* Fröd. sp. n.
14368 Saussurea
14408 Corydalis
14412 Saussurea
14413 Saussurea
14425 Saussurea
14449 Allium
14485 Saussurea
14497 Gentianella
14505 Delphinium
14520 Delphinium
14573 Senecio
14597 Gentianella
14640 Delphinium
14777 Swertia
14925 *Malus kansuensis x toringoides*, seeds
15004 *Rhododendron micranthum*

Index of Geographical Names

1 (p. 37)
Crataegus kansuensis Wilson sp. nov.

2 (p. 42)
A T'ieh-pu on his way to Drag-gam-na, photographed in K'a-cha kou. In the background spruces and *Rhododendron rufum* Batal., in flower.

3 (p. 43)
Malus toringoides Hughes, forming groves in the T'ao River Valley near Shao-ni kou, Cho-ni elevation 8500 feet.

4 (p. 43)
Juniperus distans Florin growing on limestone rocks in a defile of the Lha-gya stream,
K'a-cha kou (Valley), elevation 9000 feet.

5 (p. 46)
On the summit pass Kuang-k'e, elevation 12550 ft., foreground, looking east up a pass elevation 13150 feet which leads to the main Rock Gate or Shih-men. Kan-su, Min Shan.

6 (p. 47)
View from a pass 13150 feet, east of Kuang-k'e, looking west. The long trail visible in
the center from the Kha-gya (K'a-cha kou) valley to the summit pass Kuang-k'e 12550
feet. In the distance, center a trail is visible which ascends steeply to Ch'e-pa kou pass.

7 (p. 47)
On the summit pass Kuang-k'e, elevation 12550 feet, in midwinter, at the head of K'a-cha Kou. Kansu Min Shan.

8 (p. 48)
View from Kuang-k'e pass west. Over the dip, upper right, beyond the vertical limestone rocks, a pass leads into Ch'e-pa kou (valley). the pass is 12500 feet. Summit of Kan-su Min Shan.

9 (p. 52)
An old The-wu or T'ieh-pu from the village of sTong-ba (Tong-wa), Brag-sgam-na (Drag-gam-na). Upper or Shang T'ieh-pu.

10 (p. 52)
A young T'ieh-pu from Drag-gam-na with sword and pronged rifle.

11 (p. 52)
T'ieh-pu women of Drag-gam-na spinning yak hair while on their way to the forests for
leaf mould.

12 (p. 53)
Abies sutchuenensis Rehd. & Wils. growing at Drag-gam-na, southern slopes of the Kan-su Min Shan in the Upper The-wu country, elevation 10,000 feet. Spec. no 12977.

13 (p. 55)
Drag-gam-na or The Stone Box looking east from a pass 11700 feet called Pen-drup Khi-kha. The villages are from left to right Ti-pa, Ta-re, Nye-ri and Tong-wa, the lamasery above the last village is Lha-sam Gom-pa. The two bluffs extreme upper right are the two buttresses of the main Rock gate or Shih-men (rear).

14 (p. 57)
The preliminary Shih-men or Rock gate at the head Tsha-lu valley. The trees to the left
are *Picea purpurea* Mast. Through the narrow defile a trail leads over logs into the
broad valley which confronts the main Shih-men of the backbone of the Min Shan,
three miles instant. One of the buttresses of the rock gate is visible in the distance;
elevation at foot of cliff 11000 feet.

15 (p. 58)
The main Rock Gate or Shih-men of the Kan-su Min Shan, as seen directly south of the preliminary rock gate at an elevation of 11000 feet. The alpine meadow in foregorund is marshy and supports mainly *Cremanthodium plantagineum* Max.

16 (p. 58)
The main Rock Gate or Shih-men of the Kan-su Min Shan. The summit bluffs are 15000 feet in height. There is no passage through this gate as it is enclosed by a high rocky, circular rampart and extensive scree slopes. Photographed from an elevation of 11750 feet. The trees are *Abies sutchuenensis* Rehd. & Wils., and *Abies Faxoniana* Rehd. & Wils.

17 (p. 59)
Alpine valley with scree slopes left, extending west from in front of main Rock Gate or Shih-men whose limestone bluffs are seen here laterally from an elevation of 13,159 feet. A similar scree valley leads east, middle distance. Summit of Min Shan, Kan-su.

18 (p. 59)
Alpine valley with scree slopes leading west from in front of the main Rock Gate or
Shih-men. The high rocky bluffs to the left form the Shih-men viewed laterally. A
similar ravine leads east, opposite the valley in foreground; photographed from summit
pass, elevation 13150 feet.

19 (p. 72)
Eastern Kan-su Min Shan; view from the summit ridge between Changolo Valley and Ch'i-pu kou (valley) looking southwest from an elevation of 12700 feet, west of A-chüeh and Ta-yü kou. The main backbone of the Min Shan in the background. The steep smooth-looking slopes to the left in valley are covered with willows and *Rhododendron capitatum*. Limestone cliffs in foreground showing vertical strata.

20 (p. 78)
A limestone gate in the valley of Yi-wa kou, near the village of Na-chia and (dGah-)khu looking south. In the background *Picea Wilsonii* Mast.

21 (p. 78)
View north into second Shih-men or Rock Gate in Yi-wa kou (valley), elevation 8800
feet. For description of vegetation see The Valley of Yi-wa.

22 (p. 79)
The end of Yi-wa kou (valley) opposite the mouth of Tsha-ru Valley, right not visible in picture. The narrow part of valley, center of picture, is the beginning of Dro-tshu valley. The mountain in center is Tsha-ri-ma-mön. The crests of the mountains to the right are the border of Kan-su and Ssu-ch'uan.

23 (p. 79)
The-wu or T'ieh-pu villagers from Ra-na, Upper T'ieh-pu Land, valley of the Pai-lung
Chiang; the long prongs on their rifles are gun-rests used when shooting. They are never
without their rifles even when plowing their fields.

24 (p. 82)
At the village of Ngon-gon in the valley of the Pai-lung Chiang, upper T'ieh-pu
Country. The scare-crow strawman was to keep out a cattle epidemic from the district.
The tree in the background is *Populus cathayana* Rehd., a lama from Cho-ni in the
foreground. Elevation 7180 feet.

25 (p. 82)
Entrance to the The-wu village of Ngon-gon. The primitively carved posts represent
guardians, as does the carved bridge post left; they are to keep demons from entering
the village. The village wall is of tamped earth, the grainracks are full of bundles of
wheat, hung up to dry. White prayerflags flutter from long sticks. The hillside in the
background is covered with *Pinus tabulaeformis* Carr. Upper T'ieh-pu Land, Valley of
the Pai-lung Chiang; Cho-ni territory, S.W. Kan-su.

26 (p. 82)
Trunk of *Juniperus chinensis* L. growing in the grove at Pe-zhu on the Pai-lung Chiang
in the Upper T'ieh-pu Country. S.W. Kan-su. Elevation 7100 feet.

27 (p. 83)
The The-wu village of Wang-tsang in the valley of the Pai-lung Chiang. Wang-tsang kou (valley) to left of village. The trees back of the village are *Malus baccata* Borkh. Elevation 6400 feet. Lower T'ieh-pu land, S.W. Kan-su.

28 (p. 83)
In the mouth of Wang-ts'ang kou (valley) Lower T'ieh-pu land. The streambed is lined with *Quercus liaotungensis* Koidz., elevation 7200 feet.

29 (p. 86)
Abies chinensis Van Tiegh, growing in Wang-ts'ang Valley, Lower The-wu Land, elevation 8000 feet. Trees are from 100-150 feet in height and from 3-4 feet in diameter. Southeastern Min Shan, Kan-su.

30 (p. 87)
Populus szechuanica Schneid. var. *Rockii* Rehd. A tree growing 80 feet tall in the
forests of Wan-tsang Valley in the Lower The-wu country south of the Kan-su Min
Shan. Elevation 8000 feet. Spec. no. 14846.

31 (p. 89)
Trail in the valley of the Pai-lung Chiang around a cliff built on posts and sticks with
the river roaring a hundred feet below. Between Pe-zhu and Ni-shih-kʻa in Lower Tʻieh-
pu Land.

32 (p. 89)
The new Wang-tsang Monastery, Wang-tsang Men-chhe Gom-pa, on the left bank of
the Pai-lung Chiang, Lower T'ieh-pu Land. Elevation 6290 feet. Southern slopes of the
Kan-su Min Shan.

33 (p. 92)
In the valley of To-er (kou) or Do-ru Nang; to the left the Sa-kya Lamasery of Pe-ku
Gom-pa or Pai-ku ssu, elevation 7400 feet. The walls of the lamasery are striped red
and white. Looking down stream. In the distance, center is visible the lamasery of Ra-
gya Gom-pa.

34 (p. 92)
Cutting a trail through the snow on Yang-pu Shan, over 12300 feet. For two days our
men and The-wu Tibetans of Yang-pu cut and shovelled a path over the snow-covered
slopes to the path. The Rhododendrons visible in the snow are *Rhod. przewalskii*. The
Kan-su – Ssu-ch'uan divide.

35 (p. 95)
Looking down Ma-ya kou from a bluff elevation 8000 feet. The horizontal trail in centre of picture is south of the Pai-lung Chiang. The valley opposite To-erh kou, ist branch on the right is A-hsia kou (valley). For description of flora see Ma-ya kou chapter.

36 (p. 95)
View down Ma-ya kou (valley) from an elevation of 10600 feet, from below the summit
of Lha-mo gün-gün pass. The trees are Abies, and *Picea Wilsonii* Mast., *P. asperata*
Mast., and *P. purpurea*, with an occasional *Juniperus squamata* forma *Wilsonii*, the
latter along the margin of the small meadow in lower centre of picture.

37 (p. 95)
At the head of Ma-ya kou (Valley), eastern Min Shan, elevation 10800 feet, below Lha-
mo gün-gün pass. The trees are *Abies Faxoniana* Rehd. & Wils., the shrubs in
foreground *Rhododendron Przewalskii* Kom. Looking south.

38 (p. 97)
Looking northwest from He-ra in San-pa Kou (valley) towards Yor-wu-drag-kar, elevation 9500 feet. Lower T'ieh-pu Land southern slopes of Min Shan. The shrubs are mainly Salix, Rosa, Berberis, and scattered conifers, Picea, Abies, and Juniperus. The limestone gorge in the distance is Yor-wu-drag-kar through which the main San-pa stream flows.

39 (p. 98)
The upper or northern end of Do-ya-ya gorge. Here the limestone gives way to conglomerate, very porous yet smooth in appearance; elevation 10700 feet. Eastern Min Shan, Yellow River–Yangtze divide. Willows to right, to left on the smooth walls *Juniperus saltuaria* Rehd. & Wils.

40 (p. 120)
A lämmergeier (*Gypaetus barbatus grandis* Starr) shot on the eastern shores of the
Koko Nor, Ch'ing-hai.

41 (p. 134)
The T'ung-tzu Ho, looking downstream towards its junction with the Pa-pao Ho (Hei-kou Ho) and to the southern slopes of the Ma-lo-ho Shan about 15000 feet in height, a part of the Ch'i-lien Shan range. Elevation 9000 feet.

310 Plates

42 (p. 142)
The Tshe Chhu (River) on the Na-mo-gen Thang (Plain) at an elevation of 12050 feet. The encampment of the Rong-wo nomads in the middle distance, the spur in the background is Na-mo-ri-ön-dza-de. The nomads taking their sheep out to graze after having been in the center of their encampment for the night. Our camp was on the terrace to the left above the river.

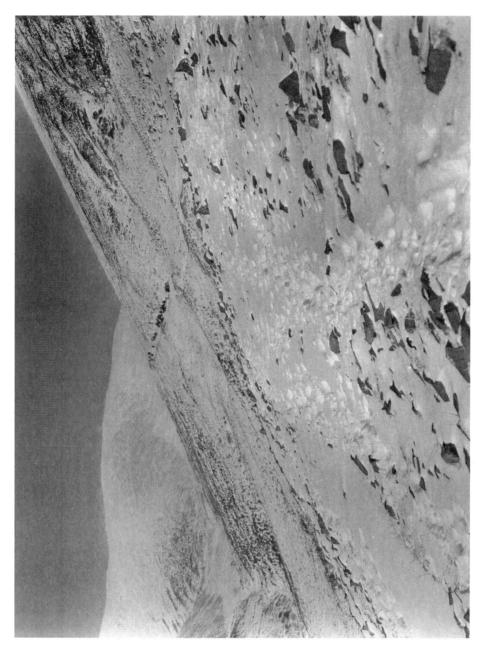

43 (p. 161)
The vanguard of our yak caravan making its way to the summit of the Khe-chhag Nye-ra (pass) 13200 feet elevation. Between La-brang and the Sog-wo A-rig encampment. May 8th 1926

44 (p. 165)
The Yellow River gorge at the mouth of the Go-chhen (left), five miles south of Tsang-gar (monastery), elevation 10200 feet. The trees are *Picea asperata* Mast. Note the fierce rapids. Looking due south up stream.

45 (p. 166)
Picea asperata Mast. at the mouth of the Go-chhen or the Great Gate where it debouches into the Yellow River south of Tsang-gar (monastery). Elevation 10300 feet.

46 (p. 166)
The Yellow River gorge near the mouth of Go-shub Valley, looking upstream, five miles south of Tsang-gar (monastery) elevation 10690 feet. The trees are *Picea asperata* Mast. It is only in protected places like this that spruces can grow.

47 (p. 166)
The Yellow River or Ma Chhu near the mouth of Go-shub Valley looking downstream
from a bluff elevation 10690 feet. The trees on the slopes are *Picea asperata* Mast.,
willows in foreground.

48 (p. 167)
A cleft in the red sandstone wall of the Great Gold Valley or Ser-chhen Nang with an ice stalactite, note man at the foot of it. On the banks of the stream *Juniperus glaucescens* Florin; elevation 10400 feet.

49 (p. 168)
Ra-gya Gom-pa (lamasery) on the Yellow River, as seen from the West looking
upstream, the small shrubs on the left hillside are *Caragana tibetica* Kom. The grass left
foreground *Stipa mongolica* Turcz.

50 (p. 168)
The Yellow River and part of Ra-gya Gom-pa (lamasery), looking west down stream.
The black dots on the plain opposite are Go-log tents. Elevation 10,000 feet. The large
buildings are the main chanting halls.

51 (p. 168)
Celebrating the feast of the Ra-gya mountain god Am-nye Khyung-ngön, north on the
slopes of the mountain of the same name. In the large tent the lamas are gathered in
prayer, on the 11th of the fourth moon. Elevation 11500 feet. In the upper left the sacred
cairn or obo.

Plates

52 (p. 168)
Ra-gya Gom-pa (lamasery) on the Yellow River or Ma Chhu as seen from above the hermit quarters (below) at the foot of mount Am-nye Khyung-ngön. The large building to the left is the printing establishment of Ra-gya, on the hillside experimental barley fields; looking upstream. The expedition quarters are across the small ravine beyond the lamasery. Photographed from an elevation of 10700 feet.

53 (p. 168)
Hermit quarters above Ra-gya Gom-pa at the foot of the cliffs (conglomerate sandstone) of Mount Khyung-ngön, elevation 10700 feet. Trees to left *Juniperus tibetica* Kom. Nettles, *Urtica dioica* L. grow below the hermit quarters, of these the lama hermits subsisted in the summer.

54 (p. 169)
Head of a blue sheep *Ovis Burrhel* common in the valley of the Yellow River on cliffs. At Ra-gya Gom-pa they are sacred and protected, they roam in herds quite tame and unafraid of man. This one was shot in Lung-mar valley, east of Ra-gya.

55 (p. 172)
View up the Yellow River from back of the summit of Am-nye Khyung-ngön, the
sacred mountain back of Ra-gya Gom-pa, showing the central cleft of the mountain, the
trees are mainly *Juniperus tibetica* Kom., and *Picea asperata* Mast. Elevation 11406
feet.

56 (p. 172)
The Yellow River or Ma Chhu above the mouth of the Lung-mar or Red Valley, elevation 10300 feet, looking upstream. The trees are mainly *Juniperus Przewalskii* Kom., and *Picea asperata* Mast. Three miles east of Ra-gya Gom-pa.

57 (p. 176)
The Yellow River looking down stream from a bluff, elevation 10900 feet on the spur
which intersects Hao-wa and Nya-rug Valleys, west of Ra-gya. Spruces (*Picea asperata*
Mast.), on the left on the northeastern slopes of the Yellow River Valley with *Juniperus
Przewalskii* Kom. on the opposite slopes. *Juniperus tibetica*, right foreground.

58 (p. 176)
The Yellow River looking down stream west of Ra-gya from a bluff on the spur which intersects Hao-wa and Nya-rug valleys, elevation 10900 feet. *Picea asperata* Mast., to left and *Juniperus Przewalskii* Kom. on the opposite slopes. In immediate foreground right, *Juniperus tibetica* Kom. A rapid can be seen in middle distance.

59 (p. 177)
The Valley of the Yellow River looking up stream, south-east from a bluff above Ta-ra-lung, between Hao-wa and Sa-khu-tu valleys, elevation 10600 feet. The valley walls to the right are forested with *Picea asperata* Mast. The wrinkled appearance of the slopes to the left is due to overgrazing.

60 (p. 177)
The Yellow River looking south, up stream, from a bluff above Sa-khu-tu ravine, elevation 10910 feet. Juniper trees in center below, in the mouth of Sa-khu-tu ravine. The narrow ridges on the left hillsides are the result of grazing by the heards of sheep of the nomads.

61 (p. 177)
The Yellow River looking down stream west-north-west from a bluff called Ta-ra-lung, elevation 10620 feet. *Picea asperata* Mast. forests on the left. *Juniperus Przewalskii* Kom., in foreground. West of Ra-gya Gom-pa.

62 (p. 177)
The Yellow River gorges as seen from a pass north of Nya-rug nang (valley), elevation
11,850 feet, looking down stream westnorthwest. The trees on the spur in foreground
are *Juniperus tibetica* Kom., spruces *Picea asperata* on the slopes to the left.

63 (p. 179)
Picea asperata Mast. forests in the valley of the Yellow River as seen from a bluff 11700 feet elevation, between Ar-tsa and Tag-so valleys, looking down stream, northwest of Ra-gya Gom-pa.

64 (p. 179)
Camp in the Tag-so Valley, elevation 10146 feet, immediately above its narrow defile
cut into the valley walls of the Yellow River; spruces, birches, willows, Lonicera,
Berberis, Ribes, etc. form the main vegetation.

65 (p. 180)
Trunk of *Picea asperata* Mast., growing in Tag-so canyon, elevation 10150 feet, northwest of Ra-gya. Tag-so is tributary of the Yellow River.

66 (p. 180)
Juniperus tibetica Kom. (no. 13946) on a grassy plot in Tag-so canyon; *Picea asperata* Mast. clings to the steep hillsides in the background. Elevation 11000 feet.

67 (p. 180)
Vegetation in Tag-so canyon; to right *Betula japonica* Sieb. var. *szechuanica* Schneid., to left *Picea asperata* Mast., in center *Salix juparica* Goerz, elevation 10150 feet.

68 (p. 180)
In the valley of Tag-so, looking upstream. *Picea asperata* Mast., covers the steep valley slopes, in middle foreground (pointed trees) *Betula japonica* Sieb. var. *szechuanica* Schneid. Shrubs to the left *Salix juparica* Goerz, other shrubs are *Sibiraea angustifolia* (Rehd.) Hao, *Lonicera syringantha* Max., and *Berberis Boschanii* Schneid. Elevation 10200 feet.

69 (p. 180)
Juniperus tibetica Kom., showing trunk and stringy bark, growing in Tag-so valley
where it forms pure stands at 10600 feet elevation. The tree is probably 500 years old.

70 (p. 189)
The Bâ valley at an elavation of 9940 feet, facing the northern slopes of the valley. The shrubs on the river-bank are *Lonicera syringantha*, back of them, *Salix Wilhelmsiana* M. v. B., *Salix juparica* Goerz, *Hippophae rhamnoides* L.; the tussocks on the hillside above are *Caragana tibetica* Kom.

71 (p. 192)
The Gyü-par Valley looking upstream, northern slopes of the Gyü-par Range, elevation
18600 feet. *Picea asperata* forest in background, the shrubs are willows, honeysuckles,
Hippophae, Ribes, etc.

72 (p. 192)
The Gyü-par Valley, northern slopes of the Gyü-par Range, looking upstream at an
elevation of 10200 feet. *Picea asperata* on the steep slopes, willows in the streambed. A
landslide exposes loess deposits with schist and shale embedded in the fine loess.

73 (p. 192)
The large bend of the Yellow River at the mouth of the Gyü-par valley, north of the
Gyü-par (Jü-par) Range, as seen from a bluff elevation 10480 feet. The rocks on the
opposite bank of the river are a deep grayish blue slate. The Gyü-par stream enters the
Yellow River, extreme lower right & through a narrow rocky defile.

74 (p. 195)
Looking down a branch of the Gyu-par Valley and the main valley towards the Yellow
River, northwest from an elevation of 11100 feet. The Yellow River flows at the foot of
the eroded steep cliffs (center distance). The trees in foreground are *Picea asperata*
Mast.

75 (p. 195)
Caragana tibetica Kom., a tussock-forming leguminous shrub growing on slate scree overlooking the Yellow River, on the northern slopes of the Gyü-par range, elevation 10400 feet.

76 (p. 195)
The Gyü-par valley looking south southeast from a bluff 11300 feet, showing spruce
forest (*Picea asperata* Mast.) on the slopes to right, the only region in which the Gyü-
par Range is forested. The triangular peak in the distance under the clouds is the second
highest of the range and is called Gyü-par sher-nying.

77 (p. 196)
One of the tallest *Picea asperata* Mast. (no. 14323) growing on the western slopes of the Gyü-par Range in the Gyü-par Valley, elevation 10200 feet. The base of the trunk is on the lower slopes of the hill. The tree is over 100 feet in height

78 (p. 199)
Tetraogallus tibetanus Przewalskii Bianchi, shot on the rocky scree slopes at an elevation of 13200 feet, near the head of Ser-chhen Valley, north-east of Ra-gya.

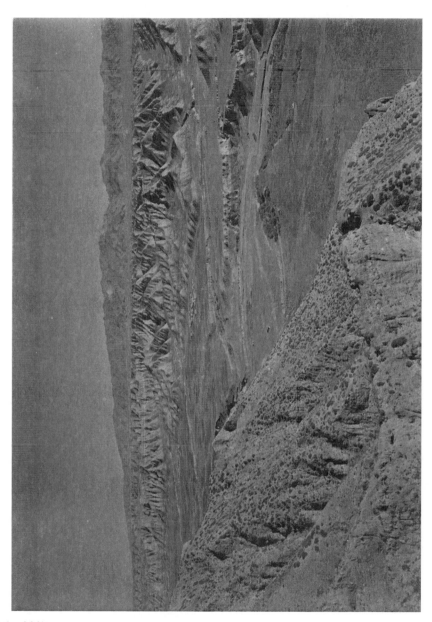

79 (p. 200)
View into the Bâ valley with the Gyu-par range in the distance, as seen from a bluff
south, elevation 10400 ft, looking north north-west. Willows along the streambed in the
valley. Near the foot of the eroded loess cliffs are two Tibetan villages called Sa-og-
rong-wa. The tussock forming plants in the immediate foreground are *Caragana
tibetica* Kom.

80 (p. 206)
Tshang-rgyur mGo-logs at their encampment west of the Yellow River, opposite Ra-
gya Gom-pa. The left one was on the point of taking a pinch of snuff, the latter consists
of the ashes of burnt yak dung. They wear one single sheepskin garment (tanned by
softening and rubbing it with butter).

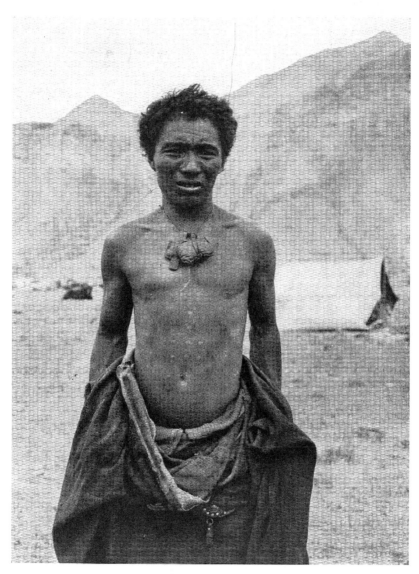

81 (p. 206)
A Tshang-gur Go-log at his encampment west of the Yellow River. The scars on his abdomen are from self-inflicted burns resorted to as counter-irritation against indigestion. Around his neck he wears an amulet containing charms.

82 (p. 206)
The chief of the Bu-tshang Go-log, his encampment is south of the Ri-mang Go-long and west of the Nga-ba tribe, south of the Yellow River. He wears a rich blue satin brocade with gold designs, trimmed with otter skins, and on his head a fox skin. Around his neck is a fine silver charm box studded with turquoise and coral.

83 (p. 207)
Ferry across Yellow River at Ra-gya Gom-pa; twelve inflated sheepskins are use to one raft, and as many as twelve persons are take across; horse and yak have to swim. Elevation at river bank 9900 feet, looking downstream. William E. Simpson my interpreter stands behind my horse.

84 (p. 207)
Tibetan ferry across the Yellow River at Ra-gya Gom-pa. Eight nomad women are
about to be ferried across on a raft of twelve sheepskins.

85 (p. 209)
In the Tsha-chhen Valley with part of our escort of Ja-zâ nomad Tibetans at an
elevation of 12500 feet. Note the grove of *Juniperus tibetica* Kom. on the valley slope.
In foreground small bushes of *Potentilla fruticosa* L. var? They were not in flower.
West of Yellow River.

86 (p. 210)
A Tibetan nomad of the Ja-zâ clan, from west of the Yellow River, their only garment is a sheepskin, their head is shaved except for a lock in the back. He wears a silver charm box, with turquoise, and a necklace of wooden beads. The right arm and shoulder is always bare.

87 (p. 210)
The Am-nye Ma-chhen Range (Ma-chhen pom-ra) as seen from the summit of Am-nye
Drug-gu, elevation 144350 feet, looking west. The valley of the Gur-zhung is below and
west of the ridge in the immediate foreground. The Tshab Chhu Valley is beyond the
ridge above the little cloud extending diagonally towards upper center of picture.
Juniperus tibetica Kom. on the slopes.

88 (p. 210)
The Yellow River gorges looking up stream from the mountains to the west of it, from a bluff 10500 feet elevation, below Am-nye Drug-gu, the mountain god of the Yön-zhi tribe. To right on the slopes *Picea asperata* Mast., middle distance *Juniperus tibetica* Kom. and willows, foreground *Rhododendron capitatum* Max. The small white flowers are *Leontopodium linearifolium* H.-M.